# Bioactive Molecules from Plants: Discovery and Pharmaceutical Applications, 3rd Edition

# Bioactive Molecules from Plants: Discovery and Pharmaceutical Applications, 3rd Edition

Guest Editor

**Javier Garcia-Pardo**

MDPI

Basel • Beijing • Wuhan • Barcelona • Belgrade • Novi Sad • Cluj • Manchester

*Guest Editor*
Javier Garcia-Pardo
Institut de Biotecnologia i de
Biomedicina (IBB)
Universitat Autónoma
de Barcelona
Barcelona
Spain

*Editorial Office*
MDPI AG
Grosspeteranlage 5
4052 Basel, Switzerland

This is a reprint of the Special Issue, published open access by the journal *Pharmaceutics* (ISSN 1999-4923), freely accessible at: https://www.mdpi.com/journal/pharmaceutics/special_issues/96Y1TTT998.

For citation purposes, cite each article independently as indicated on the article page online and as indicated below:

| Lastname, A.A.; Lastname, B.B. Article Title. *Journal Name* **Year**, *Volume Number*, Page Range. |
|---|

**ISBN 978-3-7258-7861-1 (Hbk)**
**ISBN 978-3-7258-7862-8 (PDF)**
**https://doi.org/10.3390/books978-3-7258-7862-8**

Cover image courtesy of Javier Garcia Pardo

# Contents

# About the Editor

**Javier Garcia-Pardo**

Javier Garcia-Pardo is an Associate Professor in the Department of Biochemistry and Molecular Biology at the Universitat Autònoma de Barcelona (UAB) and a researcher at the Institut de Biotecnologia i Biomedicina (IBB-UAB, Spain). His research focuses on structural biology and the molecular mechanisms underlying human diseases, with a particular interest in the identification and development of bioactive molecules from natural sources and their translation into biomedical and pharmaceutic applications. He obtained degrees in Biology and Biochemistry, followed by a master's degree in Biochemistry, Molecular Biology, and Biomedicine at UAB. He completed his PhD under the supervision of Prof. F. X. Avilés, studying the structure and function of human proteases and their inhibitors, and was awarded the Extraordinary PhD Prize. He further specialized in structural biology during an EMBO-funded stay in the laboratory of Sir Tom Blundell (University of Cambridge, UK), where he developed strong expertise in X-ray crystallography and cryo-EM. He later joined the group of Prof. Ivan Dikic at the Buchmann Institute for Molecular Life Sciences (BMLS, Frankfurt, Germany) as a senior scientist, where he worked on the development of novel compounds targeting NF-κB signaling, as well as on reticulons and bacterial protein complexes. In 2021, he returned to IBB-UAB with a Juan de la Cierva–Incorporación grant and currently leads research projects focused on the structural characterization of large protein assemblies, including bacterial effector proteins, amyloid fibrils, and membrane-associated proteins, as well as their applications in nanotechnology.

*pharmaceutics*

MDPI

*Article*

# Chemical Composition and Bioactivity Dataset Integration to Identify Antiproliferative Compounds in *Phyllanthus* Plants

**Luis Diaz [1,*], Taylor H. Díaz-Herrera [2] and Ericsson Coy-Barrera [2,*]**

[1] Bioprospecting Research Group, School of Engineering, Universidad de La Sabana, Chía 140013, Colombia
[2] Bioorganic Chemistry Laboratory, Universidad Militar Nueva Granada, Cajicá 250247, Colombia; u0500872@unimilitar.edu.co
* Correspondence: luis.diaz1@unisabana.edu.co (L.D.); inquibio@unimilitar.edu.co (E.C.-B.)

**Abstract:** Background/Objectives: *Phyllanthus* species are renowned in traditional medicine for their diverse therapeutic properties, including potential anticancer activities. This study explored the antiproliferative potential of six *Phyllanthus* species by integrating chemical composition with bioactivity assays to identify key antiproliferative compounds. Methods: The integration of liquid chromatography–mass spectrometry (LC-MS)-based chemical composition data with antiproliferative activity against three cancer cell lines—PC-3 (prostate adenocarcinoma), SiHa (cervical carcinoma), and A549 (lung carcinoma)—as well as a normal mouse fibroblast line (L929) was performed by covariate analysis. These compounds were subsequently isolated and structurally characterized using spectroscopic methods. Results: Through covariate statistics, seven $m/z$ features were found to be plausible active compounds, and after isolation, they were related to butyrolactone and arylnaphthalide lignans. Among the active isolates, an unreported compound, (+)-phyllanlathyrin **6**, was discovered in the aerial part of *Phyllanthus lathyroides*. The isolated compounds exhibited moderate to good antiproliferative activity ($IC_{50}$ < 20 μM) with selectivity to SiHa, validating the covariate-based identification approach. Conclusions: These findings highlight the potential of *Phyllanthus* species as sources of novel anticancer agents, with specific arylnaphthalide lignans showing promising cytotoxic effects that could be further developed into therapeutic leads. Additionally, this study underscores the value of combining advanced analytical techniques with bioactivity testing to uncover bioactive compounds from natural sources. The results contribute to the growing body of evidence supporting the therapeutic relevance of *Phyllanthus* species and provide a foundation for future drug development efforts targeting cancer treatment.

**Keywords:** *Phyllanthus*; antiproliferative agents; lignans; bioactivity; covariate analysis

## 1. Introduction

The genus *Phyllanthus* belongs to the family Phyllanthaceae and comprises over 2000 species distributed across tropical and subtropical regions of the world [1]. These plants have a long-standing history of use in traditional medicine, particularly in Asian, African, and South American cultures [2]. The therapeutic applications of *Phyllanthus* species are diverse, ranging from treating liver disorders, such as hepatitis and jaundice, to managing diabetes, urinary tract infections, and skin diseases [3]. Among these, *Phyllanthus niruri*, commonly known as "stonebreaker", has gained significant attention due to its claimed benefits and pharmacological properties [4]. The ethnopharmacological relevance of *Phyllanthus* species can be attributed to their rich phytochemical profile, which includes a variety of bioactive compounds such as flavonoids, tannins, alkaloids, lignans, and polyphenols. These compounds have been linked to various biological activities, including anti-inflammatory, antioxidant, antiviral, and anticancer effects [5,6]. The growing interest in natural products as sources of novel therapeutic agents has led to an increased focus on *Phyllanthus* species as potential reservoirs of bioactive molecules, particularly in the context of cancer research [5].

The phytochemical composition of *Phyllanthus* species is notably diverse, encompassing a wide array of secondary metabolites that contribute to their therapeutic potential. Among these, lignans have emerged as a key group of compounds with significant bioactivity, particularly in terms of cytotoxic and antiproliferative effects [7]. Lignans are a class of polyphenolic compounds derived from the phenylpropanoid pathway, and they are characterized by a unique structure that consists of two phenylpropanoid units linked by a carbon–carbon bond [8]. These compounds are known for their ability to modulate various biological pathways, making them promising candidates for anticancer drug development [9]. Several lignans isolated from Phyllanthus species have demonstrated potent cytotoxic and antiproliferative activities against various cancer cell lines [5]. For instance, *Phyllanthus amarus* has been shown to contain lignans such as phyllanthin, hypophyllanthin, and niranthin, which exhibit cytotoxic effects against human colorectal cancer cells [10]. These lignans have been reported to induce apoptosis, inhibit cell proliferation, and disrupt cell cycle progression, thereby highlighting their potential as anticancer agents. In addition to their cytotoxic properties, lignans from *Phyllanthus* species also possess anti-inflammatory, antioxidant, and antiviral activities, further underscoring their therapeutic relevance [3,5,6].

Among the various lignan classes, butyrolactone, aryltetralin, and arylnaphthalene(ide) lignans stand out due to their remarkable bioactivity and structural diversity [7,11]. Butyrolactone lignans, such as (+)-hinokinin, are known for their ability to inhibit topoisomerase enzymes, which are essential for DNA replication and cell division [12]. This inhibition leads to the accumulation of DNA damage in cancer cells, ultimately triggering cell death. Aryltetralin lignans, including podophyllotoxin, have been extensively studied for their antiproliferative effects, particularly in leukemia and breast cancer [13,14]. These lignans exert their effects through multiple mechanisms, including the inhibition of microtubule polymerization, induction of apoptosis, and modulation of cell cycle regulators [14]. Arylnaphthalene and arylnaphthalide lignans, another important subclass, have also garnered attention due to their potent cytotoxic activity. These compounds are characterized by a naphthalene ring system alone or conjugated with a lactone moiety, respectively, and have been shown to induce apoptosis in various cancer cell lines through the activation of caspase enzymes and the downregulation of anti-apoptotic proteins [15,16]. The structural complexity and bioactivity of arylnaphthalene lignans make them attractive targets for developing novel anticancer agents. Exploring lignan bioactivity has expanded our understanding of their therapeutic potential and provided insights into the molecular mechanisms underlying their anticancer effects.

The search for effective anticancer compounds remains a critical focus in pharmaceutical research, as cancer continues to be one of the leading causes of mortality worldwide [17]. Natural products, particularly those derived from plants, have long been a cornerstone in drug discovery due to their structural diversity and biological activity. Many successful anticancer drugs, such as etoposide, paclitaxel, and vincristine, originate from natural sources, underscoring the potential of nature as a prolific source of novel therapeutics. The importance of anticancer compounds lies in their ability to selectively target and eliminate cancer cells, often through mechanisms that involve the induction of apoptosis, inhibition of cell proliferation, and disruption of critical cellular processes such as DNA replication and repair [18]. In the context of natural product-related pharmaceutics, integrating traditional knowledge with modern drug discovery techniques has led to the identification and development of promising new anticancer agents. These compounds offer potential efficacy and often exhibit reduced toxicity compared to synthetic drugs, making them attractive candidates for therapeutic development [19]. As resistance to existing cancer therapies grows, the continued exploration of natural products is essential to uncover new molecular frameworks that can be developed into next-generation anticancer drugs [20].

Given the promising cytotoxic and antiproliferative activities exhibited by bioactive lignans obtained from *Phyllanthus* species, the present study aims further to investigate the antiproliferative potential of six *Phyllanthus* plants. The study has been focused on integrating chemical composition data obtained through liquid chromatography–mass spectrometry (LC-MS) with bioactivity datasets to identify plausible active compounds responsible for the observed antiproliferative effects. The identification and isolation of these compounds have been undertaken to enable a more detailed estimation of their antiproliferative scope. The study encompasses a comprehensive analysis of the phytochemical profiles of the six selected *Phyllanthus* species, with particular emphasis on recognizing bioactive candidates. Thus, by correlating the LC-MS-based chemical composition data with the antiproliferative activity observed against cancer cell lines, the study aims to pinpoint specific compounds that contribute to the observed activity. The subsequent isolation and structural characterization of these compounds provided valuable insights into their potential as antiproliferative agents and laid the groundwork for future natural product-based drug development efforts.

## 2. Materials and Methods

### 2.1. General Information

All reagents (MTT, Folin-Ciocalteu, DPPH, ABTS, and TPTZ), standards (trolox, gallic acid, quercetin, podophyllotoxin, streptomycin, and penicillin), and solvents (EtOH, MeOH, $CHCl_3$, $CDCl_3$, and ACN) were commercially purchased from Merck KGaA and/or Sigma-Aldrich (Darmstadt, Germany) and were used without further purification. The purity of the dry solvents was ensured as specified upon purchase. Reaction progress and product purification were monitored using thin layer chromatography (TLC) on silica gel 60 F254 plates (Merck KGaA, Darmstadt, Germany), with detection under UV light at 254 nm.

### 2.2. Plant Samples

The aerial parts of six *Phyllanthus* plants were collected in different Colombian locations. In this regard, *P. urinaria* (4°0′40″ N; 73°46′56″ W, 457 m.a.s.l.), *P. caribaeus* (4°1′7″ N; 73°47′13″ W, 457 m.a.s.l.), and *P. caroliniensis* (4°0′40″ N; 73°46′56″ W, 457 m.a.s.l.) were collected in Acacías, Meta, whereas *P. salviifolius* (5°46′31″ N; 73°3′4″ W, 457 m.a.s.l.) was collected in in Duitama, Boyacá, and *P. madeirensis* (4°46′7″ N; 74°27′35″ W, 1800 m.a.s.l.) was collected in Anolaima, Cundinamarca. Finally, *P. lathyroides* (4°31′13″ N; 74°35′19″ W, 457 m.a.s.l.) was collected in Apulo, Cundinamarca. One sample of each species was deposited in the Colombian National Herbarium (Collection code: *P. urinaria* 596082; *P. caribaeus* 596079; *P. salviifolius* 596083; *P. caroliniensis* 596080; *P. madeirensis* 595478; *P. lathyroides* 596131).

### 2.3. Ethanolic Extracts Preparation

Healthy aerial parts (200 g) from the six *Phyllanthus* plants were air-dried, crushed, and ground. The ground aerial parts were separated into five replicates and then extracted with 96% ethanol using an orbital shaker (Heidolph Instruments GmbH & Co. KG, Schwabach, Germany). The procedure spanned a week, with daily solvent removal and restoration with clean 96% ethanol. The resulting mixtures were then concentrated under reduced pressure at 40 °C using a rotary evaporator (IKA® Werke GmbH & Co. KG, Staufen, Germany). The resulting crude extracts were dried and stored at −20 °C until further analyses.

### 2.4. Antiproliferative Assay

Human cancer cell lines—prostatic adenocarcinoma (PC-3, ATCC CRL-7934), lung adenocarcinoma (A549, ATCC CCL-185), and cervical carcinoma (SiHa, ATCC HTB-35)—along with normal mouse fibroblasts (L929, ATCC CRL-6364) were cultured as monolayers in Dulbecco's Modified Eagle Medium (DMEM) supplemented with 10% (*v*/*v*) fetal bovine serum (FBS), 1% (*v*/*v*) penicillin, and 1% (*v*/*v*) streptomycin. These cultures were maintained at 37 °C in a humidified atmosphere with 5% $CO_2$. The antiproliferative effects of *Phyllanthus*-derived extracts and isolated compounds were evaluated using a previously described method [21]. Cells were seeded at $5 \times 10^3$/well in 96-well plates (100 μL) and incubated for 24 h. After incubation, the medium was replaced with 100 μL of serum-free medium containing varying concentrations of the treatments (0.5–250 μg/mL for extracts and 0.01–100 μg/mL for pure compounds). Each treatment was tested in triplicate. Controls included a PBS-containing medium as a blank, 1% (*w*/*v*) bovine serum albumin-amended medium as a negative control (100% survival), and podophyllotoxin (0.16–100 μg/mL) as a positive control. After 48 h of treatment, cell viability was determined by adding 10 μL of MTT (5 mg/mL) to each well, followed by a 3-h incubation under the same culture conditions. The resulting formazan crystals were dissolved in 100 μL of DMSO, and absorbance was measured at 570 nm using a Varioskan LUX 96-well plate reader (Thermo Fisher Scientific, Waltham, MA, USA). The antiproliferative effects were quantified as the half-maximal inhibitory concentration ($IC_{50}$) in μg/mL for extracts and μM for pure compounds. Dose–response curves were used to determine $IC_{50}$ values from non-linear regression using GraphPad 7.0 (GraphPad, San Diego, CA, USA).

### 2.5. Chemical Characterization of Test Extracts

#### 2.5.1. Total Phenolic and Flavonoid Contents

The total phenolic and flavonoid contents were evaluated in the *Phyllanthus*-derived extracts using a previously reported method [22]. Briefly, the total phenolics content (TPC) was quantified using the Folin–Ciocalteu method, with absorbance measured at 765 nm. Results were expressed using a gallic acid standard curve ($m = 7.70 \times 10^{-2}$; $b = 1.37 \times 10^{-2}$; $R^2 = 0.99$). Total flavonoid content (TFC) was determined using the aluminum chloride ($AlCl_3$) complex method, with absorbance measured at 420 nm, and the results were expressed using a quercetin standard curve ($m = 9.27 \times 10^{-2}$; $b = 1.02 \times 10^{-2}$; $R^2 = 0.99$).

#### 2.5.2. Antioxidant Capacity

The antioxidant capacity of the test extracts was measured by the FRAP (ferric reducing antioxidant power), $DPPH^{\bullet}$ (2,2-diphenyl-1-picrylhydrazyl), and $ABTS^{\bullet+}$ (2,2′-azino-bis(3-ethylbenzothiazoline-6-sulphonic acid) radical scavenging assays, measuring final absorbance at $\lambda = 593$ nm, $\lambda = 515$ nm, and 734 nm, respectively, following the previously reported method [22]. The results of $DPPH^{\bullet}$ and $ABTS^{\bullet+}$ assays were expressed as $IC_{50}$ values from inhibition percentages for each free radical using GraphPad Prism v7.0. The free radical inhibition percentage (RI) was calculated using Equation (1).

$$\%RI = \frac{(AbsC - AbsT)}{AbsC} \times 100\% \tag{1}$$

where AbsC = control absorbance value, and AbsT = treatment absorbance value.

#### 2.5.3. High-Performance Liquid Chromatography Coupled with Mass Spectrometry

Chemical characterization of the test *Phyllanthus*-derived extracts was also recorded through liquid chromatography coupled with mass spectrometry (LC-MS) using a Shimadzu Prominence HPLC system (Shimadzu Corporation, Kyoto, Japan) and a micrOTOF-Q II mass spectrometer (Bruker, Billerica, MA, USA) detector equipped with a quadrupole–time-of-flight (QToF) analyzer and electrospray ionization (ESI). Each aerial part-derived extract replicate ($n = 5$) was dissolved in absolute ethanol at a 5 mg/mL concentration, and a 20 μL aliquot was injected into the HPLC system. Separation was achieved on a Luna C18 column (4.6 mm × 150 mm, 5 μm; Phenomenex, Torrance, CA, USA) using a gradient elution method with solvent A (1% formic acid in Milli-Q $H_2O$) and solvent B (1% formic acid in acetonitrile). The elution gradient was set as follows: 5% B (0–2 min), 5% to 30% B (2–7 min), 30% B (7–10 min), 40% to 90% B (10–22 min), 90% B (22–26) min, and returning to 5% B (26–30) min. The flow rate was maintained at 0.7 mL/min, and detection was monitored at a wavelength of 270 nm. Mass spectra were simultaneously recorded using positive ion mode electrospray ionization, scanning a mass range of 100–2000 $m/z$. The mass spectrometry parameters included quadrupole energy of 7.0 eV, collision energy of

14 eV, a curved desolvation line temperature of 250 °C, a heat block temperature of 400 °C, drying gas of 8 L/min, and a nebulization gas flow rate of 1.5 L/min.

### 2.6. LC-MS-Derived Metabolite Profile Pre-Processing

Mass spectrometry-derived data were processed using MZmine 2.53 [23]. The parameters used for processing were as follows: mass detection was set to centroid mode with an MS1 noise level of $5 \times 10^2$. For the automated data analysis pipeline (ADAP) chromatogram builder, the settings included a minimum group size of 5 scans, a group intensity threshold of $4 \times 10^2$, a minimum highest intensity of $4 \times 10^2$, and an $m/z$ tolerance of 0.01. Chromatogram deconvolution parameters were set with a minimum peak height of $4 \times 10^2$, a peak duration of 0–2 min, and a baseline level of $4 \times 10^2$. Isotope grouping was performed with an $m/z$ tolerance of 0.01, a retention time tolerance of 0.1, and a maximum charge of 2. The joining aligner parameters included an $m/z$ tolerance of 0.01, a weight for $m/z$ of 70, a retention time tolerance of 0.1, and a weight for retention time of 30. Peak filtering was applied with a minimum peak in a row set to 5, an $m/z$ range of 100–2000, and enabled the reset peak number ID option. Finally, gap filling was conducted with an intensity tolerance of 10%, an $m/z$ tolerance of 0.01, and a retention time tolerance of 0.1.

### 2.7. Integration of Bioactivity and Chemical Composition Datasets

The pre-processed metabolite profile-derived data were saved as .csv files to create the feature intensity table (FIT), consisting of 2356 features across 30 samples. The resulting dataset was then autoscaled (unit variance scaling) for appropriate associations. MetaboAnalyst 5.0 (McGill University, Quebec, Canada) was used to generate an intuitive visualization of the distribution of autoscaled features [3]. The FIT was subsequently combined with the corresponding antiproliferative activity data, treated as a continuous variable, to form an integrated dataset. This matrix was then analyzed in SIMCA software (v 14.0) (Umetrics, Umeå, Sweden) for model construction using single-*Y* orthogonal partial least squares (OPLS), affording the respective scatter, variable importance in the projection (VIP) scores, and *S* plots.

### 2.8. Purification and Elucidation of Antiproliferative Candidates **1–7**

Extracts from *P. lathyroides* (*Pl*) and *P. caribaeus* (*Pcb*) (500 mg each) were initially processed using solid-phase extraction (SPE) with Strata® C18-U cartridges (55 μm, 70 Å, 500 mg, 6 mL; Phenomenex, Torrance, CA, USA). The cartridges were conditioned with methanol (6 mL) followed by water (6 mL) and methanol (6 mL). After loading the extracts previously dissolved in methanol, the cartridges were washed with water (5 mL) and then eluted with methanol (5 mL). The eluates containing the concentrated extracts were used for semi-preparative HPLC to isolate the target compounds.

Isolation was carried out on a UFLC Prominence system (Shimadzu, Columbia, MD, USA) in semi-preparative mode. This system included an LC-20AD pump, CTO-20AC column oven, SPD-20AV UV/Vis detector, SIL-10AP autosampler, and FRC-10A fraction collector, and was equipped with a reversed-phase Phenomenex Luna C18 column (250 × 10 mm, 5 μm; Phenomenex, Torrance, CA, USA) at 20 °C. The SPE-purified extracts (500 μL per injection, 80 mg/mL in MeOH) were injected ten times and separated at a flow rate of 3 mL/min using an isocratic elution method with solvents A (1% formic acid in $H_2O$) and B (1% formic acid in ACN). Six targeted peaks were collected based on single-Y OPLS-based recognition, with the following retention times and yields: 6.0–6.2 min (3.6 mg, compound **1**), 11.5–11.7 min (11.3 mg, compound **2**), 12.6–12.8 min (11.4 mg, compound **3**), 13.9–14.0 min (8.7 mg, compound **4**), 18.9–11.1 min (5.4 mg, compound **5**), 19.6–19.8 min (4.9 mg, compound **6**), and 20.3–20.4 min (3.7 mg, compound **7**). The *Pl* extract yielded **1**, **3**, **6**, and **7**, while the *Pcb* extract yielded **2**, **4**, and **5**.

The structures of the isolated compounds were determined using $^1H$ and $^{13}C$ NMR spectroscopy, including the attached proton test (APT), on an Avance 400 spectrometer (Bruker, Billerica, MA, USA) with $CDCl_3$ as the solvent (400 MHz for $^1H$; 100 MHz for

$^{13}C$). All chemical shifts are reported in δ (ppm) with tetramethylsilane (TMS) as the internal standard. The APT $^{13}C$ NMR data and optical rotation of the isolated compounds matched those reported for phyllanthusmin C (**1**) [24], (+)-acutissimalignan B (**2**) [25], phyllanthusmin A (**3**) [24], (−)-hinokinin (**4**) [26], justicidin A (**5**) [27], and justicidin P (**7**) [28]. Compound **6** was found to be an unreported compound, so its structural elucidation based on spectroscopic characteristics is presented herein.

(+)-Phyllanlathyrin (IUPAC: (+)-5-methoxy-9-(7-methoxybenzo[d][1,3]dioxol-5-yl)-8-oxo-6,8-dihydrofuro[3′,4′:6,7]naphtho[2,3-d][1,3]dioxol-6-yl acetate) (**6**): yellowish amorphous solid (mp 231–233 °C); $[\alpha]_D^{20}$ = +48.9 (c = 0.02, $CHCl_3$); $^1H$ NMR (400 MHz, $CDCl_3$) $\delta_H$ 7.76 (s, 1H), 7.58 (s, 1H), 7.33 (s, 1H), 6.86 (d, 1.4 Hz, 1H), 6.77 (d, J = 1.4 Hz, 1H), 6.06 (s, 2H), 5.92 (s, 2H), 4.03 (s, 3H), 3.86 (s, 3H), 2.18 (s, 3H). $^{13}C$ NMR (100 MHz, $CDCl_3$) $\delta_C$ 169.1 (C-9′), 167.5 (9-$\underline{C}OCH_3$), 154.4 (C-3′), 151.8 (C-4), 150.8 (C-5), 150.1 (C-5′), 138.8 (C-4′), 136.7 (C-7), 135.2 (C-1′), 133.7 (C-1), 131.7 (C-2), 126.8 (C-8′), 124.6 (C-8), 123.3 (C-7′), 109.9 (C-2′), 109.2 (C-3), 104.5 (C-6′), 101.6 (3-$OCH_2O$-5), 101.3 (3′-$OCH_2O$-5′), 98.9 (C-6), 87.6 (C-9), 60.6 (7-$OCH_3$), 56.5 (5′-$OCH_3$), 20.4 (9-$CO\underline{CH_3}$). HRESIMS (positive mode) $m/z$ 467.0966 $[M]^+$, (calcd. for $C_{24}H_{19}O_{10}$, 467.0978).

### 2.9. Statistical Analysis

Standard deviations were calculated for the total flavonoid and phenolic contents, as well as for antioxidant capacities. $IC_{50}$ values were determined using dose–response curves and regression analysis with GraphPad Prism 7.0 software. Normality was checked using the Shapiro–Wilk test. For normally distributed data, differences between treatments were analyzed with ANOVA followed by Tukey's multiple comparison test. All statistical analyses were performed using InfoStat statistical software v29.09.2020 (National University of Córdoba, Córdoba, Argentina) (significance at $\alpha$ = 0.05).

## 3. Results and Discussion

### 3.1. Chemical Characterization: Total Flavonoid and Phenolics Content and Antioxidant Capacity

The phytochemical characterization showed the metabolite content at different levels in the six extracts as well as significance differences ($p < 0.05$) (Table 1). TPC values were highest in *P. salviifolius* (228.2 mg GAE/g DE), *P. caribaeus* (215.5 mg GAE/g DE), and *P. madeirensis* (215.2 mg GAE/g DE) extracts with respect to *P. urinaria* (173.1 mg GAE/g DE), *P. lathyroides* (142.6 mg GAE/g DE), and *P. caroliniensis* (139.8 mg GAE/g DE) extracts. Total flavonoids content displayed a similar behavior since *P. salviifolius* (17.3 mg QE/g DE), *P. caroliniensis* (15.9 mg QE/g DE), *P. urinaria* (14.2 mg QE/g DE), and *P. caribaeus* (13.3 mg QE/g DE) extracts showed the highest content with respect to *P. lathyroides* (9.2 mg QE/g DE) and *P. madeirensis* (9.2 mg QE/g DE) extracts.

**Table 1.** Characterization of *Phyllanthus*-derived extracts.

| Extract | TPC (mg GAE/g DE) | TFC (mg QE/g DE) | Antioxidant Capacity | | |
|---|---|---|---|---|---|
| | | | FRAP (TE µM/mg DE) | DPPH ($IC_{50}$ µg/mL) | ABTS ($IC_{50}$ µg/mL) |
| *P. urinaria* | 173.1 ± 20.6 $^b$ | 14.2 ± 0.2 $^c$ | 34.8 ± 0.9 $^b$ | 19.9 ± 1.7 $^{ab}$ | 8.5 ± 1.1 $^{ab}$ |
| *P. caribaeus* | 215.5 ± 5.9 $^a$ | 13.3 ± 0.7 $^c$ | 28.4 ± 0.2 $^c$ | 17.6 ± 1.6 $^a$ | 7.3 ± 0.7 $^a$ |
| *P. salviifolius* | 228.2 ± 5.4 $^a$ | 17.3 ± 0.1 $^a$ | 38.3 ± 2.7 $^b$ | 21.9 ± 4.6 $^{ab}$ | 7.6 ± 0.9 $^a$ |
| *P. caroliniensis* | 139.8 ± 6.2 $^c$ | 15.9 ± 0.3 $^b$ | 27.1 ± 0.5 $^c$ | 24.3 ± 1.7 $^b$ | 10.5 ± 0.8 $^b$ |
| *P. lathyroides* | 142.6 ± 5.6 $^c$ | 9.2 ± 0.4 $^d$ | 24.2 ± 0.7 $^d$ | 24.9 ± 1.5 $^b$ | 11.2 ± 0.7 $^b$ |
| *P. madeirensis* | 215.2 ± 6.1 $^a$ | 9.8 ± 0.5 $^d$ | 45.7 ± 1.1 $^a$ | 25.3 ± 1.2 $^b$ | 6.9 ± 0.6 $^a$ |

GAE = Gallic Acid Equivalent; QE = Quercetin Equivalent; DE = Dry Extract; TE = Trolox Equivalent; TPC = Total phenolics content; TFC = Total flavonoid content; FRAP = Ferric-reducing antioxidant power; DPPH = 2,2-diphenyl-1-picrylhydrazyl radical scavenging assay; ABTS = 2,2′-azino-*bis*(3-ethylbenzothiazoline-6-sulphonic acid radical scavenging assay. Values are expressed as mean ± standard deviation (SD), $n$ = 3. Distinct lowercase superscript letters imply significant differences consistent with Tukey's test ($p < 0.05$).

In addition, the five test extracts also showed antioxidant capacity by three assays at different levels (Table 1). The FRAP assay showed the highest antioxidant capacity for *P. madeirensis* (45.7 TEAC μM/mg DE), *P. salviifolius* (38.3 TEAC μM/mg DE), and *P. urinaria* (34.8 TEAC μM/mg DE) extracts. On the other hand, the antioxidant capacity for scavenging DPPH radical showed that *P. caribaeus* ($IC_{50}$ = 17.6 μg/mL) and *P. urinaria* ($IC_{50}$ = 19.9 μg/mL) extracts had the best $IC_{50}$ values for this assay. In contrast, the ABTS assay showed the best $IC_{50}$ for the *P. salviifolius* ($IC_{50}$ = 7.3 μg/mL) and *P. caribaeus* ($IC_{50}$ = 7.6 μg/mL) extracts (Table 1).

Antioxidant capacity and TPC and TFC have been reported in some *Phyllanthus* species. *P. amarus* has exhibited a phenolic content of 13.68 to 13.90 mg GAE/g DE and antioxidant capacity with DPPH radical scavenging activity ($IC_{50}$ of 0.344 mg/mL) and FRAP (6.23 to 16.2 mg GAE/g sample) [29]. Liu et al. [30] and Bansal et al. [31,32] reported a phenolic content of 62 to 513 mg GAE/g extract and 250 to 262 mg GAE/L, respectively. Additionally, the antioxidant capacity for this species with $IC_{50}$ values between 4.2 and 142 μg/mL has been measured. Fang et al. [33] reported the antioxidant capacity for isolated compounds from *P. urinaria* with $IC_{50}$ values ca. 20 μM.

### 3.2. Characterization of *Phyllanthus* Aerial Part-Derived Extracts Based on LC-ESI-MS Data

A reverse phase LC-ESI-MS was employed to determine the chemical composition of the six extracts derived from *Phyllanthus* species. The mass-to-charge features recorded for each extract were extracted from the MS raw data and compiled into a feature intensity table (FIT), identifying 414 distinct features. This finding indicated a relevant metabolite diversity among the test extracts. While some features were shared across multiple extracts, others were unique to specific ones. This diversity was further visualized through a heat map illustrating a global LC-MS-based metabolite distribution using the top 100 differential metabolites according to the VIP scores (Figure 1).

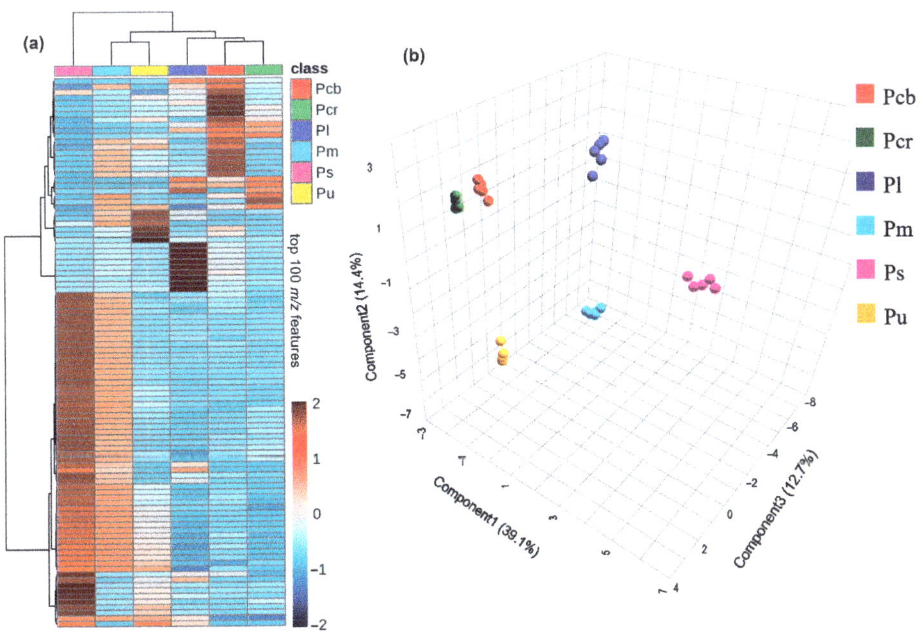


**Figure 1.** LC-MS-based differentiation of *Phyllanthus* aerial part-derived extracts from six plants and five technical replicates per plant. *Pcb = P. caribaeus; Pcr = P. caroliniensis; Pl = P. lathyroides;*

*Pm* = *P. madeirensis; Ps* = *P. salviifolius; Pu* = *P. urinaria*. (**a**) Heat map depicting the chemical distribution based on the intensity of the top 100 most contrasting *m*/*z* features detected across *Phyllanthus* aerial part-derived extracts. Each column represents the average levels (*n* = 5) per plant extract, and each colored cell corresponds to the autoscaled intensity of a detected *m/z* feature, with the color scale ranging from dark red (high intensity) to dark blue (low intensity). (**b**) Three-dimensional score plot derived from sparse partial least square discriminant analysis (sPLS-DA), showing the separation of samples based on the three first principal components. Plant extract grouping was used as a categorical variable, explaining 66.2% of the variance.

The extracts were grouped into *Phyllanthus* species (*n* = 6) and the feature intensities were autoscaled to highlight differential metabolites. The heat map, with a color scale ranging from dark red (=2), indicating high feature intensity, to dark blue (=−2), indicating low feature intensity, revealed distinct metabolic profiles of the *Phyllanthus*-derived extracts (Figure 1a). In addition, a hierarchical clustering analysis (HCA) of the autoscaled data showed that the presence or abundance of specific metabolites clearly distinguished the *Phyllanthus* extracts. *P. salviifolius* was separated from the rest of the *Phyllanthus* plants, exhibiting a higher number of differential metabolites, possibly by the presence of flavonoids and other phenolics, which are very common in *Phyllanthus* plants [3]. The other five plants exhibited profiles that share components but were differentiated by the presence of particular metabolites, as expected, and deserve further investigation in future studies.

Finally, a sparse partial least squares discriminant analysis (sPLS-DA) further highlighted the chemical differentiation between the *Phyllanthus* chemical composition, with the first three principal components (PC1, PC2, and PC3) explaining 66.2% of the variance. The separation between the species was evident in the three-dimensional score plot (PC1 × PC2 × PC3) (Figure 1b), involving discrimination between *P. urinaria, P. salviifolius*, and *P. madeirensis* from *P. caribaeus, P. caroliniensis*, and *P. lathyroides* along C2, while *P. lathyroides* was separated from *P. caribaeus* and *P. caroliniensis* along C3.

### 3.3. Antiproliferative Activity of Phyllanthus Aerial Part-Derived Extracts

Several previous studies have explored the bioactivity of *Phyllanthus* plants against cancer cell lines [3,5]. However, *P. urinaria* is the only plant assessed for anticancer properties among the tested extracts. The antiproliferative activity of ethanolic aerial part extracts from the six *Phyllanthus* plants was evaluated against three cancer cell lines—PC-3, SiHa, and A549—as well as a normal cell line, L929. The results in Table 2 showed that the extracts exhibited varying degrees of antiproliferative activity, suggesting differential content and properties of antiproliferative compounds across the extracts.

**Table 2.** Antiproliferative activity against cancer and normal cell lines of *Phyllanthus* aerial part-derived ethanolic extracts.

| | PC-3 [b] | | SiHa [b] | | A549 [b] | | L929 [c] | |
|---|---|---|---|---|---|---|---|---|
| **Samples [a]** | **$IC_{50}$ [c]** | **CI [d]** | **$IC_{50}$ [c]** | **CI [d]** | **$IC_{50}$ [c]** | **CI [d]** | **$IC_{50}$ [c]** | **CI [d]** |
| *P. urinaria* | 77.2 | 72.3–80.5 | 101 | 96.3–107 | 96.3 | 91.3–102 | >250 | - |
| *P. caribaeus* | 112 | 107–117 | 49.8 | 45.7–52.8 | 188 | 179–196 | 235 | 227–244 |
| *P. salviifolius* | 95.3 | 90.7–101 | 131 | 126–135 | 133 | 122–139 | 207 | 201–213 |
| *P. caroliniensis* | 127 | 121–133 | 67.5 | 64.3–69.2 | >250 | - | >250 | - |
| *P. lathyroides* | 89.6 | 83.3–92.6 | 22.1 | 20.2–24.6 | 105 | 97.3–111 | 225 | 219–232 |
| *P. madeirensis* | 235 | 230–241 | >250 | - | >250 | - | >250 | - |

[a] Aerial part-derived extracts from *Phyllanthus* plants; [b] test human cancer cell lines: PC-3 (prostate adenocarcinoma), SiHa (cervical carcinoma), A549 (lung carcinoma); [c] normal cell line: L929 (murine fibroblasts), values expressed as half-maximal inhibitory concentration ($IC_{50}$) in μg/mL; [d] CI = $IC_{50}$ confidence interval (95% confidence) after non-linear regression.

The $IC_{50}$ values ranged from 22.1 to 250 μg/mL. Notably, the extracts exhibited low activity against A549 ($IC_{50}$ > 96 μg/mL) and showed lower activity against fibroblasts

($IC_{50}$ > 207 µg/mL). In contrast, SiHa cells were the most sensitive cell line to the extracts, with an average $IC_{50}$ of 74.3 µg/mL–except *P. madeirensis* ($IC_{50}$ > 250 µg/mL)–followed by PC-3 cells, which had an average $IC_{50}$ of 123 µg/mL. The *P. lathyroides*-derived extract was the most potent, with an $IC_{50}$ of 22.1 µg/mL against SiHa cells, and exhibited good selectivity as it showed low activity against fibroblasts ($IC_{50}$ = 225 µg/mL). In this regard, tested extracts demonstrated the lowest activity against A549 and moderate-to-low activity against PC-3 cells compared to that exhibited against SiHA. This pattern indicated that *Phyllanthus* aerial part extracts exhibit selectivity towards PC-3 and, particularly, SiHa cancer cell lines.

### 3.4. Recognition of Antiproliferative Candidates from Phyllanthus Plants Using Chemical Composition and Bioactivity Dataset Integration

The identification of active metabolites produced by *Phyllanthus* aerial parts was achieved by integrating antiproliferative activity and LC-MS-based chemical composition datasets through multiple-covariate statistical analysis. A single-Y orthogonal partial least squares (OPLS) modeling was employed to associate these datasets. Since the SiHa cell line was selectively affected by extracts according to the $IC_{50}$ values (Table 1), it was utilized as the antiproliferative activity dataset, serving as the continuous Y variable. The OPLS regression incorporated $t_1$ (a predictive score) and $to_1$ (an orthogonal component). This incorporation effectively differentiated the aerial part extracts based on antiproliferative activity (Y-data) and chemical composition (X-data). This model demonstrated strong fit ($R^2X$ = 0.875, $R^2Y$ = 0.795) and predictability ($Q^2Y$ = 0.705), covering 51.3% of the variance in antiproliferative activity along $t_1$ and 43.2% in chemical composition along $to_1$. The discrimination mode among *Phyllanthus* aerial part extracts was revealed in the OPLS-derived scores plot (Figure 2a), where antiproliferative activity-influencing metabolite profiles were visualized through a color gradient, ranging from red (250 µg/mL) to aquamarine (0 µg/mL). The most active extracts ($IC_{50}$ < 70 µg/mL) were grouped on the II and III quadrants, though involving low spreading, while the least active extracts were positioned on the I and IV quadrants with high variability among replicates ($n$ = 5). This pattern advised that certain metabolites in those extracts with the highest activity would be associated with the resulting antiproliferative activity against the SiHa cell line.

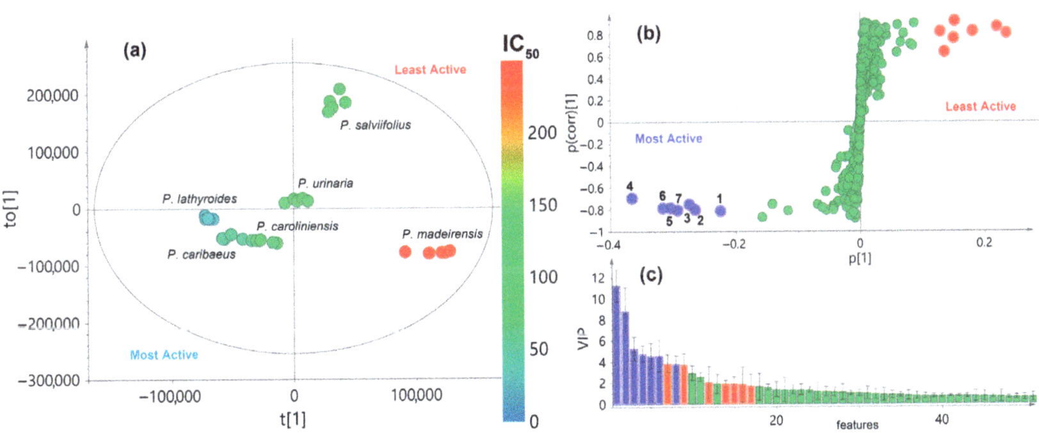


**Figure 2.** Integration of chemical and bioactivity datasets (i.e., feature-based chemical composition and antiproliferative activity) for *Phyllanthus* aerial part extracts using single-*Y* orthogonal partial least squares (OPLS) analysis. The $IC_{50}$ values, represented as the continuous *Y*-variable, are depicted on a color scale (red = 250 µg/mL; aquamarine = 0 µg/mL). (**a**) Scores plot. (**b**) OPLS-derived *S*-plot. (**c**) OPLS-derived variable importance in the projection (VIP) plot.

To further pinpoint bioactive compounds, the OPLS-DA-derived loadings were analyzed using an *S*-plot conversion ($p_1 \times p_{(corr)1}$), which categorized the statistical weight of differential compounds. The *S*-plot led to the visualization of the successful association of X-data with $t_1$ through covariance and correlation under centering scaling (Figure 2b). This analysis highlighted the most significant metabolite distinctions between the least active ($p1 > 0$) and most active ($p1 < 0$) *Phyllanthus* aerial part-derived extracts, represented by those compounds distantly situated in the *S*-plot boundaries, indicating reliable metabolite discrimination. Hence, metabolite-based features **1–7** emerged as the maximum linking ($p_{(corr)1} < -0.5$, $p1 < -0.2$) with plausible bioactives, while the other seven features were strongly associated with the lowest bioactivity. The VIP plot (Figure 2c) confirmed the importance of these metabolites in the integrative discrimination process, identifying them as potential bioactive candidates (VIP scores > 3). These metabolites exhibited *m*/*z* and rt values in the ranges of 355.1165–513.1385 and 6.3–19.8 min, respectively, with specific rt/*m*/*z* pairs, pinpointed as 513.1385/6.3 (**1**), 357.1351/11.4 (**2**), 381.0984/12.3 (**3**), 355.1165/13.7 (**4**), 395.1144/18.2 (**5**), 467.0966/18.9 (**6**), and 425.1247/19.8 (**7**). Based on their LC and MS performance, these metabolite-related features were classified as compounds with middle polarity and low molecular weight. Feature **4** revealed the most substantial impact on the model due to its high classifying magnitude ($p1 < -0.35$) and was particularly abundant in *P. caribaeus* and *P. caroliniensis* extracts. Compounds 1 and **2** exhibited the highest reliability with their differential p(corr)1 values, indicating their presence across various most active *Phyllanthus* extracts. In this regard, all seven differential metabolites (**1–7**) were considered potential antiproliferative candidates, likely contributing to the observed activity against the SiHa cell line in the studied extracts.

This analysis demonstrated the successful application of single-Y OPLS for bioactive identification by integrating antiproliferative activity/chemical composition datasets. Principal component analysis (PCA) was not used as the primary multivariate statistical method since the use of supervised statistical methods like OPLS or PLS effectively maximized the covariate performance of differential features (independent variables) depending on antiproliferative activity (dependent variable), a task not as effectively achieved by unsupervised methods such as PCA [34]. Additionally, the fundamental benefit of associating datasets based on chemical composition lies in using LC-MS-based profiles as independent variables combined with the bioactivity of natural origin mixtures as the dependent variable [35]. Since single-Y OPLS requires a continuous variable, it is more convenient for multiple-covariate dataset integration than categorical variables, whose use can result in significant information loss [36,37]. Additionally, this method can detect unstable metabolites, which is crucial during extract fractionation [38]. However, a limitation of this approach is directing the analysis to false positives. This situation can be produced by other extract constituents synergistically or antagonistically acting against the target cell [39]. Therefore, the present study carefully performed the MS-guided isolation and studied their antiproliferative activity on cell lines of compounds **1–7** to validate the observed correlations.

### *3.5. Isolation and Identification of Antiproliferative Candidates*

Semi-preparative HPLC separations were carried out to purify compounds **1–7** from the tested active plant extracts (i.e., *P. lathyroides* and *P. caribaeus*). Compounds **1**, **3**, **6**, and **7** were isolated from the *P. lathyroides* extract, while compounds **2**, **4**, and **5** were obtained from the *P. caribaeus* extract. Once isolated, compounds **1–7** were identified and structurally elucidated through comprehensive NMR and MS spectroscopy analysis. These metabolites were identified as lignans, namely phyllanthusmin C (**1**) [24], (+)-acutissimalignan B (**2**) [25], phyllanthusmin A (**3**) [24], (−)-hinokinin (**4**) [26], justicidin A (**5**) [27], and justicidin P (**7**) [28], with their $^{13}C$ NMR data corresponding to those reported in the literature. The structures of compounds **1–5** and **7** are presented in Figure 3. Contrarily, compound **6** did not match the previously reported data, so further elucidation was required. Hence, compound **6** was then characterized and structurally elucidated using various spectroscopic techniques. High-resolution mass spectrometry (HRMS) (Supplementary Material, Figure S1) established its molecular formula as $[C_{24}H_{19}O_{10}]^+$ with a quasimolecular ion peak at $m/z$ 467.0966 $[M+H]^+$ (error = 2.57 ppm). Its $^1$H NMR spectrum, recorded in $CDCl_3$, revealed characteristic signals indicative of a polysubstituted polyaromatic moiety [40]. Specifically, the $^1$H NMR spectrum (Figure S2) displayed ten distinct singlet signals, and the $^{13}$C NMR spectrum presented twenty-four signals, which, upon detailed analysis using HMQC and APT experiments (Figure S3), corresponded to fourteen quaternary carbons, two methylene carbons, five methine carbons, and three methyl groups. Comparison of the $^1$H and $^{13}$C NMR data with literature values suggested a similarity to justicidin-like arylnaphthalide lignan [41], involving a very comparable NMR profile to that of justicidin P [28], isolated in this study as well. However, compound **6** differed from justicidin P by a singlet at $\delta_H$ 2.18 (3H), corresponding to a methyl group forming an acetyl group, and two doublets ($J$ = 1.4 Hz) at $\delta_H$ 6.86 (1H) and 6.77 (1H) related to *meta*-positioned hydrogens attached to the aryl group, instead of a multiplet (3H) at 6.86–6.66 in justicidin P. A singlet at $\delta_H$ 6.08 (2H) was also attributed to a further methylenedioxy group at C4,C5. The aromatic region also exhibited signals characteristic of hydrogens forming an *ortho* system involving a tetrasubstituted aromatic ring, with two singlets at $\delta_H$ 7.76 (1H) and 7.58 (1H), corresponding to those protons at positions 3 and 6 of the arylnaphthalide lignan moiety. A singlet at $\delta_H$ 7.33 (1H) was assigned to the proton at position C9, geminal to the acetyl group. The $^{13}$C NMR spectrum provided additional structural details, including signals for aromatic carbons within the range of $\delta_C$ 98 to 155, consistent with the benzomethylenedioxy rings, signals for the methoxy groups at $\delta_C$ 60.6 and 56.5, and the carbonyl carbons at $\delta_C$ 169.5 (lactone) and 167.5 (acetyl). The HMQC experiment demonstrated the respective connectivity between hydrogens and their corresponding carbons and the HMBC experiment revealed heteronuclear long-range correlations, which were relevant in corroborating the above-mentioned structural assignments for **6** (Figure 3).

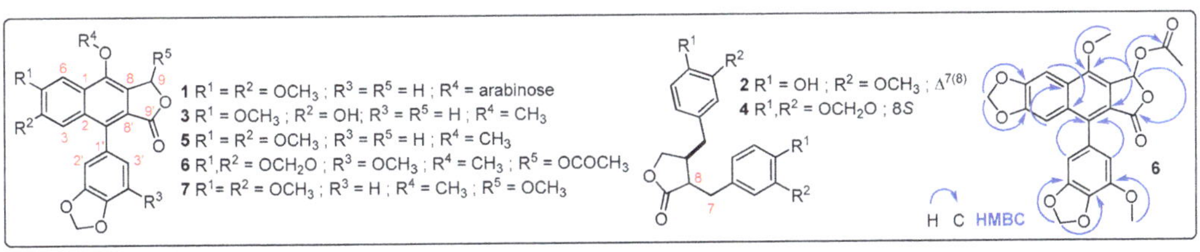


**Figure 3.** Structures of isolated compounds after statistical pattern recognition from *Phyllanthus* aerial parts. HMBC correlations in compound **6**.

The hydrogens of the two methoxy groups correlated with the attached carbons in the arylnaphthalide moiety at $\delta_C$ 150.1 and 136.7, as well as the methylenedioxy-attached hydrogens, which correlated to the bearing carbons at $\delta_C$ 151.8/150.8 and 138.8/151.8, respectively. The signal at $\delta_H$ 7.33 (H-9) showed significant correlations with $\delta_C$ 136.7 (C-7), 167.5 (acetyl), 126.8 (C-8′), and 169.1 (C-9′), indicating the attachment of the lactone moiety to the naphthalene core structure. Additionally, correlations between H-2′ and H-6′ established the connection of the methoxymethylenedioxyaryl moiety to the naphthalide (Figure 3). In addition, compound **6** was found to be optically active, exhibiting a dextrorotatory orientation ($[\alpha]_D^{20}$ = +48.9), likely due to the presence of a chiral carbon at C-9 bearing an acetyl group. However, determining the absolute configuration of this stereogenic center was not possible and will be addressed in a future study. Thus, compound **6** was identified as (+)-phyllanlathyrin (IUPAC: (+)-5-methoxy-9-(7-methoxybenzo[*d*][1,3]dioxol-5-yl)-8-oxo-6,8-dihydrofuro [3′,4′:6,7]naphtho [2,3-*d*][1,3]dioxol-6-yl acetate) (**6**), representing an unreported arylnaphthalide lignan isolated from *P. lathyroides* aerial parts.

The purified compounds **1–7** were further evaluated for their antiproliferative activity against the four cell lines to corroborate their OPLS-based bioactive differentiation. The resulting $IC_{50}$ values are presented in Table 3. The confidence intervals (CI) for the $IC_{50}$ values are also provided, indicating the reliability of these measurements. As anticipated, the SiHa cell line was the most susceptible to treatment with these compounds due to the focus on the LC-MS-based covariate pattern recognition, although some compounds exhibited good activity against PC-3 (i.e., **2** and **4**). Remarkably, lignans **1** and **6** demonstrated the most potent antiproliferative effect on SiHa cells, with an $IC_{50}$ of 2.35 and 3.62 μM, respectively, although this outcome was less potent than the positive control, podophyllotoxin ($IC_{50}$ = 1.85 μM). In addition, compound **1** exhibited moderate antiproliferative activity against PC-3 ($IC_{50}$ = 18.3 μM) and SiHa ($IC_{50}$ = 2.55 μM), with a notable decrease in efficacy against A549 ($IC_{50}$ = 62.5 μM). It shows no significant cytotoxicity toward the fibroblast cell line L929c ($IC_{50}$ > 100 μM). As mentioned, compound **2** was highly effective against PC-3 ($IC_{50}$ = 6.55 μM) but much less so against SiHa ($IC_{50}$ = 56.8 μM) and A549 ($IC_{50}$ = 78.6 μM), with no observed cytotoxicity toward L929. Lignan **3** demonstrated balanced activity across all cancer cell lines, showing moderate $IC_{50}$ values for PC-3 (16.3 μM) and SiHa (8.61 μM), though it was less effective against A549 (93.2 μM). It also exhibits some activity against L929 ($IC_{50}$ = 92.3 μM). On the other hand, compound **4** showed potent activity against PC-3 ($IC_{50}$ = 4.55 μM) and moderate activity against A549 ($IC_{50}$ = 12.3 μM) but is less effective against SiHa ($IC_{50}$ = 61.2 μM) and somewhat active against L929 ($IC_{50}$ = 76.3 μM), while lignan **5** displayed moderate efficacy against PC-3 ($IC_{50}$ = 17.6 μM) and A549 ($IC_{50}$ = 50.3 μM), with better activity against SiHa ($IC_{50}$ = 6.53 μM) and low activity against L929 (IC50 > 100 μM). Novel compound **6** demonstrated moderate activity against PC-3 ($IC_{50}$ = 15.3 μM) and significant efficacy against SiHa ($IC_{50}$ = 3.62 μM), with intermediate results for A549 ($IC_{50}$ = 26.5 μM) and some activity against fibroblasts ($IC_{50}$ = 86.3 μM). Similar to compound **6**, compound **7** showed moderate activity against PC-3 ($IC_{50}$ = 15.2 μM) and SiHa ($IC_{50}$ = 5.23 μM), with moderate results for A549 ($IC_{50}$ = 45.3 μM) and the lowest activity against L929 ($IC_{50}$ > 100 μM).

**Table 3.** Antiproliferative activity against cancer cell lines and fibroblasts of compounds **1–7** isolated from *Phyllanthus* aerial part extracts.

| | PC-3 [b] | | SiHa [b] | | A549 [b] | | L929 [c] | |
|---|---|---|---|---|---|---|---|---|
| **Compounds** [a] | $IC_{50}$ [c] | CI [d] | $IC_{50}$ [c] | CI [d] | $IC_{50}$ [c] | CI [d] | $IC_{50}$ [c] | CI [d] |
| **1** | 18.3 | 17.3–19.2 | 2.55 | 2.46–2.67 | 62.5 | 60.2–64.1 | >100 | - |
| **2** | 6.55 | 6.05–6.98 | 56.8 | 55.9–57.7 | 78.6 | 75.9–80.6 | >100 | - |
| **3** | 16.3 | 15.5–17.3 | 8.61 | 8.21–8.77 | 93.2 | 90.6–96.1 | 92.3 | 90.0–94.4 |
| **4** | 4.55 | 4.38–4.66 | 61.2 | 58.9–63.5 | 12.3 | 11.5–13.0 | 76.3 | 74.9–77.2 |
| **5** | 17.6 | 16.8–18.2 | 6.53 | 6.01–6.89 | 50.3 | 47.8–52.6 | >100 | - |
| **6** | 15.3 | 14.9–16.9 | 3.62 | 3.48–3.71 | 26.5 | 25.1–27.9 | 86.3 | 85.0–88.2 |
| **7** | 15.2 | 14.6–15.9 | 5.23 | 4.98–5.36 | 45.3 | 44.6–46.4 | >100 | - |
| **ppt** [e] | 2.22 | 2.03–2.37 | 1.85 | 1.78–1.96 | 4.82 | 4.62–5.13 | 2.53 | 2.44–2.61 |

[a] Isolated compounds **1–7** from *Phyllathus* aerial parts (Figure 3); [b] test cancer cell lines: PC-3 (prostate adenocarcinoma), SiHa (cervical carcinoma), A549 (lung carcinoma)); [c] normal cell line: L929 (murine fibroblasts), values expressed as half-maximal inhibitory concentration ($IC_{50}$) in μM; [d] CI = $IC_{50}$ confidence interval (95% confidence) after non-linear regression. [e] ppt = podophyllotoxin used as positive control.

Hinokinin (**4**) is a well-known cytotoxic compound [42], and it is highlighted due to its strong activity against PC-3 and A549. Its moderate activity against L929 has suggested potential cytotoxicity concerns in non-cancerous cells. Similarly, (+)-acutissimalignan B (**2**) is another recognized cytotoxic lignan, previously isolated from the aerial part of *P. acutissima* [25], and it was highly selective for PC-3, with minimal activity against SiHa and A549, making it a possible lead compound for prostate cancer treatment with reduced off-target effects. Reports on the antiproliferative properties and mechanisms of action of (−)-hinokinin have been well-characterized. Cunha et al. [43] described a reduction in the G1 and S phases and an arrest in the G2/M phase, along with the downregulation of genes such as CCNA1 (Cyclin A1), CCND1 (Cyclin D1), and CCNE1 (Cyclin E1), and the upregulation of CDKN1A and CDKN1B (Cyclin Dependent Kinase Inhibitors 1A and 1B) in MFC-10A, MCF-7, and SKBR-3 cell lines. These findings suggest an action mechanism closely associated with cell division. Additionally, stimulation of apoptotic processes was observed, mediated by the expression of pro-apoptotic genes such as PUMA, NOXA, and the caspases CASP3 and CASP8.

Similarly, arylnaphthalene lignans, such as justicidin B, have demonstrated action mechanisms against proliferative cell lines. In glomerular mesangial cells (MCs), justicidin B exhibited inhibitory effects on factors involved in mitogen-induced processes, reduced cell proliferation induced by platelet-derived growth factor subunit B (PDGF-BB), and decreased levels of proliferating cell nuclear antigen (PCNA), which is associated with the S/G2 and G1 stages [44]. Noteworthy cell morphology changes, including elongation, shrinkage, and apoptosis, were observed, along with the downregulation of anti-apoptotic, apoptotic, and DNA repair proteins, such as Bcl-2, CASP3, and PARP-1 (poly(ADP-ribose) polymerase 1), in HeLa cells [45]. Furthermore, phytosterols like (-)-β-sitosterol-3-*O*-β-D-(6-*O*-palmitoyl) glucopyranoside, isolated from *P. songboiensis* extracts, have been shown to stimulate the production of IFN-γ in natural killer (NK) cells, contributing to the reduction of malignant cell transformation [46].

Phyllanthusmin C (**1**) and the novel (+)-phyllanlathyrin (**6**) show broad-spectrum activity across cancer cell lines, with less cytotoxicity toward fibroblasts. The secondary metabolites characterized in this study are likely involved in anti-apoptotic processes, the inhibition of DNA repair mechanisms, and/or the arrest of cell division processes. This fact makes them potential candidates for treating multiple cancer types. However, the positive control showed superior efficacy across all cell lines. This performance indicates that while the tested compounds **1**–**7** have potential, they require optimization to achieve comparable effectiveness. In this context, these findings provide valuable insights into the cytotoxic profiles of the covariate-based recognized lignans, highlighting compound **6** as a particularly promising candidate for further investigation. Future studies should focus on optimizing these compounds' structures to enhance selectivity and reduce off-target effects. Additionally, the potential for combination therapies should be explored to maximize therapeutic efficacy. Additionally, it is essential to note that there may be other active compounds that were not detected due to potential antagonistic interactions among different extract components. Therefore, more in-depth integrative analyses are recommended to identify any missed bioactives, even those with lower activity but possible synergistic roles that could enhance the overall antiproliferative effect. Lastly, since various concentrations of compounds **1**–**7** were employed and they were distinct from those of the original extracts, the antiproliferative activity of the isolates cannot be compared with extracts. Consequently, further studies are needed to determine if **1**–**7** are the sole antiproliferative agents in the tested *P. lathyroides* or *P. caribaeus* aerial part-derived extracts.

## 4. Conclusions

The aerial part-derived extracts from the test plants were assessed for their antiproliferative effects against three cancer cell lines (PC-3, SiHa, and A549), demonstrating notable selectivity towards SiHa, with reduced toxicity to fibroblasts. This study represents the first investigation into the inhibitory potential of five *Phyllanthus*-derived extracts (excluding *P. urinaria*, previously reported) against these specific cancer cell lines, highlighting their preferential activity against SiHa. Additionally, integrating LC-MS-based chemical composition data with antiproliferative activity datasets provided a robust framework for identifying seven bioactive compounds in *Phyllanthus* species, underscoring the effectiveness of this approach in natural product research. Thus, the antiproliferative properties of these seven compounds, isolated from two *Phyllanthus* plants, validated the OPLS-based bioactive recognition, with compounds **1** and **6** showing the most potent activity against SiHa, indicating their potential as lead compounds for further development in cervical cancer therapies. Furthermore, this covariate-based integration led to the identification, isolation, and structural elucidation of a novel arylnaphthalide lignan, (+)-phyllanlathyrin (**6**), which exhibited selective activity against SiHa cells. However, our findings also allowed for the possibility of other active compounds being overlooked due to potential antagonistic effects within the extracts, suggesting the need for deeper integrative analyses. Overall, this study identified promising candidates for cancer treatment and emphasized the importance of further structural optimization and exploring combination therapies to enhance efficacy and reduce toxicity. These findings pave the way for future research aimed at developing novel anticancer agents from *Phyllanthus*-derived arylnaphthalene lignans.

**Supplementary Materials:** The following supporting information can be downloaded at: https://www.mdpi.com/article/10.3390/pharmaceutics16111381/s1, Figure S1: HRMS spectrum of (+)-phyllanlathyrin (6); Figure S2: $^1$H NMR spectrum of (+)-phyllanlathyrin (**6**) (400 MHz, $CDCl_3$); Figure S3: APT experiment of (+)-phyllanlathyrin (**6**) (100 MHz, $CDCl_3$).

**Author Contributions:** Conceptualization, L.D. and E.C.-B.; methodology and investigation, L.D., T.H.D.-H. and E.C.-B.; formal analysis, L.D., T.H.D.-H. and E.C.-B.; resources, L.D. and E.C.-B.; writing—original draft preparation, T.H.D.-H. and E.C-B.; writing—review and editing, L.D., T.H.D.-H. and E.C.-B.; supervision, project administration, and funding acquisition, L.D. and E.C.-B. All authors have read and agreed to the published version of the manuscript.

**Funding:** This research was funded by Vicerrectoría de Investigaciones at Universidad Militar Nueva Granada (UMNG) through the project EXT-CIAS-3854, validity 2023, and General Research Directorate at Universidad de La Sabana (project INGPHD-56-2023).

**Institutional Review Board Statement:** This study did not require ethical approval since no animals or humans were involved. Plant accessions were gathered according to the current Colombian ethical legislation (legal decree 1376-2014 and collection permit 1198-2014) and specific regulations related to non-commercial, scientific studies on exotic plants (legal ruling 1348-2014).

**Informed Consent Statement:** Not applicable.

**Data Availability Statement:** Data are available from authors upon reasonable request.

**Acknowledgments:** The authors thank UMNG and Universidad de la Sabana for their financial support.

**Conflicts of Interest:** The authors declare no conflicts of interest.

## References

1. Kiran, K.R.; Swathy, P.S.; Paul, B.; Shama Prasada, K.; Radhakrishna Rao, M.; Joshi, M.B.; Rai, P.S.; Satyamoorthy, K.; Muthusamy, A. Untargeted Metabolomics and DNA Barcoding for Discrimination of *Phyllanthus* Species. *J. Ethnopharmacol.* **2021**, *273*, 113928. [CrossRef] [PubMed]
2. Calixto, J.B.; Santos, A.R.S.; Filho, V.C.; Yunes, R.A. A Review of the Plants of the Genus *Phyllanthus*: Their Chemistry, Pharmacology, and Therapeutic Potential. *Med. Res. Rev.* **1998**, *18*, 225–258. [CrossRef]
3. Nisar, M.F.; He, J.; Ahmed, A.; Yang, Y.; Li, M.; Wan, C. Chemical Components and Biological Activities of the Genus *Phyllanthus*: A Review of the Recent Literature. *Molecules* **2018**, *23*, 2567. [CrossRef]
4. Bagalkotkar, G.; Sagineedu, S.R.; Saad, M.S.; Stanslas, J. Phytochemicals from *Phyllanthus niruri* Linn. and Their Pharmacological Properties: A Review. *J. Pharm. Pharmacol.* **2006**, *58*, 1559–1570. [CrossRef]
5. Seyed, M.A. A Comprehensive Review on *Phyllanthus* Derived Natural Products as Potential Chemotherapeutic and Immunomodulators for a Wide Range of Human Diseases. *Biocatal. Agric. Biotechnol.* **2019**, *17*, 529–537. [CrossRef]
6. Díaz-Herrera, T.H.; Caldas Martínez, M.L.; Coy-Barrera, E. *Phyllanthus*-Derived Naturally-Occurring Products: An Overview of Their Effects against Viruses in Cell Models. *Curr. Bioact. Compd.* **2024**, *20*, 1–26. [CrossRef]
7. Nawfetrias, W.; Devy, L.; Esyanti, R.R.; Faizal, A. Phyllanthus Lignans: A Review of Biological Activity and Elicitation. *Horticulturae* **2024**, *10*, 195. [CrossRef]
8. Álvarez-Caballero, J.M.; Coy-Barrera, E. Lignans. In *Antioxidants Effects in Health—The Bright and the Dark Side;* Nabavi, S.M., Silva, A.T.S., Eds.; Elsevier: London, UK, 2022; pp. 387–416, ISBN 978-0-12-819096-8.
9. Zálešák, F.; Bon, D.J.-Y.D.; Pospíšil, J. Lignans and Neolignans: Plant Secondary Metabolites as a Reservoir of Biologically Active Substances. *Pharmacol. Res.* **2019**, *146*, 104284. [CrossRef]
10. Shimaa, M.I.A.; Ibrahim, J.; Mohd, N.A.; Mohamed, S.A.; Kok, C.M. Lignans and Polyphenols of *Phyllanthus amarus* Schumach and Thonn Induce Apoptosis in HCT116 Human Colon Cancer Cells through Caspases-Dependent Pathway. *Curr. Pharm. Biotechnol.* **2021**, *22*, 262–273.
11. Patel, J.R.; Tripathi, P.; Sharma, V.; Chauhan, N.S.; Dixit, V.K. *Phyllanthus amarus*: Ethnomedicinal Uses, Phytochemistry and Pharmacology: A Review. *J. Ethnopharmacol.* **2011**, *138*, 286–313. [CrossRef]

12. Solyomváry, A.; Beni, S.; Boldizsar, I. Dibenzylbutyrolactone Lignans—A Review of Their Structural Diversity, Biosynthesis, Occurrence, Identification and Importance. *Mini-Rev. Med. Chem.* **2017**, *17*, 1053–1074. [CrossRef] [PubMed]
13. Shen, S.; Tong, Y.; Luo, Y.; Huang, L.; Gao, W. Biosynthesis, Total Synthesis, and Pharmacological Activities of Aryltetralin-Type Lignan Podophyllotoxin and Its Derivatives. *Nat. Prod. Rep.* **2022**, *39*, 1856–1875. [CrossRef] [PubMed]
14. Motyka, S.; Jafernik, K.; Ekiert, H.; Sharifi-Rad, J.; Calina, D.; Al-Omari, B.; Szopa, A.; Cho, W.C. Podophyllotoxin and Its Derivatives: Potential Anticancer Agents of Natural Origin in Cancer Chemotherapy. *Biomed. Pharmacother.* **2023**, *158*, 114145. [CrossRef] [PubMed]
15. Luo, J.; Hu, Y.; Kong, W.; Yang, M. Evaluation and Structure-Activity Relationship Analysis of a New Series of Arylnaphthalene Lignans as Potential Anti-Tumor Agents. *PLoS ONE* **2014**, *9*, e93516. [CrossRef]
16. Park, S.; Kim, S.; Shin, D. Arylnaphthalene Lactones: Structures and Pharmacological Potentials. *Phytochem. Rev.* **2021**, *20*, 1033–1054. [CrossRef]
17. Trajanoska, K.; Bhérer, C.; Taliun, D.; Zhou, S.; Richards, J.B.; Mooser, V. From Target Discovery to Clinical Drug Development with Human Genetics. *Nature* **2023**, *620*, 737–745. [CrossRef]
18. Ye, J.; Wu, J.; Liu, B. Therapeutic Strategies of Dual-Target Small Molecules to Overcome Drug Resistance in Cancer Therapy. *Biochim. Biophys. Acta-Rev. Cancer* **2023**, *1878*, 188866. [CrossRef]
19. Chaachouay, N.; Zidane, L. Plant-Derived Natural Products: A Source for Drug Discovery and Development. *Drugs Drug Candidates* **2024**, *3*, 184–207. [CrossRef]
20. Gielecińska, A.; Kciuk, M.; Mujwar, S.; Celik, I.; Kołat, D.; Kałuzińska-Kołat, Ż.; Kontek, R. Substances of Natural Origin in Medicine: Plants vs. Cancer. *Cells* **2023**, *12*, 986. [CrossRef]
21. Balachandran, C.; Duraipandiyan, V.; Arun, Y.; Sangeetha, B.; Emi, N.; Al-Dhabi, N.A.; Ignacimuthu, S.; Inaguma, Y.; Okamoto, A.; Perumal, P.T. Isolation and Characterization of 2-Hydroxy-9,10-Anthraquinone from *Streptomyces olivochromogenes* (ERINLG-261) with Antimicrobial and Antiproliferative Properties. *Rev. Bras. Farmacogn.* **2016**, *26*, 285–295. [CrossRef]
22. Buitrago, D.; Buitrago-Villanueva, I.; Barbosa-Cornelio, R.; Coy-Barrera, E. Comparative Examination of Antioxidant Capacity and Fingerprinting of Unfractionated Extracts from Different Plant Parts of Quinoa (*Chenopodium quinoa*) Grown under Greenhouse Conditions. *Antioxidants* **2019**, *8*, 238. [CrossRef] [PubMed]
23. Pluskal, T.; Castillo, S.; Villar-Briones, A.; Orešič, M. MZmine 2: Modular Framework for Processing, Visualizing, and Analyzing Mass Spectrometry-Based Molecular Profile Data. *BMC Bioinform.* **2010**, *11*, 395. [CrossRef] [PubMed]
24. Wu, W.-H.; Kang, Y.-P.; Wang, N.-H.; Jou, H.-J.; Wang, T.-A. Sesame Ingestion Affects Sex Hormones, Antioxidant Status, and Blood Lipids in Postmenopausal Women. *J. Nutr.* **2006**, *136*, 1270–1275. [CrossRef] [PubMed]
25. Tuchinda, P.; Kornsakulkarn, J.; Pohmakotr, M.; Kongsaeree, P.; Prabpai, S.; Yoosook, C.; Kasisit, J.; Napaswad, C.; Sophasan, S.; Reutrakul, V. Dichapetalin-Type Triterpenoids and Lignans from the Aerial Parts of *Phyllanthus acutissima*. *J. Nat. Prod.* **2008**, *71*, 655–663. [CrossRef]
26. Desai, D.C.; Jacob, J.; Almeida, A.; Kshirsagar, R.; Manju, S.L. Isolation, Structural Elucidation and Anti-Inflammatory Activity of Astragalin, (−)-Hinokinin, Aristolactam I and Aristolochic Acids (I & II) from *Aristolochia indica*. *Nat. Prod. Res.* **2014**, *28*, 1413–1417. [CrossRef]
27. Fukamiya, N.; Lee, K.-H. Antitumor Agents, 81. Justicidin-A and Diphyllin, Two Cytotoxic Principles from *Justicia procumbens*. *J. Nat. Prod.* **1986**, *49*, 348–350. [CrossRef]
28. Wang, C.L.J.; Ripka, W.C. Total Synthesis of (±)-Justicidin P. A New Lignan Lactone from *Justicia extensa*. *J. Org. Chem.* **1983**, *48*, 2555–2557. [CrossRef]
29. Lim, Y.Y.; Murtijaya, J. Antioxidant Properties of *Phyllanthus amarus* Extracts as Affected by Different Drying Methods. *LWT-Food Sci. Technol.* **2007**, *40*, 1664–1669. [CrossRef]
30. Liu, X.; Cui, C.; Zhao, M.; Wang, J.; Luo, W.; Yang, B.; Jiang, Y. Identification of Phenolics in the Fruit of Emblica (*Phyllanthus emblica* L.) and Their Antioxidant Activities. *Food Chem.* **2008**, *109*, 909–915. [CrossRef]
31. Bansal, V.; Sharma, A.; Ghanshyam, C.; Singla, M.L. Rapid HPLC Method Development for Determination of Vitamin C, Phenolic Acids, Hydroxycinnamic Acid and Flavonoids in *Emblica officinalis* Juice. *J. Liq. Chromatogr. Relat. Technol.* **2015**, *38*, 619–624. [CrossRef]
32. Bansal, V.; Sharma, A.; Ghanshyam, C.; Singla, M.L. Coupling of Chromatographic Analyses with Pretreatment for the Determination of Bioactive Compounds in *Emblica officinalis* Juice. *Anal. Methods* **2014**, *6*, 410–418. [CrossRef]
33. Fang, S.; Koteswara Rao, Y.; Tzeng, Y. Anti-Oxidant and Inflammatory Mediator's Growth Inhibitory Effects of Compounds Isolated from Phyllanthus Urinaria. *J. Ethnopharmacol.* **2008**, *116*, 333–340. [CrossRef]
34. Worley, B.; Powers, R. Multivariate Analysis in Metabolomics. *Curr. Metabolomics* **2013**, *1*, 92–107. [CrossRef] [PubMed]
35. Zhu, C.-S.; Lin, Z.-J.; Xiao, M.-L.; Niu, H.-J.; Zhang, B. The Spectrum-Effect Relationship—A Rational Approach to Screening Effective Compounds, Reflecting the Internal Quality of Chinese Herbal Medicine. *Chin. J. Nat. Med.* **2016**, *14*, 177–184. [CrossRef] [PubMed]
36. MacCallum, R.C.; Zhang, S.; Preacher, K.J.; Rucker, D.D. On the Practice of Dichotomization of Quantitative Variables. *Psychol. Methods* **2002**, *7*, 19–40. [CrossRef] [PubMed]
37. Altman, D.G.; Royston, P. The Cost of Dichotomising Continuous Variables. *BMJ* **2006**, *332*, 1080. [CrossRef]

38. Caesar, L.K.; Kellogg, J.J.; Kvalheim, O.M.; Cech, N.B. Opportunities and Limitations for Untargeted Mass Spectrometry Metabolomics to Identify Biologically Active Constituents in Complex Natural Product Mixtures. *J. Nat. Prod.* **2019**, *82*, 469–484. [CrossRef]
39. Britton, E.R.; Kellogg, J.J.; Kvalheim, O.M.; Cech, N.B. Biochemometrics to Identify Synergists and Additives from Botanical Medicines: A Case Study with Hydrastis Canadensis (Goldenseal). *J. Nat. Prod.* **2018**, *81*, 484–493. [CrossRef]
40. Liu, G.; Wu, J.; Si, J.; Wang, J.; Yang, M. Complete Assignments of $^{1}$H and $^{13}$C NMR Data for Three New Arylnaphthalene Lignan from *Justicia procumbens*. *Magn. Reson. Chem.* **2008**, *46*, 283–286. [CrossRef]
41. da Silva, R.; Maia Ruas, M.; Marcos Donate, P. Complete Assignments of $^{1}$H and $^{13}$C NMR Spectral Data for Arylnaphthalene Lignan Lactones. *Magn. Reson. Chem.* **2007**, *45*, 902–904. [CrossRef]
42. Marcotullio, M.C.; Pelosi, A.; Curini, M. Hinokinin, an Emerging Bioactive Lignan. *Molecules* **2014**, *19*, 14862–14878. [CrossRef] [PubMed]
43. Cunha, N.L.; Teixeira, G.M.; Martins, T.D.; Souza, A.R.; Oliveira, P.F.; Símaro, G.V.; Rezende, K.C.S.; Gonçalves, N.D.S.; Souza, D.G.; Tavares, D.C.; et al. (−)-Hinokinin Induces G2/M Arrest and Contributes to the Antiproliferative Effects of Doxorubicin in Breast Cancer Cells. *Planta Medica* **2016**, *82*, 530–538. [CrossRef] [PubMed]
44. Ai, Z.; Zhou, S.; Wu, S.; Hong, Z.; Zhang, Y.; Wang, M.; Yang, Y.; Wu, H. Justicidin B Inhibits PDGF-BB-Induced Proliferation and ECM Accumulation in Mesangial Cells via Nrf2/HO-1 and Akt/MTOR Signaling Pathway. *Nat. Prod. Commun.* **2020**, *15*, 1934578X20923821. [CrossRef]
45. Tajuddeen, N.; Muyisa, S.; Maneenet, J.; Nguyen, H.H.; Naidoo-Maharaj, D.; Maharaj, V.; Awale, S.; Bringmann, G. Justicidin B and Related Lignans from Two South African *Monsonia* Species with Potent Activity against HeLa Cervical Cancer Cells. *Phytochem. Lett.* **2024**, *60*, 234–238. [CrossRef]
46. Ren, Y.; Yuan, C.; Deng, Y.; Kanagasabai, R.; Ninh, T.N.; Tu, V.T.; Chai, H.-B.; Soejarto, D.D.; Fuchs, J.R.; Yalowich, J.C.; et al. Cytotoxic and Natural Killer Cell Stimulatory Constituents of *Phyllanthus songboiensis*. *Phytochemistry* **2015**, *111*, 132–140. [CrossRef]

*pharmaceutics*

MDPI

*Article*

# Targeting 7KCh-Induced Cell Death Response Mediated by p38, P2X7 and GSDME in Retinal Pigment Epithelium Cells with Sterculic Acid

**Ana Pariente [1], Rafael Peláez [1], Rodrigo Ochoa [1,2], Álvaro Pérez-Sala [1], Ángela Villanueva-Martínez [1], Miriam Bobadilla [1,*] and Ignacio M. Larráyoz [1,3,*]**

1 Biomarkers and Molecular Signaling Group, Neurodegeneration Area, Center for Biomedical Research of La Rioja (CIBIR), Piqueras 98, 26006 Logroño, Spain; apariente@riojasalud.es (A.P.); rpelaez@riojasalud.es (R.P.); rochoa.iacs@aragon.es (R.O.); aperez@riojasalud.es (Á.P.-S.); avmartinez@riojasalud.es (Á.V.-M.)

2 Proteomics Research Core Facility, Aragonese Institute of Health Sciences (IACS), San Juan Bosco 13, 50009 Zaragoza, Spain

3 Biomarkers, Artificial Intelligence and Signaling (BIAS), Department of Nursing, University of La Rioja, Duquesa de la Victoria 88, 26006 Logroño, Spain

* Correspondence: mbobadilla@riojasalud.es (M.B.); ignacio.larrayoz@unirioja.es (I.M.L.)

**Abstract:** Age-related macular degeneration (AMD) is the main cause of blindness in developed countries. AMD is characterized by the formation of drusen, which are lipidic deposits, between retinal pigment epithelium (RPE) and the choroid. One of the main molecules accumulated in drusen is 7-Ketocholesterol (7KCh), an oxidized-cholesterol derivative. It is known that 7KCh induces inflammatory and cytotoxic responses in different cell types and the study of its mechanism of action is interesting in order to understand the development of AMD. Sterculic acid (SA) counteracts 7KCh response in RPE cells and could represent an alternative to improve currently used AMD treatments, which are not efficient enough. In the present study, we determine that 7KCh induces a complex cell death signaling characterized by the activation of necrosis and an alternative pyroptosis mediated by P2X7, p38 and GSDME, a new mechanism not yet related to the response to 7KCh until now. On the other hand, SA treatment can successfully attenuate the activation of both necrosis and pyroptosis, highlighting its therapeutic potential for the treatment of AMD.

**Keywords:** 7-ketocholesterol; sterculic acid; AMD; retina; cell death; P2X7; GSDME

## 1. Introduction

Age-related macular degeneration (AMD) is the leading cause of legal blindness in people older than 65 years in developed countries. This disease affects the macula, is related to aging and is characterized by a gradual loss of central vision [1–4]. There are mainly two types of AMD, called dry AMD and wet AMD, of which dry AMD is the milder and more common form of the disease [1,5,6]. As a consequence of aging, waste material not digested by retinal pigment epithelium cells (RPE) accumulates around it, forming yellowish specific extracellular deposits called drusen, which is the first hallmark of the disease [6–10]. Given the importance of the RPE in drusen formation, currently, much of the research focuses on this part of the retina [11]. The increase in the number and size of drusen causes the molecules inside them, mainly lipid compounds, cholesterol derivatives and unfolded-oxidized proteins, to start promoting different inflammatory and cytotoxic responses that damage the RPE, triggering the development of dry AMD [3,8,10,12,13]. Due to its slow progress, dry AMD does not usually provoke the complete loss of vision [3].

Wet AMD, by contrast, is the most severe form of AMD and it is characterized by the abnormal growth of choroidal vessels from the macula through the RPE in a process called choroidal neovascularization (CNV) [1,3,6,11]. At some point, during this process, blood and fluids will leak, damaging the macula and causing the irreversible loss of central vision [5,6]. Anti-VEGF therapies are the only treatment currently available for AMD patients, which are not exempt from serious side effects. In addition, there are patients who do not respond well or can develop resistance, causing a decrease in the efficacy of these treatments [1,5,6,14–16]. For these reasons, there is an urgent need to search for new therapeutic targets that allow the development of new therapies.

One of these alternative therapeutic targets is 7-Ketocholesterol (7KCh), an oxidized form of cholesterol or oxysterol strongly connected with the development of AMD, as it forms and accumulates in drusen as a consequence of aging [17–22]. It is known that 7KCh can trigger oxidative stress, inflammation and cell death responses in different cell types, including retinal cells, although its response seems to depend on the cell type and the dose used [13,23–34]. The investigation of the mechanism of action of 7KCh in retinal cells could provide a new perspective on the molecular mechanisms that can lead to the development of AMD.

On the other hand, sterculic acid (SA), a natural cyclopropane fatty (Figure 1) acid obtained from the plant *Sterculia foetida*, has been described as the most potent functional antagonist of 7KCh in retinal cells [23,24,33–35], being able to protect RPE cells against 7KCh deleterious effects at low concentrations, as well as being able to reduce laser-induced choroidal neovascularization in a rat model [23,24,34]. These results suggest that SA could be a good alternative to the treatment of ocular diseases, such as AMD [35,36]. SA is known to inhibit Stearoyl coenzyme-A desaturase 1 (SCD1) activity, an enzyme from de novo lipogenesis, both in vitro and in vivo [37–39]. However, its protective effect against the 7KCh response is independent of the inhibition of SCD1 [35,36].

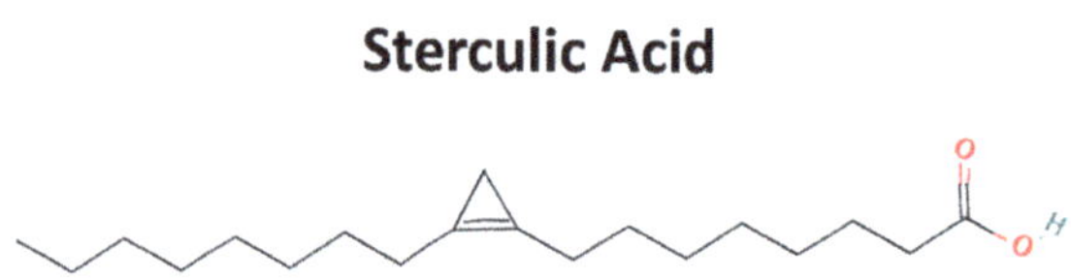


**Figure 1.** Chemical structure of sterculic acid (image obtained from Pubchem, https://pubchem.ncbi.nlm.nih.gov/compound/Sterculic-acid#section=2D-Structure, accessed on 26 October 2023, CID: 12921).

In our previous work, we performed genome-wide transcriptomic analysis in RPE cells in order to unravel the molecular pathways induced by 7KCh in these cells, as well as the mechanisms by which SA is able to protect against this response. We found that 7KCh exposure modulates the expression of several genes associated with lipid metabolism, reticulum stress, inflammation and cell death and activates signaling pathways related to these genes, such as unfolded protein response (UPR), the classical MAPK pathway and the JNK and p38 pathways. Co-treatment with SA significantly reversed gene expression alterations as well as attenuated the activation of these pathways [34]. In the present work, we continued with the study of the signaling induced by 7KCh in RPE cells, focusing on the cell death pathways activated by this oxysterol. Similarly, we have evaluated the protective effect of SA on the 7KCh-induced cytotoxic response.

## 2. Materials and Methods

### 2.1. Cell Lines and Culture

mRPE cells were obtained as a gift from Dr. SP Becerra, from the National Eye Institute (NIH, Bethesda, MD, USA). These cells came from the eyes of 3- to 5-year-old Rhesus monkeys, and were maintained for two weeks in a monolayer until biochemical and physiological markers characteristic of differentiated tissue were expressed [40]. DMEM/F12 1:1 medium (Hyclone-Thermo Scientific, Waltham, MA, USA) was used to grow the cells. Cell culture medium was supplemented with 5% fetal bovine serum (Invitrogen, Alcobendas, Madrid, Spain), 1.5% of sodium pyruvate, 1% of penicillin/streptomycin and 1% of non-essential amino-acids (Hyclone-Thermo Scientific, Waltham, MA, USA). Cultured cells were kept in an atmosphere of 37 °C, 5% $CO_2$ and 85% humidity.

### 2.2. Cell Treatments

Cells were seeded in P100 plates for DNA, RNA and protein purification at a density of $1.2 \times 10^6$ cells/plate or in 12-well plates for cell viability assays at a density of 100,000 cells/plate. Cells were allowed to attach and 100% confluency was reached after 24 h after seeding. Then, the culture media was replaced with serum-free media for 24 h and, after that, treatments were added. Cells were treated with 15–20 μM 7KCh (Sigma-Aldrich, Madrid, Spain) alone or with 10 μM SA (chemically synthesized by the Center for Applied Chemistry and Biotechnology, PPQF, University of Alcalá, Alcalá de Henares, Spain), 1–10 μg/mL og Cycloheximide (CHX, Sigma-Aldrich, Madrid, Spain), 5–25 μM of Caspase 3 inhibitor Z-DEVD-FMK (Merck, Darmstadt, Germany), 3–10 μM of RIPK1 inhibitor Necrostatin-1 (Nec-1, Cayman Chemical, Ann Arbor, MI, USA), 0.5–2 μM of MLKL inhibitor Necrosulfonamide (NSA, Tocris, Bristol, UK), 5 μM of P2X7 inhibitor A-839977 (Santa Cruz Biotechnology, Dallas, TX, USA), 25–75 μM of pannexin-1 inhibitor Carbenoxolone (CBX, Sigma-Aldrich, Madrid, Spain), 20 μM of JNK inhibitor SP600125 (StressMarq Bioscience Inc., Victoria, BC, Canada) or 40 μM of p38 inhibitor SB203580 (Sigma-Aldrich, Madrid, Spain). Then, 7KCh was prepared in β-cyclodextrin (Sigma-Aldrich, Madrid, Spain) as previously described [13,24]; SA, Z-DEVD-FMK, Nec-1, NSA, A-839977, CBX, SP600125 and SB203580 were dissolved in DMSO (Sigma-Aldrich, Madrid, Spain) and CHX was dissolved in absolute ethanol.

### 2.3. Cell Viability Assays

CellTitter 96 Aqueous One Solution Reagent MTS assay (Promega, Madison, WI, USA) was used to validate the toxicity or protective effect of the different treatments following the manufacturer's guidelines. Cells were washed two times with 1X PBS and then MTS reagent was added at a ratio of 1:10 in culture medium. Plates were incubated at 37 °C and absorbance was measured at baseline and after 4 h of incubation using a Biotek Synergy H4 multi-mode plate reader (BioTek Instruments, Covina, CA, USA). The results were presented as a percentage of viability with respect to the control cells.

### 2.4. DNA Laddering

Cell extracts were obtained from cells in culture, collecting both floating cells and cells adhered to the plate. Once harvested, cells were centrifuged (1000 rpm, 5 min) to remove residual media and washed with 1× PBS at 4 °C. The resulting pellet was resuspended in 250 μL of Tris-EDTA (tris (hydroxymethyl) aminomethane-ethylenediaminetetraacetic acid)/Triton X-100 buffer (10 mM Tris pH 8, 1 mM EDTA and 0.2% *v*/*v* Triton X-100, Sigma-Aldrich, Madrid, Spain) to extract the DNA from the cells by lysis. Then, extracts were incubated for 10 min on ice and centrifuged for 15 min at 4 °C and 13,000× *g*. The supernatant was then incubated with 10 mg/mL RNAse-A (Macherey-Nagel, Dueren, Germany) at 37 °C for 1 h, and with 20 μM proteinase K (Abm, Richmond, BC, Canada) and 10% SDS (sodium dodecyl sulfate, Sigma-Aldrich, Madrid, Spain) at 50 °C for 1 h in order to remove RNA and protein debris. To precipitate the DNA, 0.1 volume of 5 M NaCl (Sigma-Aldrich, Madrid, Spain) and one volume of isopropanol (Sigma-Aldrich, Madrid, Spain) at −20 °C were added to the samples, which were incubated on ice for 10 min. Afterward, the samples were centrifuged for 15 min at 4 °C and 13,000× *g*, and the resulting pellet was resuspended in 12 μL of Tris-EDTA buffer (10 mM Tris pH 8, 1 mM EDTA). Finally, the DNA concentration was calculated using a NanoDrop 1000 spectrophotometer (Thermo Fisher Scientific, Madrid, Spain) using Tris-EDTA buffer as a blank.

The DNA laddering pattern of bands was visualized by running 1 μg of DNA from each sample mixed with loading buffer (6× Loading buffer, Thermo Fisher Scientific, Madrid, Spain) on 2% agarose gel electrophoresis in TAE buffer (Tris-Acetate-EDTA, Alfa Aesar, Haverhill, MA, USA) with GelRed® (10,000× GelRed Nucleid Acid Stain, Biotium, Fremont, CA, USA) as the nucleic intercalating agent. Gels were run for 3 h at 50 V and bands were visualized under UV light in a Gel-Doc EQ System Universal Hood II transilluminator (Bio-Rad, Hercules, CA, USA). Images were taken using the Quantity One 4.6.3. program (Bio-Rad, Hercules, CA, USA).

### 2.5. In Vivo Immunofluorescence

DNA fluorescent staining was used to study the morphology of the cell nucleus in order to identify the different types of cell death depending on the state of the DNA. Cells were stained in vivo with 1:2000 Hoechst 33342 (Sigma-Aldrich, Madrid, Spain), 1:2000 propidium iodide (Sigma-Aldrich, Alcobendas, Madrid, Spain) or 1:10,000 YO-PRO1 (Invitrogen, Alcobendas, Madrid, Spain). The fluorochromes were directly added to the medium, to avoid losing dead cells floating in the medium, and incubated for 20 min at 37 °C. Cells were visualized and photographed at various times using a TSC SP5 confocal microscope (Leica, Wetzlar, Germany) in the case of Hoechst and propidium iodide co-staining, or the Incucyte® Live-Cell Analysis System (Sartorius, Göttingen, Germany) in the case of propidium iodide and YO-PRO1 co-staining. Image analysis was carried out using the Image J 1.52h (Fiji, National Institute of Health, Bethesda, MD, USA) and Incucyte® 2022A (Sartorius, Göttingen, Germany) programs, respectively. The excitation/emission wavelengths were 350/461 nm for Hoechst, 535/625 nm for propidium iodide and 488/515 nm for YO-PRO1.

### 2.6. RNA Purification and Quantitative Real-Time PCR (qRT-PCR)

Total RNA was isolated from cell cultures in 1 mL of TRIzol (Invitrogen, Madrid, Spain) and purified using the RNAsy mini kit (Qiagen, Valencia, CA, USA) and DNAse I (Qiagen, Valencia, CA, USA) following the manufacturer's instructions. Then, reverse-transcription was carried out with 1 μg of total RNA, random primers and the SuperScript III kit (Invitrogen, Madrid, Spain) to obtain the first-strand cDNA. Then, cDNA was mixed with forward and reverse primers to a concentration of 0.3 μM (Table 1) and SYBR Green PCR Master Mix. Then, a 7300 Real-Time PCR System (Applied Biosystems, Madrid, Spain) was used to take quantitative measurements in the following cycling conditions: initial denaturation at 95 °C for 10 min, followed by 40 cycles of 95 °C for 15 s and 60 °C for 1 min. In order to validate the amplicon specificity, a dissociation curve was implemented from 60 °C to 95 °C. The interpolation of the Ct value from the corresponding standard line was used to calculate the gene expression levels using *18S* rRNA expression to normalize the results.

**Table 1.** Primers used in this work in mRPE cells.

| Gene | Oligonucleotide Sequence |
|---|---|
| *18S*—Forward | 5′-ATGCTCTTAGCTGAGTGTCCCG-3′ |
| *18S*—Reverse | 5′-ATTCCTAGCTGCGGTATCCAGG-3′ |
| *CASP1*—Forward | 5′-CCATTCCCCTCCTACCCTGA-3′ |
| *CASP1*—Reverse | 5′-TGCTTCGTCTTCCTTTTTGGC-3′ |
| *GSDMD*—Forward | 5′-CGAAGATCACGGGTGGGG-3′ |
| *GSDMD*—Reverse | 5′-GCAGGATTTTGTGCTCTGGC-3′ |
| *IL1B*—Forward | 5′-ACACTCCGGGATGCACAGC-3′ |
| *IL1B*—Reverse | 5′-CCCAAGGCCACAGGTATTTT-3′ |
| *IL18*—Forward | 5′-ACCAAGGAAATCGGCCCCTA-3′ |
| *IL18*—Reverse | 5′-CCATACCTCTAGGCTGGCTATC-3′ |
| *NLRP3*—Forward | 5′-GGCTGGAGCTGTTGAAATGG-3′ |
| *NLRP3*—Reverse | 5′-CCCTTTGCACGAAGTCCTCC-3′ |

### 2.7. Western Blotting

Protein extracts were obtained using RIPA buffer (Thermo Scientific, Madrid, Spain) containing protease (EDTA-free complete, Roche, Basilea, Switzerland) and phosphatase (PhosStop, Roche, Basilea, Switzerland) inhibitors from scraped cells in culture. The centrifugation of homogenates for 30 min at 15,000× g was performed to collect supernatants and the protein concentration was determined using a BCA kit (Pierce, Rockford, IL, USA) and a Biotek Synergy H4 multi-mode plate reader (BioTek Instruments, Covina, CA, USA), following the manufacturer's instructions. Samples were prepared by mixing 20 μg of protein with 4× sample buffer (Invitrogen, Madrid, Spain) and 10× sample-reducing agent (Invitrogen, Madrid, Spain), and were then heated at 70 °C for 10 min. Then, 4–12% SDS-polyacrylamide gels (Invitrogen, Madrid, Spain) were used to run both the samples and the molecular weight marker SeeBlue plus 2 prestained standard (Invitrogen, Madrid, Spain). Then, protein transfer was performed onto 0.2 μm polyvinylidene difluoride (PVDF) membranes (iBlot system, Invitrogen, Madrid, Spain) and blocking was carried out at room temperature for 2 h with 5% non-fat dry milk in 1× TBS (tris buffered saline) and 0.1% Tween (Sigma-Aldrich, Madrid, Spain).

For protein identification, membranes were incubated overnight at 4 °C with different primary antibodies: α-p-p38 1:5000 (#4511, Cell Signaling, Danvers, MA, USA), α-p38 1:5000 (#8690, Cell Signaling, Danvers, MA, USA), α-GSDMD 1:1000 (ab210070, Abcam, Cambridge, UK), α-GSDME 1:1000 (ab215191, Abcam, Cambridge, UK) or α-cleaved-Caspase 3 1:1000 (#9664, Cell Signaling, Danvers, MA, USA). The detection of α-Actin 1:10,000 (#A5441, Sigma-Aldrich, Madrid, Spain) was utilized to standardize protein levels. To visualize immunoreactivity, membranes were incubated with α-rabbit (#7074, Cell Signaling, Danvers, MA, USA) or α-mouse (715-035-1514, Jackson Immunoresearch Lab. West Grobe, PA, USA) and chemiluminescence Clarity Western ECL kit (Bio-Rad, Hercules, CA, USA), and exposed to X-ray films (Amersham Hyperfilm ECL, GE Healthcare, Buckinghamshire, UK) or the ChemiDoc Imaging System (Bio-Rad, Hercules, CA, USA).

### 2.8. Statistical Analysis

GraphPad Prism 6 was used to perform statistical analysis. Data were presented as means ± SEM (standard error of the means) and were considered statically significant when $p \leq 0.05$. Normally distributed data were analyzed by ANOVA testing followed by Sidak or Tukey's post hoc test, following program recommendations. Student's *t*-test was used for comparisons of two conditions.

## 3. Results

### 3.1. 7KCh-Induced Cell Death Is Not Mediated by Apoptosis in mRPE Cells

We treated mRPE cells with 15–20 μM 7KCh with or without 10 μM SA for 24 h (Figure 2) in order to validate the toxicity effect of 7KCh and the protection of SA, as previously described [34,35]. Once again, a dose-dependent decrease in cell viability was observed with 7KCh treatment, as well as a decrease in this effect when adding SA. With the aim to find out the type of cell death induced by 7KCh in mRPE cells, we evaluated the participation of apoptosis in this cytotoxic response, as in our previous work, we showed an alteration in the expression of several genes involved in this pathway, such as *APAF1*, *GADD45A*, *PMAIP1* or *TP53I3* [34]. First, cells were treated with the Caspase 3 inhibitor Z-DEVD-FMK in the concentration range of 5–25 μM and 15 μM 7KCh (Figure 3A). No protective effect was observed over 7KCh-induced cell death. Since apoptosis involves the synthesis of specific proteins to be activated [41,42], such as PUMA or NOXA, the next approach was to inhibit protein synthesis with CHX. Treatment with 1–10 μg/mL CHX for 24 h could not protect mRPE cells against the toxicity of 15 μM 7KCh (Figure 3B). Lastly, we evaluated whether treatment with 15–20 μM 7KCh with or without 10 μM SA caused DNA laddering fragmentation, characteristic of apoptosis, in mRPE cells. However, the ladder fragmentation pattern observed in the positive control (1% $v/v$ ethanol) did not occur in 7KCh-treated cells (Figure 3C), indicating that apoptosis is not induced in response to 7KCh in mRPE cells.

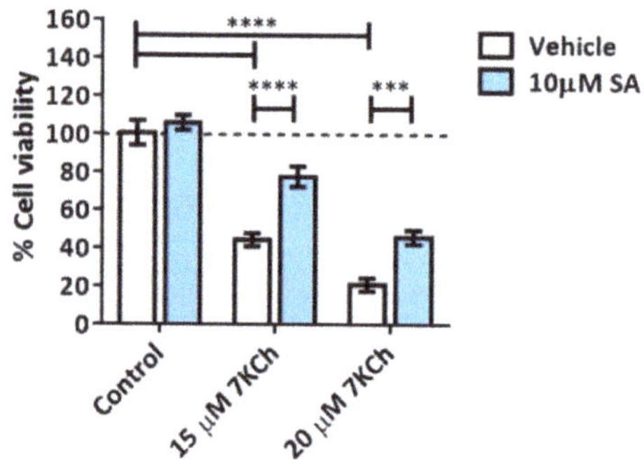


**Figure 2.** Cell toxicity induced by 7KCh and protective effect exerted by SA in mRPE cells. Cell viability was determined by MTS assay in mRPE cells treated with 15–20 μM 7KCh and 10 μM SA for 24 h. Data are expressed over the control and presented as a mean ± SEM of at least three different experiments. The dashed line is a guidance mark of the control value. The ANOVA test was used for statistical analysis, followed by the Sidak post hoc test. *** $p < 0.001$; **** $p < 0.0001$.

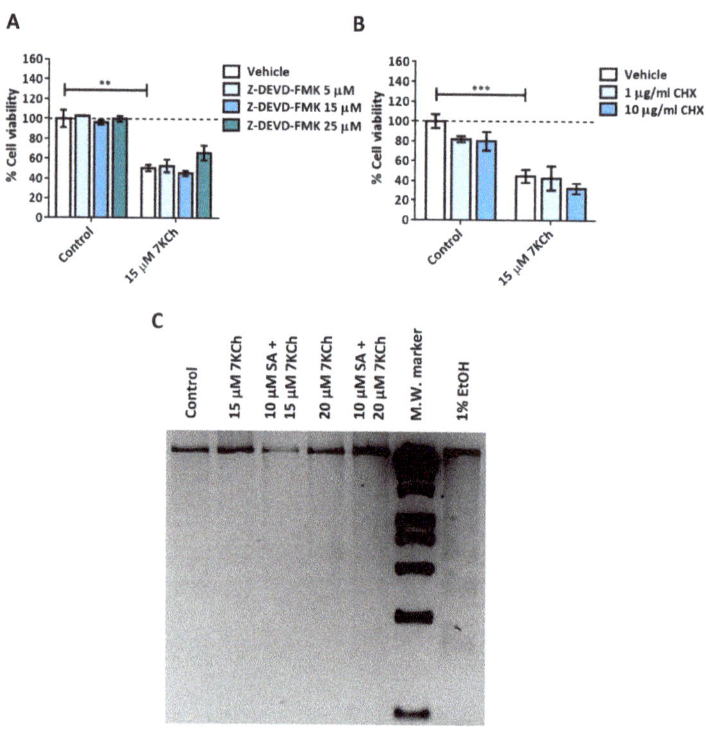


**Figure 3.** Participation of apoptosis in the toxic response induced by 7KCh in retinal cells. (**A**) Cell viability was determined by MTS assay in mRPE cells treated with 15 μM 7KCh and 5–25 μM Z-DEVD-FMK for 24 h. (**B**) Cell viability was determined by MTS assay in mRPE cells treated with 15 μM 7KCh and 1–10 μg/mL CHX for 24 h. (**C**) DNA laddering was detected by electrophoresis in mRPE cells treated with 15–20 μM 7KCh and 10 μM SA for 24 h. CHX and Z-DEVD-FMK were added with pretreatment for 2 h with respect to 7KCh. M.W., molecular weight. Data are expressed over the control and represented as mean ± SEM of at least three different experiments. The dashed line is a guidance mark of the control value. The ANOVA test was used for statistical analysis, followed by the Sidak post hoc test. ** $p < 0.01$; *** $p < 0.001$.

### 3.2. Activation of Necrosis in Response to 7KCh in mRPE Cells

We then studied the role of necrosis in the toxicity mechanisms of 7KCh in mRPE cells. The control and 15 μM 7KCh-treated cells were stained with Hoechst and propidium iodide to assess for the presence of necrotic nuclei (Figure 4A). The percentage of mRPE cells exposed to 7KCh positive for propidium iodide was approximately 55% with respect to the total number of cells observed quantified with Hoechst staining (Figure 4B). In addition, the nucleus of these cells presented a ring shape (white arrows), with the genetic material being released from the nucleus (Figure 4A), as occurs in necrosis. In order to verify the implication of necrosis processes in the 7KCh-induced response, mRPE cells were also treated with two necroptosis inhibitors: 3–10 μM Necrostatin-1 (Nec-1) as the RIPK1 inhibitor (Figure 4C), and 0.5–2 μM Necrosulfonamide (NSA) as the MLKL inhibitor (Figure 4D). Both inhibitors could successfully attenuate 15 μM 7KCh-induced toxicity. Taken together, these results indicate that at least a significant part of the 7KCh-induced cell death in mRPE cells is mediated by necrotic processes.

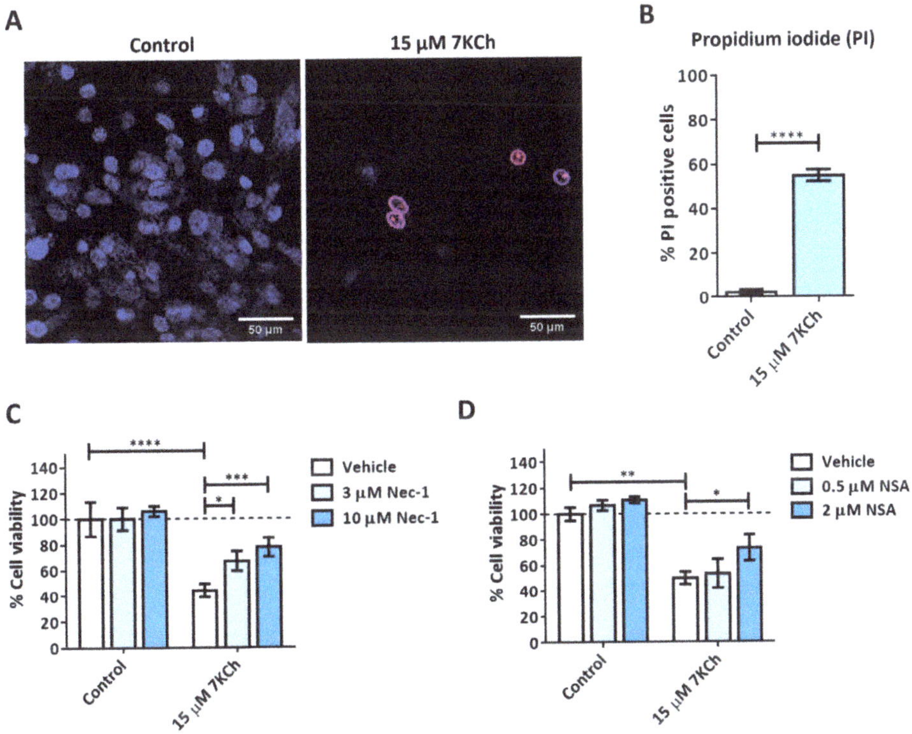


**Figure 4.** 7KCh-induced necrosis in mRPE cells. (**A**) Representative images obtained in vivo by confocal microscopy in the control and 7KCh-exposed mRPE cells for 24 h stained with Hoechst (blue) and propidium iodide (IP, red). White arrow indicates an example of necrotic core. (**B**) Quantification of the percentage of nuclei stained with propidium iodide. Data expressed as the mean ± SEM of seven independent fields. (**C**) Cell viability determined by MTS assay in mRPE cells trated with 15 μM 7KCh and 3–10 μM Nec-1 or (**D**) 0.5–2 μM NSA. Data are expressed over the control and represented as mean ± SEM of at least three different experiments. Dashed line is a guidance mark of the control value. The ANOVA test was used for statistical analysis, followed by the Sidak post hoc test. * $p < 0.05$; ** $p < 0.01$; *** $p < 0.001$; **** $p < 0.0001$.

### 3.3. The P2X7 Receptor Is Involved Is the Cell Death Response to 7KCh in mRPE Cells and Participates in p38 Activation and in Necrotic Processes

A pannexin-1-independent involvement of the P2X7 purinoceptor in the 7KCh-initiated necrosis response has been previously described in ARPE-19 cells [22], so we decided to check whether the same occurs in the mRPE cell line. Cells were treated for 24 h with the P2X7 inhibitor A-839977 at a concentration of 5 μM or with the pannexin-1 inhibitor Carbenoxolone (CBX) at concentrations of 25–75 μM in combination with 15–20 μM 7KCh, and cell viability was evaluated with MTS assay. The addition of A-839977 (Figure 5A), but not CBX (Figure 5B), protected mRPE cells against 7KCh cell death, indicating the participation of P2X7 in the cytotoxic response induced by 7KCh, but with the independence of pannexin 1.

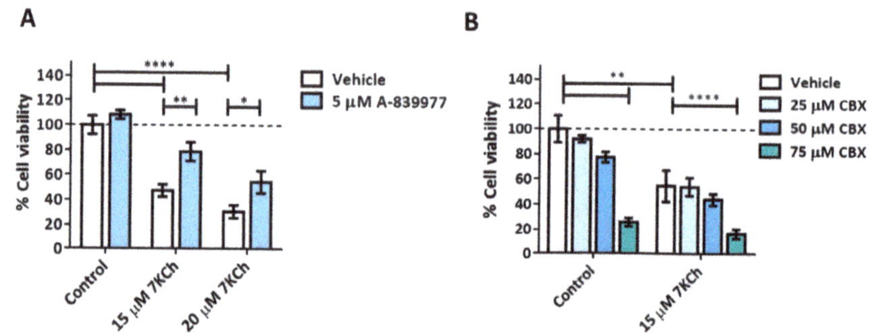


**Figure 5.** Effect of P2X7 inhibition on 7KCh-induced cell death. (**A**) Cell viability was determined by MTS assay in mRPE cells treated with 15–20 μM 7KCh and 5 μM A-839977 for 24 h. (**B**) Cell viability was determined by MTS assay in mRPE cells treated with 15 μM 7KCh and 25–75 μM CBX. Data are expressed over the control and represented as mean ± SEM of at least three different experiments. The dashed line is a guidance mark of the control value. The ANOVA test was used for statistical analysis, followed by the Sidak post hoc test. * $p < 0.05$; ** $p < 0.01$; **** $p < 0.0001$.

To further study the role of P2X7 in cell death induced by 7KCh, YO-PRO1 was used to label the cells, as this fluorochrome is capable of entering through the pores formed by this receptor when activated [43]. Cells were treated with 5 μM A-839977 and 15 μM 7KCh, and then simultaneously stained with propidium iodide and YO-PRO1 (Figure 6A). Cells positive for propidium iodide (red), YO-PRO1 (green) or both fluorochromes (merge) were quantified with respect to the total number of cells observed in the brightfield at 6 h, 12 h and 24 h. In mRPE cells treated with 7KCh, an increase in the number of nuclei positive for propidium iodide and/or YO-PRO1 was observed (Figure 6B). Moreover, the internalization of YO-PRO1 occurred before the internalization of propidium iodide since, at all times, the evaluated percentage of cells stained with YO-PRO1 was higher than that of cells stained with propidium iodide, although the difference narrowed over time. Approximately 8% of cells stained with propidium iodide and 20% with YO-PRO1 were observed at 6 h; 20% of cells were stained with propidium iodide and 33% with YO-PRO1 at 12 h; and 45% of cells were positive for propidium iodide and 60% for YO-PRO1 at 24 h. In addition, practically all cells positive for propidium iodide were also positive for YO-PRO1 (Figure 6B, merge). P2X7 inhibition with A-839977 significantly reduced the number of cells stained with both propidium iodide and YO-PRO1 at 12 h and 24 h (also at 6 h in the case of YO-PRO1). This indicates that in response to 7KCh, the P2X7 receptor is activated before necrosis begins and, in turn, participates in its induction. Furthermore, the percentage of cells solely stained with YO-PRO1 suggests the involvement of at least a second type of cell death, in addition to necrosis, in response to 7KCh-induced toxicity in mRPE cells.

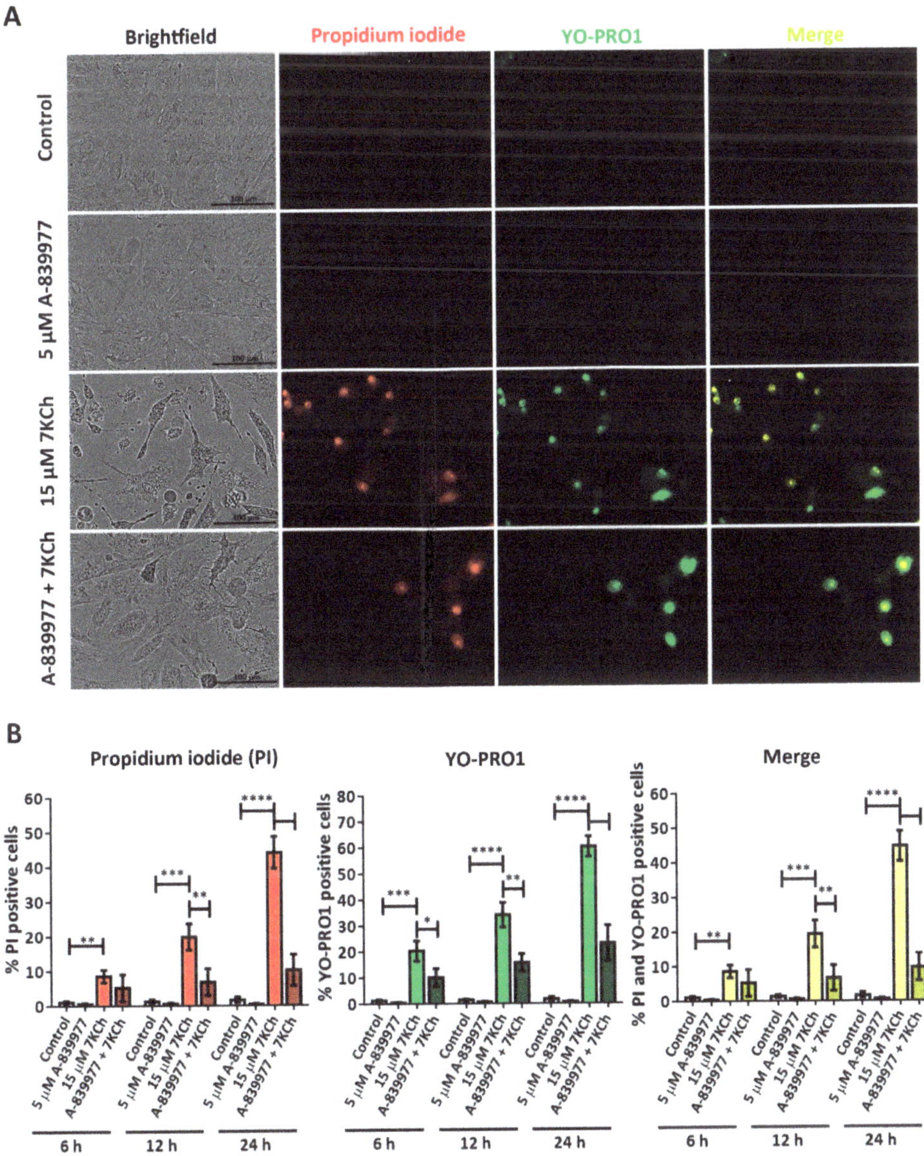


**Figure 6.** Involvement of the P2X7 receptor in 7KCh-induced toxicity in mRPE cells stained with propidium iodide (red) and YO-PRO 1 (green). (**A**) Representative images obtained in vivo in brightfield and fluorescence of mRPE cells treated with 15 µM 7KCh and/or 5 µM A-839977 for 24 h using Incucyte® equipment. Merge (yellow) represents the combination of propidium iodide and YO-PRO1 fluorescence. (**B**) Percentage of cells stained with propidium iodide (IP), YO-PRO1 and both fluorochromes (merge) at 6 h, 12 h and 24 h of treatment with respect to the total cell number quantified in brightfield. Data are expressed as the mean ± SEM of four independent fields. The ANOVA test was used for statistical analysis, followed by Tukey's post hoc test. * $p < 0.05$; ** $p < 0.01$; *** $p < 0.001$; **** $p < 0.0001$.

### 3.4. SA Prevents the Activation of Necrosis and P2X7 Induced by 7KCh in mRPE Cells

To verify the ability of SA to inhibit necrosis and P2X7 activation, mRPE cells were treated with 10 µM and 15 µM 7KCh and simultaneously stained with propidium iodide and YO-PRO1 (Figure 7A). The total number of cells labeled with each one of the fluorochromes was counted, again, at 6 h, 12 h and 24 h with respect to the total number of cells observed in brightfield (Figure 7B). The percentages of cells positive for propidium iodide and YO-PRO1 with 7KCh treatment were similar to those observed in Figure 6. The addition of 10 µM SA significantly decreased the entry of propidium iodide in evaluated mRPE cells at all times, as well as the internalization of YO-PRO1 after 12 h (Figure 7B). These data show that SA can attenuate both 7KCh-induced necrosis and P2X7 activation in mRPE cells due to the decreased internalization of propidium iodide and YO-PRO1, respectively.

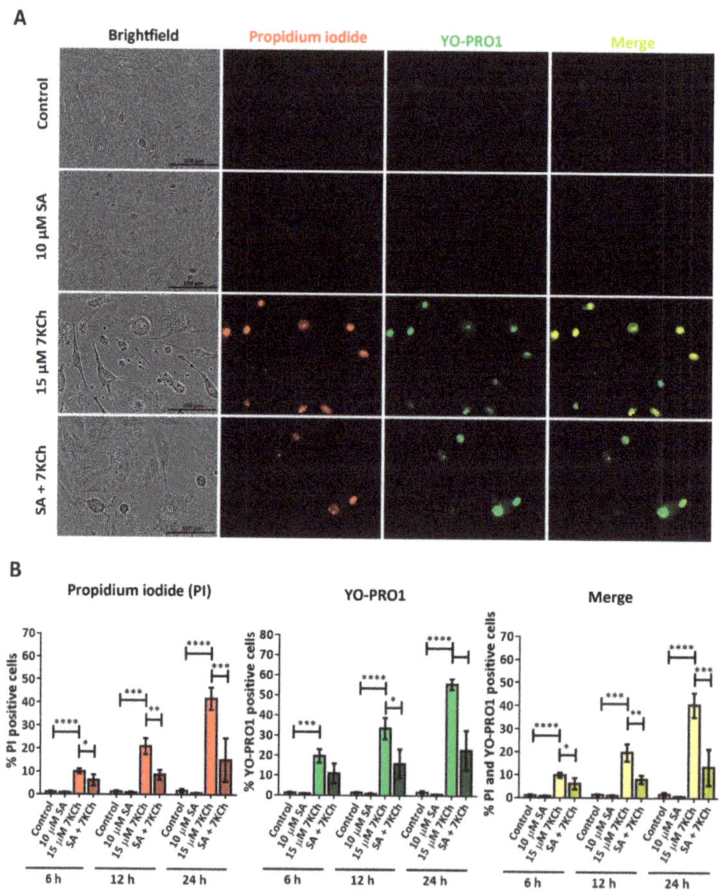


**Figure 7.** SA attenuation of necrosis and P2X7 activation induced in response to 7KCh in mRPE cells stained with propidium iodide (red) and YO-PRO1 (green). (**A**) Representative images obtained in vivo in brightfield and fluorescence of mRPE cells treated with 15 µM 7KCh and/or 10 µM SA for 24 h using Incucyte® equipment. Merge (yellow) represents the combination of propidium iodide and YO-PRO1 fluorescence. (**B**) Percentage of cells stained with propidium iodide (IP), YO-PRO1 and both fluorochromes (merge) at 6 h, 12 h and 24 h of treatment with respect to the total cell number quantified in brightfield. Data are expressed as the mean ± SEM of four independent fields. The ANOVA test was used for statistical analysis, followed by Tukey's post hoc test. * $p < 0.05$; ** $p < 0.01$; *** $p < 0.001$; **** $p < 0.0001$.

3.5. *7KCh-Induced p38 Phosphorylation Is Mediated by P2X7 Activation*

As in our previous work, the participation of p38 in the induction of cell death in response to 7KCh in mRPE cells was observed [34]. Therefore, we decided to assess the effect of P2X7 inhibition on p38 activation. The p38 phosphorylation status was evaluated by Western blot in mRPE cells treated with 20 µM 7KCh in combination with 10 µM SA and 5 µM A-839977 for 24 h (Figure 8A). A significant decrease in p38 phosphorylation levels was observed with the addition of both SA and A-839977 compared to 7KCh alone (Figure 8B). Although the effect of P2X7 inhibition on p38 activation was less than with SA treatment, these results indicate that at least an important part of p38 phosphorylation is mediated by the activation of P2X7.

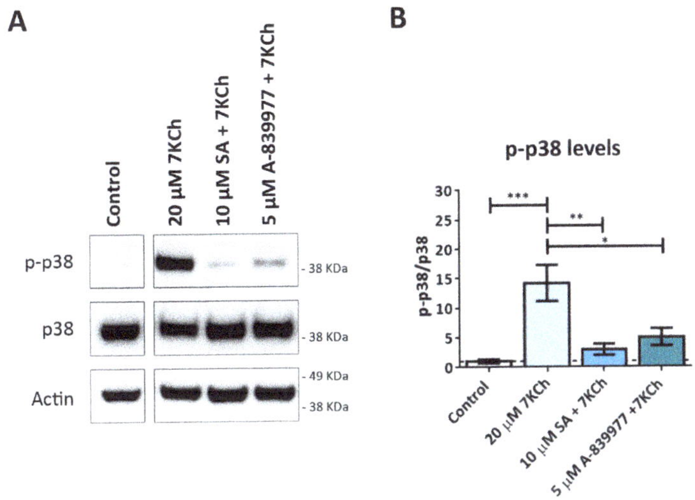


**Figure 8.** Effect of P2X7 inhibition on 7KCh-induced p38 phosphorylation. (**A**) Western blot estimation of phosphorylated p38 (p-p38) levels in mRPE cells exposed to 20 µM 7KCh alone or in combination with 10 µM SA or 5 µM A-839977 for 24 h. (**B**) Quantification of p-p38 levels over p38 levels. Data are represented as mean ± SEM of three different experiments. The dashed line is a guidance mark of the control value. The ANOVA test was used for statistical analysis, followed by Tukey's post hoc test. * $p < 0.05$; ** $p < 0.01$; *** $p < 0.001$.

3.6. *7KCh Treatment Does Not Induce Classical Pyroptosis Mediated by GSDMD and NLRP3 in mRPE Cells*

The last type of cell death to be evaluated was pyroptosis, with which the activation of P2X7 has been also associated [44]. To validate the involvement of this signaling pathway, the expression levels of *CASP1*, *NLRP3*, *GSDMD*, *IL1B* and *IL18* were evaluated in mRPE cells treated with 15 µM 7KCh by RT-qPCR (Figure 9A). An increase in the expression levels of *CASP1* and *IL1B* was observed. No change was detected in *GSDMD* or *IL18* expression levels, and, in the case of *NLRP3*, its expression could not even be quantified. On the other hand, the inhibition of Caspase 1 was not able to attenuate either the inflammation or the toxicity induced by 15 µM 7KCh in mRPE cells. Furthermore, we could not detect GSDMD activation by Western blot (Figure 9B) in mRPE cells treated with 7KCh. In fact, the band corresponding to the complete protein of around 53 KDa could not be detected in any of the mRPE cell samples. This band could be observed in the human RPE cell line ARPE-19, used as the positive control. Instead, two smaller bands, possibly non-specific, were observed in the mRPE cell samples. Taken together, these results suggest that classical pyroptosis is not activated in the 7KCh-induced cytotoxic response in mRPE cells.

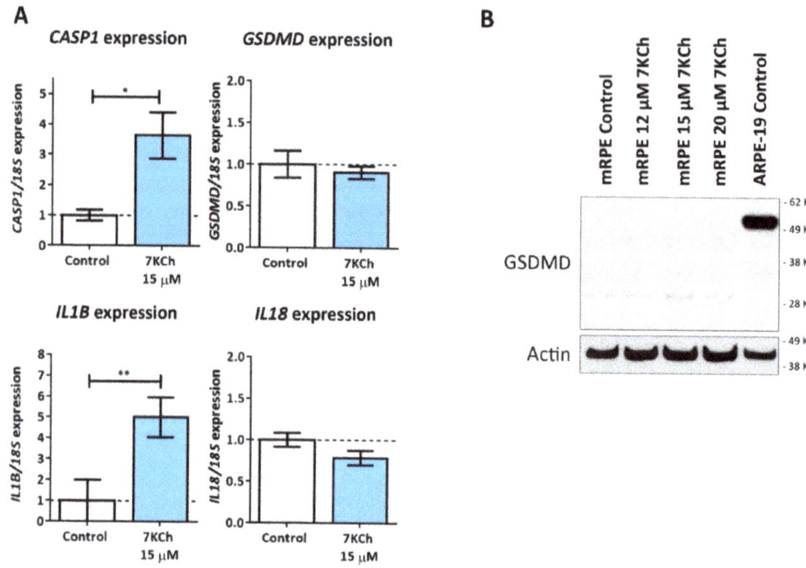


**Figure 9.** Classic pyroptosis in the cytotoxic response induced by 7KCh in mRPE cells. (**A**) Quantification of RT-qPCR of *CASP1, GSDMD, IL1B* and *IL18* expression levels in mRPE cells treated with 15 μM 7KCh for 24 h, normalized with respect to *18S* expression. (**B**) Western blot detection of GSDMD in mRPE cells exposed to 12–20 μM 7KCh. Statistical analysis in (**A**) was carried out using Student's *t*-test. Data are expressed over the control and represented as mean ± SEM of at least three different experiments. The dashed line is a guidance mark of the control value. * $p < 0.05$; ** $p < 0.01$.

### *3.7. 7KCh Induces an Alternative Pyroptosis Mechanism Mediated by the Activation of P2X7, p38 and GSDME*

The data obtained in our previous work from the transcriptome of mRPE cells showed an increase in the expression levels of the DFNA5 gene in cells treated with 15 μM 7KCh [34]. This gene codes for the GSDME protein, belonging to the same family of proteins as GSDMD and with a similar function, but with a difference in that it is activated by Caspase 3 [45]. The ability of 7KCh to promote the processing and activation of GSDME (N-GSDME) and Caspase 3 (C-Caspase 3) was verified by Western blot in mRPE cells treated with 20 μM alone or in combination with 10 μM SA for 6 h, 12 h and 24 h, or with the P2X7 inhibitor A-839977 at a concentration of 5 μM after 24 h of treatment (Figure 10A). The band corresponding to the active form of GSDME, of approximately 35 KDa, was observed after 12 h of treatment, with 20 μM 7KCh, reaching its highest levels of activation at 24 h. Caspase 3 activation was not detected until 24 h of exposition to 7KCh and, in addition, a long exposure of about 17–19 KDa was necessary to be able to visualize this band, showing a minimal activation of Caspase 3 despite the evident activation of GSMDE. Moreover, Caspase 3 inhibition could not prevent GSDME activation (Figure 10B), suggesting that there must be an alternative protease to Caspase 3 capable of promoting the cleavage of the N-terminus of GSDME. The addition of 10 μM SA reduced the activation of GSDME both at 12 h and 24 h, as well as Caspase 3 cleavage at 24 h. Treatment with 5 μM A-839977 could also prevent GSMDE and Caspase 3 activation at 24 h, validating the role of P2X7 in 7KCh-induced GSDME-mediated pyroptosis in mRPE cells.

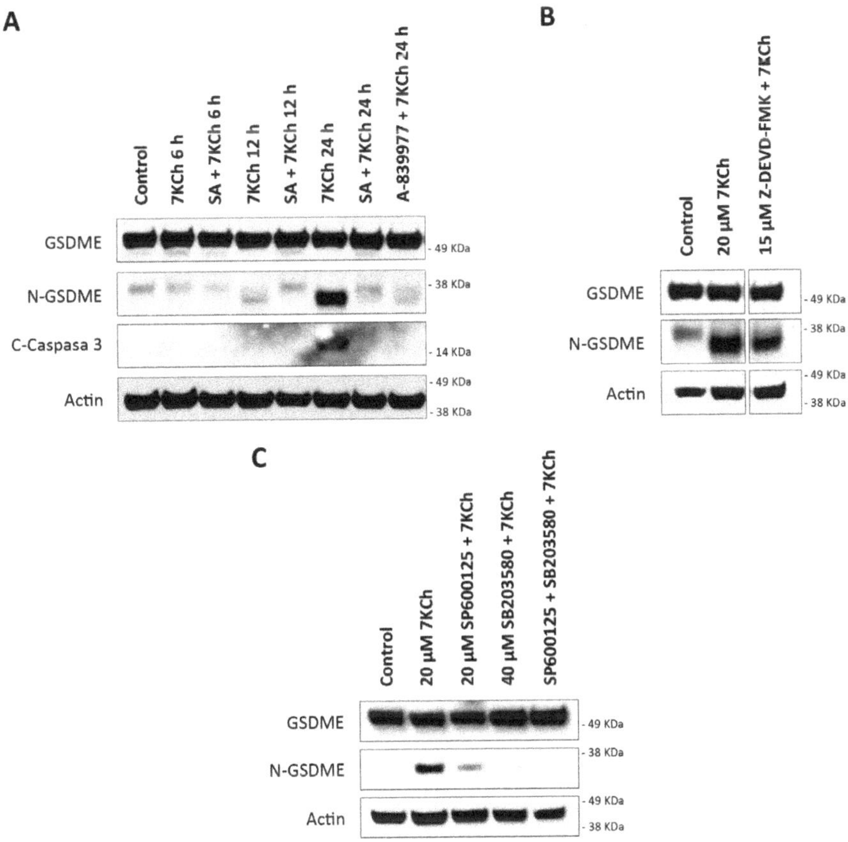


**Figure 10.** GSDME activation in mRPE cells exposed to 7KCh. (**A**) Western blot detection of GSDME cleavage (N-GSDME) and Caspase 3 cleavage (C-Caspase 3) in mRPE cells treated with 20 µM 7KCh alone or in combination with 10 µM SA at 6–24 h, and in combination with 5 µM A-839977 at 24 h. (**B**) Western blot detection of N-GSDME in mRPE cells treated with 20 µM 7KCh alone or in combination with 15 µM Z-DEVD-FMK for 24 h. (**C**) Western blot detection of N-GSDME in mRPE cells treated with 20 µM 7KCh alone or with 20 µM SP600125 and/or 40 µM SB203580 for 24 h. SP600125 and SB203580 were added with pretreatment for 2 h with respect to 7KCh.

On the other hand, the transcription of *DFNA5* is promoted by p53 [45]. Among the kinases capable of activating p53 by phosphorylation are both JNK and p38, previously described by our research group as being activated in response to 7KCh in mRPE cells [34]. The effect of JNK inhibition with 20 µM SP600125 and p38 inhibition with 40 µM SB203580 on GSMDE activation induced by 20 µM 7KCh was evaluated by Western blot (Figure 10C). Only treatment with SB203580, or with a combination of SP600125 with SB203580, could prevent GSDME activation. With SP600125 treatment, however, GSDME activation was still observed, albeit slightly less than with 7KCh treatment alone. This suggests that the p38-mediated phosphorylation of p53 could be responsible for the increased levels of *DFNA5* transcription and, taken together, these results show an alternative pyroptotic mechanism in response to 7KCh in mRPE cells mediated by the activation of P2X7, p38 and GSDME.

## 4. Discussion

In this work, we described a new mechanism of cell death induced by 7KCh in mRPE cells mediated by necrosis and alternative pyroptosis associated with GSDME. Once again, SA treatment was able to successfully attenuate this signaling, highlighting the importance of SA as a functional antagonist of 7KCh.

First of all, we evaluated the role of apoptosis in the cytotoxic response of 7KCh. Apoptosis and oxyapoptophagy are the main cell death mechanisms described for 7KCh, and, in general, for oxysterols [25,29,46–56]. Oxyapoptophagy is a type of cell death characteristic of oxysterols that combines the activation of apoptosis and autophagy [25,29,49–52,54,56]. In our previous work, we showed alterations in the expression of several genes associated with apoptosis with 7KCh-treatment in mRPE cells, such as *APAF1*, *GADD45A*, *PMAIP1* and *TP53I3* [34], suggesting that apoptosis is induced in response to 7KCh in these cells. However, the data obtained in the present work did not support the participation of apoptosis, and consequently, oxyapoptophagy in the cytotoxic response of 7KCh in retinal cells. To begin with, Caspase 3 inhibition with Z-DEVD-FMK could not attenuate 7KCh-induced cell death in mRPE cells (Figure 3B) and, despite the cleavage of Caspase 3 observed by Western blot (Figure 10A), long exposures of the films (more than 1 h of exposition) were needed in order to visualize the band, which also appeared when cell death had been already triggered (Figure 10A). This suggests that, although Caspase 3 is cleavage, its activation is not determinative for the induction of cell death in response to 7KCh. Moreover, when protein synthesis activation, which is needed for apoptosis activation [41,42], was inhibited with CHX, no change in cell viability was observed either (Figure 3B), nor could the DNA laddering fragmentation pattern characteristic of apoptosis be detected (Figure 3C). Therefore, although there might be some induction of apoptosis in response to 7KCh, its activation is late and not decisive for triggering cell death induced by this oxysterol in mRPE cells.

Given the absence of apoptosis, we next studied the role of necrosis in the response to 7KCh in retinal cells. Unlike apoptosis, there have not been many studies that describe an induction of necrosis in response to 7KCh [22,48,57], and, in some cases, it has been ruled out as a mechanism of cell death [51]. Treatment with 7KCh in mRPE cells increased the internalization and the number of nuclei stained with propidium iodide, which showed a necrotic morphology (Figure 4A). In addition, the percentage of nuclei stained with propidium iodide with 15 μM 7KCh treatment was approximately 50% (Figure 4B), coinciding with the percentage of cell death observed in mRPE cells by MTS assay with the same dose of 7KCh (Figure 2). Moreover, treatment with Nec-1 and NSA, two necroptosis inhibitors, could attenuate 7KCh-induced toxicity in mRPE cells (Figure 4C,D). This indicates that at least an important part of the cytotoxicity caused by 7KCh in mRPE cells is mediated by necrotic processes. On the other hand, this necrosis seems to be associated with P2X7 purinoceptor activation, since its inhibition with A-839977 protected mRPE cells from 7KCh-induced toxicity (Figure 5A) and significantly decreased the internalization of propidium iodide (Figure 6). P2X7-like purinoceptors are ATP-activated receptors that act as ion channels and form pores in the membrane, allowing the entry of several molecules such as the fluorescent probe YO-PRO1, which can be used to study the aperture of these receptors [43]. We observed, with exposure to 7KCh, an increase in the number of positive nuclei for YO-PRO1 after 6 h of treatment, and practically all of the labeling with propidium iodide colocalized with YO-PRO1 staining (Figures 6 and 7), validating P2X7 implication in necrosis activation induced by 7KCh in mRPE cells. SA treatment successfully reduced the internalization of both fluorochromes, implying an inhibition of both necrosis induction and P2X7 opening (Figure 7), and highlighting the therapeutic potential of this molecule.

P2X7 receptors are associated with the activation of inflammation through inflammasome formation and with various types of cell death, including apoptosis, necrosis and pyroptosis [22,43,44,58,59]. P2X7 involvement in 7KCh-induced necrosis has been previously described in ARPE-19 cells, which was independent of pannexin 1 activation [22], supporting our results. We also checked the effect of P2X7 inhibition on p38 activation, as we had previously described an implication of this protein in the induction of cell death promoted by 7KCh in mRPE cells [34]. A-839977 significantly reduced p38 phosphorylation levels induced in response to 7KCh (Figure 8), as well as SA treatment, in agreement with our previous work [34]. This indicates that an important part of p38 activation is dependent on the induction of P2X7 receptors, consistent with previous reports [43,60,61].

The last type of cell death to be evaluated was pyroptosis, since P2X7 activation has also been related to these mechanisms and pyroptotic cells are also labeled with propidium iodide [43,44,59]. Pyroptosis is a mechanism of inflammatory cell death and the canonical or classical pathway is characterized by the activation of NLRP3 inflammasome, which is responsible for the activation of Caspase 1. Once activated, Caspase 1 promotes the cleavage of the GSDMD N-terminus end, which forms pores in the membrane through which IL-1β and IL-18 are released, followed by the entire cytosolic content [42,62,63]. Although the induction of the NLRP3 inflammasome in response to 7KCh has been reported several times [18,19,24], no direct association between 7KCh and pyroptosis has been established so far. In the genome-wide analysis study carried out in our previous work, we described an increase in the gene expression of *CASP1* in 7KCh-treated cells [34]. In the present work, we confirmed this gene expression alteration by qRT-PCR and observed a modulation of *IL1B* expression in response to 7KCh (Figure 9A). However, no changes in the expression levels of *GSDMD* and *IL18* were seen, and the expression of *NLRP3* could not even be detected. Interestingly, despite detecting the expression of *GSDMD*, the band corresponding to the protein was not observed by Western blot in mRPE cells (Figure 9B). In the transcriptome analysis of our previous work, the expression of *GSDMD* could also be detected in all of the samples tested, but with a relatively low number of counts compared to the number of counts quantified in ARPE-19 cells (approximately 80 counts in ARPE-19 cells and 25 counts in mRPE cells). This could explain why, in mRPE cells, the GSDMD protein is not observed, since the levels of GSDMD would be too low to be detected, but can be observed in ARPE-19 cells, used as the positive control. Taken together, these results indicate that the canonical pyroptosis pathway is not involved in the response to 7KCh.

Transcriptome sequencing data also showed a positive regulation of *DFNA5*, the gene that codes GSDME, with 7KCh [34]. GSDME belongs to the same family of proteins as GSDMD and also acts as an effector protein for pyroptosis. The cleavage of the N-terminus end of GSDME in a ~35 KDa fragment is induced by the action of Caspase 3 instead of Caspase 1, which means that GSDME can be activated independently of the NLRP3 inflammasome [45,62–66]. Both GSDME and Caspase 3 cleavage were observed in mRPE cells treated with 20 μM 7KCh (Figure 10A). However, as mentioned above, long exposures were required for the detection of Caspase 3, whereas this was not the case for the visualization of active GSDME. Furthermore, contrary to what might be expected, it was observed that while GSDME activation was already detected after 12 h of exposure to 20 μM 7KCh, Caspase 3 cleavage was not visualized until 24 h. Similarly, Caspase 3 inhibition could not prevent GSDME cleavage (Figure 10B), suggesting that an alternative protein to Caspase 3 must exist, being capable of promoting GSDME activation. In this way, it seems to be GSDME itself that increases Caspase 3 activation in a positive feedback loop, since it has been described that GSDME can also promote Caspase 3 induction [45,65].

Regarding the cleavage of GSDME, the appearance, on some occasions, of two intermediate bands in mRPE cells is striking (Figure 10A): a lower band than that corresponding to the complete protein, and another of about 38 KDa higher than that corresponding to active GSDME (~35 KDa). These intermediate bands can be even seen in control cells, suggesting some basal GSDME processing under normal conditions, but without the protein becoming activated. In those treatments where there is no cell and GSDME has not been activated yet, such as the control or treatment with 20 μM 7KCh at 6 h, these intermediate bands can be seen. When GSDME begins to activate, as in 20 μM 7KCh after 12 h of treatment, the upper intermediate band disappears, although the ~38 KDa band can still be seen. In contrast, when GSDME is fully activated, such as during 24 h of treatment with 20 μM 7KCh, the ~35 KDa band is most strongly detected, while the two intermediate bands have disappeared, indicating that GSDME is fully cleaved and active. The presence of several intermediate bands of GSDME has also been visualized in several previous studies [67–69], although, for the moment, there has been no work that confirms that they are intermediate cleavages.

*DFNA5* is associated with the development of sensorineural deafness and, in tumor cells, it has been shown to act as a tumor suppressor gene. In different tissues and cell types, it has been described that *DFNA5* transcription is induced by p53 through a binding site for this protein located in intron 1 of the gene [45,70,71]. The regulation and activation of p53 is complex, since it presents several phosphorylation, acetylation, methylation and ubiquitination sites, and it is a target of multiple enzymes. Among proteins capable of phosphorylate p53, we can find both JNK and p38 [45,72]. The inhibition of p38, but not JNK, attenuated GSDME activation in mRPE cells (Figure 10C), indirectly suggesting that the activation of p53 through p38 might be responsible for the increased expression of the gene that codes for GSDME. These results are in agreement with our previous work, where we showed that p38, but not JNK, was implicated in cell death induced by 7KCh in mRPE cells [34]. P2X7 inhibition not only decreased p38 phosphorylation, as mentioned above (Figure 8), but also reduced GSDME activation (Figure 10A). This confirms the involvement of P2X7 in pyroptosis and suggests that 7KCh activates an alternative pyroptotic mechanism mediated by P2X7, p38 and GSDME in mRPE cells not previously described. On the other hand, treatment with 10 μM SA, in addition to decreasing the activation of both P2X7 (Figure 7) and p38 (Figure 6), was able to prevent GSDME cleavage at 12 h and 24 h of treatment, as well as Caspase 3 cleavage at 24 h (Figure 10A). These results indicate that SA acts at various levels of the 7KCh-induced response and, once again, highlights its therapeutic potential, postulating it as the most potent functional antagonist of 7KCh in retinal cells.

GSDME activation could explain why, although there is a little induction, apoptosis is not fully triggered, since when Caspase 3 activates GSDME, the initial apoptotic response culminates in pyroptosis instead of apoptosis [45]. In addition, pyroptosis is sometimes considered a secondary necrosis or one of the types of regulated necrosis, and some studies have described that GSDME can induce necrosis [45,66,67,73–75]. Thus, the necrosis and pyroptosis observed in mRPE cells in response to 7KCh might be related. The activation of P2X7 would first induce the pyroptotic response, which would end up triggering necrosis. This could explain why, with 7KCh-treatment, the internalization of YO-PRO1 is observed before propidium iodide labeling, and there is always a percentage of cells only stained with YO-PRO1 that, as cell death progresses, are also stained with propidium iodide (Figures 6 and 7).

7KCh is one of the main molecules accumulated in drusen and its relationship with the development of AMD is well established [17–20,22–24,30,34,35,76]. In the current work, we showed that 7KCh induces a cell death mechanism in retinal cells characterized by the induction of necrosis and pyroptosis, and mediated by the activation of P2X7, p38 and GSDME. There have been several studies that have associated P2X7 activation with the pathogenesis of AMD, both in vivo and in vitro [22,44,59]. Recent publications have also described the induction of GSDME in retinal cells exposed to all-trans-retinal, another molecule associated with the development of AMD [65,77]. Thus, the characterization of the response induced by 7KCh in retinal cells offers new opportunities to search for alternative therapeutic targets for this disease. On the other hand, we have demonstrated, once again, the ability of SA to counteract the harmful effect of 7KCh on mRPE cells, in agreement with previous studies [23,24,34–36], being able to act at various levels of the response induced by this oxysterol. For these reasons, SA is currently the molecule that works best to attenuate the effect of 7KCh in retinal cells and it is presented as a new opportunity to complement or improve the therapies used to treat AMD. Although there is still a long way to go and further research is needed, the present work is a step forward in the study of the mechanism of action of 7KCh in the pathogenesis of AMD and in the therapeutic potential of SA to target the effect of this oxysterol.

**Author Contributions:** Conceptualization, A.P., Á.P.-S., R.P. and I.M.L.; methodology, A.P. and I.M.L.; software, Á.P.-S.; validation, A.P., Á.P.-S., R.O. and R.P.; formal analysis, A.P., Á.P.-S., R.P. and I.M.L.; investigation, A.P., R.O., M.B., Á.V.-M. and R.P.; resources, I.M.L.; data curation, A.P., Á.P.-S., R.P. and I.M.L.; writing—original draft preparation, A.P., Á.P.-S., R.O., M.B., Á.V.-M., R.P. and I.M.L.; writing—review and editing, A.P., M.B. and I.M.L.; visualization, A.P.; supervision, I.M.L.; project administration, I.M.L.; funding acquisition, I.M.L. All authors have read and agreed to the published version of the manuscript.

**Funding:** This research was funded, in part, by a grant (PI19/01805) from the Instituto de Salud Carlos III, co-funded by European Regional Development Fund (ERDF) "A way to build Europe" and by Fundación Rioja Salud. I.M.L. was supported by a Miguel Servet contract (CPII20/00029) from the Instituto de Salud Carlos III, co-funded by European Social fund (ESF) "Investing in your future". M.B. was supported by "Retos Colaboración" project, GLIOPROG561 (RTC2019-007399-1) from the Ministry of Science and Innovation (Spain). Á.V.-M. was supported by Agencia de Desarrollo Económico de la Rioja (ADER) engagements.

**Institutional Review Board Statement:** Not applicable.

**Informed Consent Statement:** Not applicable.

**Data Availability Statement:** Not applicable.

**Conflicts of Interest:** I.M.L. holds a patent on "Methods of treatment using sterculic acid". The funders had no role in the design of the study; in the collection, analyses, or interpretation of data; in the writing of the manuscript, or in the decision to publish the results.

## Abbreviations

| | |
|---|---|
| 7KCh | 7-Ketocholesterol |
| AMD | age-related macular degeneration |
| CBX | carbenoxolone |
| CHX | cicloheximide |
| CNV | choroidal neovascularization |
| mRPE | monkey retinal pigment epithelium cells |
| Nec-1 | necrostatine-1 |
| NSA | necrosulfonamide |
| PVDF | polyvinylidene difluoride |
| qRT-PCR | quantitative real-time PCR |
| RPE | retinal pigment epithelium |
| SA | sterculic acid |
| SCD1 | stearoyl coenzyme-A desaturase 1 |
| SDS | sodium dodecyl sulphate |
| SEM | standard error of the means |
| TBS | tris buffered saline |
| UPR | unfolded protein response |

## References

1. Deng, Y.; Qiao, L.; Du, M.; Qu, C.; Wan, L.; Li, J.; Huang, L. Age-related macular degeneration: Epidemiology, genetics, pathophysiology, diagnosis, and targeted therapy. *Genes Dis.* **2022**, *9*, 62–79. [CrossRef]
2. Mehrzadi, S.; Hemati, K.; Reiter, R.J.; Hosseinzadeh, A. Mitochondrial dysfunction in age-related macular degeneration: Melatonin as a potential treatment. *Expert Opin. Ther. Targets* **2020**, *24*, 359–378. [CrossRef] [PubMed]
3. Taylor, A. Introduction to the issue regarding research regarding age related macular degeneration. *Mol. Asp. Med.* **2012**, *33*, 291–294. [CrossRef] [PubMed]
4. Xu, Q.; Cao, S.; Rajapakse, S.; Matsubara, J.A. Understanding AMD by analogy: Systematic review of lipid-related common pathogenic mechanisms in AMD, AD, AS and GN. *Lipids Health Dis.* **2018**, *17*, 3. [CrossRef] [PubMed]
5. Stahl, A. The Diagnosis and Treatment of Age-Related Macular Degeneration. *Dtsch. Arztebl. Int.* **2020**, *117*, 513–520. [CrossRef]
6. Thomas, C.J.; Mirza, R.G.; Gill, M.K. Age-Related Macular Degeneration. *Med. Clin. N. Am.* **2021**, *105*, 473–491. [CrossRef]
7. Curcio, C.A. Antecedents of Soft Drusen, the Specific Deposits of Age-Related Macular Degeneration, in the Biology of Human Macula. *Investig. Ophthalmol. Vis. Sci.* **2018**, *59*, AMD182–AMD194. [CrossRef]
8. Curcio, C.A. Soft Drusen in Age-Related Macular Degeneration: Biology and Targeting Via the Oil Spill Strategies. *Investig. Ophthalmol. Vis. Sci.* **2018**, *59*, AMD160–AMD181. [CrossRef]

9. Spaide, R.F.; Ooto, S.; Curcio, C.A. Subretinal drusenoid deposits AKA pseudodrusen. *Surv. Ophthalmol.* **2018**, *63*, 782–815. [CrossRef]
10. Curcio, C.A.; Johnson, M.; Rudolf, M.; Huang, J.D. The oil spill in ageing Bruch membrane. *Br. J. Ophthalmol.* **2011**, *95*, 1638–1645. [CrossRef]
11. Abokyi, S.; To, C.H.; Lam, T.T.; Tse, D.Y. Central Role of Oxidative Stress in Age-Related Macular Degeneration: Evidence from a Review of the Molecular Mechanisms and Animal Models. *Oxidative Med. Cell. Longev.* **2020**, *2020*, 7901270. [CrossRef] [PubMed]
12. Garcia-Garcia, J.; Usategui-Martin, R.; Sanabria, M.R.; Fernandez-Perez, E.; Telleria, J.J.; Coco-Martin, R.M. Pathophysiology of Age-Related Macular Degeneration. Implications for Treatment. *Ophthalmic Res.* **2022**, *65*, 615–636. [CrossRef] [PubMed]
13. Moreira, E.F.; Larrayoz, I.M.; Lee, J.W.; Rodriguez, I.R. 7-Ketocholesterol is present in lipid deposits in the primate retina: Potential implication in the induction of VEGF and CNV formation. *Investig. Ophthalmol. Vis. Sci.* **2009**, *50*, 523–532. [CrossRef]
14. Bobadilla, M.; Pariente, A.; Oca, A.I.; Pelaez, R.; Perez-Sala, A.; Larrayoz, I.M. Biomarkers as Predictive Factors of Anti-VEGF Response. *Biomedicines* **2022**, *10*, 1003. [CrossRef] [PubMed]
15. Oca, A.I.; Perez-Sala, A.; Pariente, A.; Ochoa, R.; Velilla, S.; Pelaez, R.; Larrayoz, I.M. Predictive Biomarkers of Age-Related Macular Degeneration Response to Anti-VEGF Treatment. *J. Pers. Med.* **2021**, *11*, 1329. [CrossRef]
16. Mettu, P.S.; Allingham, M.J.; Cousins, S.W. Incomplete response to Anti-VEGF therapy in neovascular AMD: Exploring disease mechanisms and therapeutic opportunities. *Prog. Retin. Eye Res.* **2021**, *82*, 100906. [CrossRef]
17. Poli, G.; Biasi, F.; Leonarduzzi, G. Oxysterols in the pathogenesis of major chronic diseases. *Redox Biol.* **2013**, *1*, 125–130. [CrossRef]
18. Indaram, M.; Ma, W.; Zhao, L.; Fariss, R.N.; Rodriguez, I.R.; Wong, W.T. 7-Ketocholesterol increases retinal microglial migration, activation, and angiogenicity: A potential pathogenic mechanism underlying age-related macular degeneration. *Sci. Rep.* **2015**, *5*, 9144. [CrossRef]
19. Shi, G.; Chen, S.; Wandu, W.S.; Ogbeifun, O.; Nugent, L.F.; Maminishkis, A.; Hinshaw, S.J.; Rodriguez, I.R.; Gery, I. Inflammasomes Induced by 7-Ketocholesterol and Other Stimuli in RPE and in Bone Marrow-Derived Cells Differ Markedly in Their Production of IL-1beta and IL-18. *Investig. Ophthalmol. Vis. Sci.* **2015**, *56*, 1658–1664. [CrossRef]
20. Amaral, J.; Lee, J.W.; Chou, J.; Campos, M.M.; Rodriguez, I.R. 7-Ketocholesterol induces inflammation and angiogenesis in vivo: A novel rat model. *PLoS ONE* **2013**, *8*, e56099. [CrossRef]
21. Rodriguez, I.R.; Clark, M.E.; Lee, J.W.; Curcio, C.A. 7-ketocholesterol accumulates in ocular tissues as a consequence of aging and is present in high levels in drusen. *Exp. Eye Res.* **2014**, *128*, 151–155. [CrossRef] [PubMed]
22. Olivier, E.; Dutot, M.; Regazzetti, A.; Leguillier, T.; Dargere, D.; Auzeil, N.; Laprevote, O.; Rat, P. P2X7-pannexin-1 and amyloid beta-induced oxysterol input in human retinal cell: Role in age-related macular degeneration? *Biochimie* **2016**, *127*, 70–78. [CrossRef] [PubMed]
23. Huang, J.D.; Amaral, J.; Lee, J.W.; Larrayoz, I.M.; Rodriguez, I.R. Sterculic acid antagonizes 7-ketocholesterol-mediated inflammation and inhibits choroidal neovascularization. *Biochim. Biophys. Acta* **2012**, *1821*, 637–646. [CrossRef] [PubMed]
24. Huang, J.D.; Amaral, J.; Lee, J.W.; Rodriguez, I.R. 7-Ketocholesterol-induced inflammation signals mostly through the TLR4 receptor both in vitro and in vivo. *PLoS ONE* **2014**, *9*, e100985. [CrossRef]
25. Leoni, V.; Nury, T.; Vejux, A.; Zarrouk, A.; Caccia, C.; Debbabi, M.; Fromont, A.; Sghaier, R.; Moreau, T.; Lizard, G. Mitochondrial dysfunctions in 7-ketocholesterol-treated 158N oligodendrocytes without or with alpha-tocopherol: Impacts on the cellular profil of tricarboxylic cycle-associated organic acids, long chain saturated and unsaturated fatty acids, oxysterols, cholesterol and cholesterol precursors. *J. Steroid Biochem. Mol. Biol.* **2017**, *169*, 96–110. [CrossRef]
26. Sul, O.J.; Li, G.; Kim, J.E.; Kim, E.S.; Choi, H.S. 7-ketocholesterol enhances autophagy via the ROS-TFEB signaling pathway in osteoclasts. *J. Nutr. Biochem.* **2021**, *96*, 108783. [CrossRef]
27. Tani, M.; Kamata, Y.; Deushi, M.; Osaka, M.; Yoshida, M. 7-Ketocholesterol enhances leukocyte adhesion to endothelial cells via p38MAPK pathway. *PLoS ONE* **2018**, *13*, e0200499. [CrossRef]
28. Yang, C.; Xie, L.; Gu, Q.; Qiu, Q.; Wu, X.; Yin, L. 7-Ketocholesterol disturbs RPE cells phagocytosis of the outer segment of photoreceptor and induces inflammation through ERK signaling pathway. *Exp. Eye Res.* **2019**, *189*, 107849. [CrossRef]
29. Nury, T.; Sghaier, R.; Zarrouk, A.; Menetrier, F.; Uzun, T.; Leoni, V.; Caccia, C.; Meddeb, W.; Namsi, A.; Sassi, K.; et al. Induction of peroxisomal changes in oligodendrocytes treated with 7-ketocholesterol: Attenuation by alpha-tocopherol. *Biochimie* **2018**, *153*, 181–202. [CrossRef]
30. Wang, H.; Ramshekar, A.; Kunz, E.; Hartnett, M.E. 7-ketocholesterol induces endothelial-mesenchymal transition and promotes fibrosis: Implications in neovascular age-related macular degeneration and treatment. *Angiogenesis* **2021**, *24*, 583–595. [CrossRef]
31. Vejux, A.; Abed-Vieillard, D.; Hajji, K.; Zarrouk, A.; Mackrill, J.J.; Ghosh, S.; Nury, T.; Yammine, A.; Zaibi, M.; Mihoubi, W.; et al. 7-Ketocholesterol and 7beta-hydroxycholesterol: In vitro and animal models used to characterize their activities and to identify molecules preventing their toxicity. *Biochem. Pharmacol.* **2020**, *173*, 113648. [CrossRef] [PubMed]
32. Brahmi, F.; Vejux, A.; Sghaier, R.; Zarrouk, A.; Nury, T.; Meddeb, W.; Rezig, L.; Namsi, A.; Sassi, K.; Yammine, A.; et al. Prevention of 7-ketocholesterol-induced side effects by natural compounds. *Crit. Rev. Food Sci. Nutr.* **2019**, *59*, 3179–3198. [CrossRef] [PubMed]
33. Pariente, A.; Pelaez, R.; Perez-Sala, A.; Larrayoz, I.M. Inflammatory and cell death mechanisms induced by 7-ketocholesterol in the retina. Implications for age-related macular degeneration. *Exp. Eye Res.* **2019**, *187*, 107746. [CrossRef] [PubMed]

34. Pariente, A.; Perez-Sala, A.; Ochoa, R.; Bobadilla, M.; Villanueva-Martinez, A.; Pelaez, R.; Larrayoz, I.M. Identification of 7-Ketocholesterol-Modulated Pathways and Sterculic Acid Protective Effect in Retinal Pigmented Epithelium Cells by Using Genome-Wide Transcriptomic Analysis. *Int. J. Mol. Sci.* **2023**, *24*, 7459. [CrossRef] [PubMed]
35. Pariente, A.; Perez-Sala, A.; Ochoa, R.; Pelaez, R.; Larrayoz, I.M. Genome-Wide Transcriptomic Analysis Identifies Pathways Regulated by Sterculic Acid in Retinal Pigmented Epithelium Cells. *Cells* **2020**, *9*, 1187. [CrossRef] [PubMed]
36. William, S.; Duncan, T.; Redmond, T.M. Pretreatment of human retinal pigment epithelial cells with sterculic acid forestalls fenretinide-induced apoptosis. *Sci. Rep.* **2022**, *12*, 22442. [CrossRef]
37. Herrera-Meza, M.S.; Mendoza-Lopez, M.R.; Garcia-Barradas, O.; Sanchez-Otero, M.G.; Silva-Hernandez, E.R.; Angulo, J.O.; Oliart-Ros, R.M. Dietary anhydrous milk fat naturally enriched with conjugated linoleic acid and vaccenic acid modify cardiovascular risk biomarkers in spontaneously hypertensive rats. *Int. J. Food Sci. Nutr.* **2013**, *64*, 575–586. [CrossRef]
38. Pelaez, R.; Pariente, A.; Perez-Sala, A.; Larrayoz, I.M. Sterculic Acid: The Mechanisms of Action beyond Stearoyl-CoA Desaturase Inhibition and Therapeutic Opportunities in Human Diseases. *Cells* **2020**, *9*, 140. [CrossRef]
39. Ortinau, L.C.; Nickelson, K.J.; Stromsdorfer, K.L.; Naik, C.Y.; Pickering, R.T.; Haynes, R.A.; Fritsche, K.L.; Perfield, J.W., 2nd. Sterculic oil, a natural inhibitor of SCD1, improves the metabolic state of obese OLETF rats. *Obesity* **2013**, *21*, 344–352. [CrossRef]
40. Becerra, S.P.; Fariss, R.N.; Wu, Y.Q.; Montuenga, L.M.; Wong, P.; Pfeffer, B.A. Pigment epithelium-derived factor in the monkey retinal pigment epithelium and interphotoreceptor matrix: Apical secretion and distribution. *Exp. Eye Res.* **2004**, *78*, 223–234. [CrossRef]
41. Nikoletopoulou, V.; Markaki, M.; Palikaras, K.; Tavernarakis, N. Crosstalk between apoptosis, necrosis and autophagy. *Biochim. Biophys. Acta* **2013**, *1833*, 3448–3459. [CrossRef] [PubMed]
42. Tang, D.; Kang, R.; Berghe, T.V.; Vandenabeele, P.; Kroemer, G. The molecular machinery of regulated cell death. *Cell Res.* **2019**, *29*, 347–364. [CrossRef] [PubMed]
43. Kopp, R.; Krautloher, A.; Ramirez-Fernandez, A.; Nicke, A. P2X7 Interactions and Signaling—Making Head or Tail of It. *Front. Mol. Neurosci.* **2019**, *12*, 183. [CrossRef] [PubMed]
44. Kerur, N.; Hirano, Y.; Tarallo, V.; Fowler, B.J.; Bastos-Carvalho, A.; Yasuma, T.; Yasuma, R.; Kim, Y.; Hinton, D.R.; Kirschning, C.J.; et al. TLR-independent and P2X7-dependent signaling mediate Alu RNA-induced NLRP3 inflammasome activation in geographic atrophy. *Investig. Ophthalmol. Vis. Sci.* **2013**, *54*, 7395–7401. [CrossRef]
45. Jiang, M.; Qi, L.; Li, L.; Li, Y. The caspase-3/GSDME signal pathway as a switch between apoptosis and pyroptosis in cancer. *Cell Death Discov.* **2020**, *6*, 112. [CrossRef]
46. Chang, M.C.; Chen, Y.J.; Liou, E.J.; Tseng, W.Y.; Chan, C.P.; Lin, H.J.; Liao, W.C.; Chang, Y.C.; Jeng, P.Y.; Jeng, J.H. 7-Ketocholesterol induces ATM/ATR, Chk1/Chk2, PI3K/Akt signalings, cytotoxicity and IL-8 production in endothelial cells. *Oncotarget* **2016**, *7*, 74473–74483. [CrossRef] [PubMed]
47. Kim, A.; Nam, Y.J.; Lee, C.S. Taxifolin reduces the cholesterol oxidation product-induced neuronal apoptosis by suppressing the Akt and NF-kappaB activation-mediated cell death. *Brain Res. Bull.* **2017**, *134*, 63–71. [CrossRef]
48. Koh, S.S.; Ooi, S.C.; Lui, N.M.; Qiong, C.; Ho, L.T.; Cheah, I.K.; Halliwell, B.; Herr, D.R.; Ong, W.Y. Effect of Ergothioneine on 7-Ketocholesterol-Induced Endothelial Injury. *Neuromolecular Med.* **2021**, *23*, 184–198. [CrossRef]
49. Nury, T.; Zarrouk, A.; Mackrill, J.J.; Samadi, M.; Durand, P.; Riedinger, J.M.; Doria, M.; Vejux, A.; Limagne, E.; Delmas, D.; et al. Induction of oxiapoptophagy on 158N murine oligodendrocytes treated by 7-ketocholesterol-, 7beta-hydroxycholesterol-, or 24(S)-hydroxycholesterol: Protective effects of alpha-tocopherol and docosahexaenoic acid (DHA; C22:6 n-3). *Steroids* **2015**, *99*, 194–203. [CrossRef]
50. Nury, T.; Zarrouk, A.; Vejux, A.; Doria, M.; Riedinger, J.M.; Delage-Mourroux, R.; Lizard, G. Induction of oxiapoptophagy, a mixed mode of cell death associated with oxidative stress, apoptosis and autophagy, on 7-ketocholesterol-treated 158N murine oligodendrocytes: Impairment by alpha-tocopherol. *Biochem. Biophys. Res. Commun.* **2014**, *446*, 714–719. [CrossRef]
51. Paz, J.L.; Levy, D.; Oliveira, B.A.; de Melo, T.C.; de Freitas, F.A.; Reichert, C.O.; Rodrigues, A.; Pereira, J.; Bydlowski, S.P. 7-Ketocholesterol Promotes Oxiapoptophagy in Bone Marrow Mesenchymal Stem Cell from Patients with Acute Myeloid Leukemia. *Cells* **2019**, *8*, 482. [CrossRef]
52. Ragot, K.; Delmas, D.; Athias, A.; Nury, T.; Baarine, M.; Lizard, G. alpha-Tocopherol impairs 7-ketocholesterol-induced caspase-3-dependent apoptosis involving GSK-3 activation and Mcl-1 degradation on 158N murine oligodendrocytes. *Chem. Phys. Lipids* **2011**, *164*, 469–478. [CrossRef] [PubMed]
53. Ragot, K.; Mackrill, J.J.; Zarrouk, A.; Nury, T.; Aires, V.; Jacquin, A.; Athias, A.; Pais de Barros, J.P.; Vejux, A.; Riedinger, J.M.; et al. Absence of correlation between oxysterol accumulation in lipid raft microdomains, calcium increase, and apoptosis induction on 158N murine oligodendrocytes. *Biochem. Pharmacol.* **2013**, *86*, 67–79. [CrossRef] [PubMed]
54. Soh, S.; Ong, W.Y. Effect of Withanolide A on 7-Ketocholesterol Induced Cytotoxicity in hCMEC/D3 Brain Endothelial Cells. *Cells* **2022**, *11*, 457. [CrossRef] [PubMed]
55. Xiao, Q.; Che, X.; Cai, B.; Tao, Z.; Zhang, H.; Shao, Q.; Pu, J. Macrophage autophagy regulates mitochondria-mediated apoptosis and inhibits necrotic core formation in vulnerable plaques. *J. Cell. Mol. Med.* **2020**, *24*, 260–275. [CrossRef]
56. Yammine, A.; Zarrouk, A.; Nury, T.; Vejux, A.; Latruffe, N.; Vervandier-Fasseur, D.; Samadi, M.; Mackrill, J.J.; Greige-Gerges, H.; Auezova, L.; et al. Prevention by Dietary Polyphenols (Resveratrol, Quercetin, Apigenin) Against 7-Ketocholesterol-Induced Oxiapoptophagy in Neuronal N2a Cells: Potential Interest for the Treatment of Neurodegenerative and Age-Related Diseases. *Cells* **2020**, *9*, 2346. [CrossRef]

57. Ghelli, A.; Porcelli, A.M.; Zanna, C.; Rugolo, M. 7-Ketocholesterol and staurosporine induce opposite changes in intracellular pH, associated with distinct types of cell death in ECV304 cells. *Arch. Biochem. Biophys.* **2002**, *402*, 208–217. [CrossRef]
58. Adinolfi, E.; Giuliani, A.L.; De Marchi, E.; Pegoraro, A.; Orioli, E.; Di Virgilio, F. The P2X7 receptor: A main player in inflammation. *Biochem. Pharmacol.* **2018**, *151*, 234–244. [CrossRef]
59. Yang, D. Targeting the P2X7 Receptor in Age-Related Macular Degeneration. *Vision* **2017**, *1*, 11. [CrossRef]
60. Bianco, F.; Perrotta, C.; Novellino, L.; Francolini, M.; Riganti, L.; Menna, E.; Saglietti, L.; Schuchman, E.H.; Furlan, R.; Clementi, E.; et al. Acid sphingomyelinase activity triggers microparticle release from glial cells. *EMBO J.* **2009**, *28*, 1043–1054. [CrossRef]
61. Green, J.P.; Souilhol, C.; Xanthis, I.; Martinez-Campesino, L.; Bowden, N.P.; Evans, P.C.; Wilson, H.L. Atheroprone flow activates inflammation via endothelial ATP-dependent P2X7-p38 signalling. *Cardiovasc. Res.* **2018**, *114*, 324–335. [CrossRef] [PubMed]
62. Feng, S.; Fox, D.; Man, S.M. Mechanisms of Gasdermin Family Members in Inflammasome Signaling and Cell Death. *J. Mol. Biol.* **2018**, *430*, 3068–3080. [CrossRef] [PubMed]
63. Orning, P.; Lien, E.; Fitzgerald, K.A. Gasdermins and their role in immunity and inflammation. *J. Exp. Med.* **2019**, *216*, 2453–2465. [CrossRef] [PubMed]
64. Li, Y.Q.; Peng, J.J.; Peng, J.; Luo, X.J. The deafness gene GSDME: Its involvement in cell apoptosis, secondary necrosis, and cancers. *Naunyn Schmiedebergs Arch. Pharmacol.* **2019**, *392*, 1043–1048. [CrossRef] [PubMed]
65. Liao, X.X.; Dai, Y.Z.; Zhao, Y.Z.; Nie, K. Gasdermin E: A Prospective Target for Therapy of Diseases. *Front. Pharmacol.* **2022**, *13*, 855828. [CrossRef] [PubMed]
66. De Schutter, E.; Croes, L.; Ibrahim, J.; Pauwels, P.; Op de Beeck, K.; Vandenabeele, P.; Van Camp, G. GSDME and its role in cancer: From behind the scenes to the front of the stage. *Int. J. Cancer* **2021**, *148*, 2872–2883. [CrossRef]
67. Wen, S.; Wang, Z.H.; Zhang, C.X.; Yang, Y.; Fan, Q.L. Caspase-3 Promotes Diabetic Kidney Disease Through Gasdermin E-Mediated Progression to Secondary Necrosis During Apoptosis. *Diabetes Metab. Syndr. Obes.* **2020**, *13*, 313–323. [CrossRef]
68. Zhao, W.; Zhang, L.; Zhang, Y.; Jiang, Z.; Lu, H.; Xie, Y.; Han, W.; He, J.; Shi, Z.; Yang, H.; et al. The CDK inhibitor AT7519 inhibits human glioblastoma cell growth by inducing apoptosis, pyroptosis and cell cycle arrest. *Cell Death Dis.* **2023**, *14*, 11. [CrossRef]
69. Zhou, B.; Ryder, C.B.; Dubyak, G.R.; Abbott, D.W. Gasdermins and pannexin-1 mediate pathways of chemotherapy-induced cell lysis in hematopoietic malignancies. *Sci. Signal* **2022**, *15*, eabl6781. [CrossRef]
70. Masuda, Y.; Futamura, M.; Kamino, H.; Nakamura, Y.; Kitamura, N.; Ohnishi, S.; Miyamoto, Y.; Ichikawa, H.; Ohta, T.; Ohki, M.; et al. The potential role of DFNA5, a hearing impairment gene, in p53-mediated cellular response to DNA damage. *J. Hum. Genet.* **2006**, *51*, 652–664. [CrossRef]
71. Op de Beeck, K.; Van Camp, G.; Thys, S.; Cools, N.; Callebaut, I.; Vrijens, K.; Van Nassauw, L.; Van Tendeloo, V.F.; Timmermans, J.P.; Van Laer, L. The DFNA5 gene, responsible for hearing loss and involved in cancer, encodes a novel apoptosis-inducing protein. *Eur. J. Hum. Genet.* **2011**, *19*, 965–973. [CrossRef] [PubMed]
72. Jenkins, L.M.; Durell, S.R.; Mazur, S.J.; Appella, E. p53 N-terminal phosphorylation: A defining layer of complex regulation. *Carcinogenesis* **2012**, *33*, 1441–1449. [CrossRef] [PubMed]
73. Chen, K.W.; Demarco, B.; Broz, P. Beyond inflammasomes: Emerging function of gasdermins during apoptosis and NETosis. *EMBO J.* **2020**, *39*, e103397. [CrossRef] [PubMed]
74. Mazlo, A.; Tang, Y.; Jenei, V.; Brauman, J.; Yousef, H.; Bacsi, A.; Koncz, G. Resolution Potential of Necrotic Cell Death Pathways. *Int. J. Mol. Sci.* **2022**, *24*, 16. [CrossRef] [PubMed]
75. Tixeira, R.; Shi, B.; Parkes, M.A.F.; Hodge, A.L.; Caruso, S.; Hulett, M.D.; Baxter, A.A.; Phan, T.K.; Poon, I.K.H. Gasdermin E Does Not Limit Apoptotic Cell Disassembly by Promoting Early Onset of Secondary Necrosis in Jurkat T Cells and THP-1 Monocytes. *Front. Immunol.* **2018**, *9*, 2842. [CrossRef]
76. Larrayoz, I.M.; Huang, J.D.; Lee, J.W.; Pascual, I.; Rodriguez, I.R. 7-ketocholesterol-induced inflammation: Involvement of multiple kinase signaling pathways via NFkappaB but independently of reactive oxygen species formation. *Investig. Ophthalmol. Vis. Sci.* **2010**, *51*, 4942–4955. [CrossRef]
77. Cai, B.; Liao, C.; He, D.; Chen, J.; Han, J.; Lu, J.; Qin, K.; Liang, W.; Wu, X.; Liu, Z.; et al. Gasdermin E mediates photoreceptor damage by all-trans-retinal in the mouse retina. *J. Biol. Chem.* **2022**, *298*, 101553. [CrossRef]

*pharmaceutics*

MDPI

*Article*

# The Neuroprotective Effects of Flavonoid Fisetin against Corticosterone-Induced Cell Death through Modulation of ERK, p38, and PI3K/Akt/FOXO3a-Dependent Pathways in PC12 Cells

**Pei-Rong Chang [1,2,†], Je-Wen Liou [2,†], Pei-Yi Chen [1,3], Wan-Yun Gao [4], Chia-Ling Wu [3], Ming-Jiuan Wu [5] and Jui-Hung Yen [1,4,*]**

1 Department of Molecular Biology and Human Genetics, Tzu Chi University, Hualien 970374, Taiwan; 880802candy@gmail.com (P.-R.C.); pyc571@gmail.com (P.-Y.C.)
2 Department of Biochemistry, School of Medicine, Tzu Chi University, Hualien 970374, Taiwan; jwliou@mail.tcu.edu.tw
3 Laboratory of Medical Genetics, Genetic Counseling Center, Hualien Tzu Chi Hospital, Buddhist Tzu Chi Medical Foundation, Hualien 970374, Taiwan; 106726102@gms.tcu.edu.tw
4 Institute of Medical Sciences, Tzu Chi University, Hualien 970374, Taiwan; 108353105@gms.tcu.edu.tw
5 Department of Biotechnology, Chia-Nan University of Pharmacy and Science, Tainan 717301, Taiwan; mingjiuanwu@gmail.com
* Correspondence: imyenjh@gms.tcu.edu.tw or imyenjh@hotmail.com; Tel.: +886-3-856-5301 (ext. 2683)
† These authors contributed equally to this work.

**Abstract:** The overactive hypothalamic–pituitary–adrenal (HPA) axis is believed to trigger the overproduction of corticosterone, leading to neurotoxicity in the brain. Fisetin is a flavonoid commonly found in fruits and vegetables. It has been suggested to possess various biological activities, including antioxidant, anti-inflammatory, and neuroprotective effects. This study aims to explore the potential neuroprotective properties of fisetin against corticosterone-induced cell death and its underlying molecular mechanism in PC12 cells. Our results indicate that fisetin, at concentrations ranging from 5 to 40 μM, significantly protected PC12 cells against corticosterone-induced cell death. Fisetin effectively reduced the corticosterone-mediated generation of reactive oxygen species (ROS) in PC12 cells. Fisetin treatments also showed potential in inhibiting the corticosterone-induced apoptosis of PC12 cells. Moreover, inhibitors targeting MAPK/ERK kinase 1/2 (MEK1/2), p38 MAPK, and phosphatidylinositol 3-kinase (PI3K) were found to significantly block the increase in cell viability induced by fisetin in corticosterone-treated cells. Consistently, fisetin enhanced the phosphorylation levels of ERK, p38, Akt, and c-AMP response element-binding protein (CREB) in PC12 cells. Additionally, it was found that the diminished levels of p-CREB and p-ERK by corticosterone can be restored by fisetin treatment. Furthermore, the investigation of crosstalk between ERK and CREB revealed that p-CREB activation by fisetin occurred through the ERK-independent pathway. Moreover, we demonstrated that fisetin effectively counteracted the corticosterone-induced nuclear accumulation of FOXO3a, an apoptosis-triggering transcription factor, and concurrently promoted FOXO3a phosphorylation and its subsequent cytoplasmic localization through the PI3K/Akt pathway. In conclusion, our findings indicate that fisetin exerts its neuroprotective effect against corticosterone-induced cell death by modulating ERK, p38, and the PI3K/Akt/FOXO3a-dependent pathways in PC12 cells. Fisetin emerges as a promising phytochemical for neuroprotection.

**Keywords:** hypothalamic–pituitary–adrenal axis; corticosterone; fisetin; ERK; PI3K/Akt; FOXO3a

## 1. Introduction

Stress can trigger the release of corticosteroids, which exert widespread effects in the central nervous system. Prolonged exposure to stress and elevated corticosteroid levels may lead to neuropathology, cognitive impairment, psychiatric conditions, and neurodegenerative disorders [1]. The hypothalamic–pituitary–adrenal (HPA) axis, a hypothetical construct comprising the hypothalamus, pituitary gland, and adrenal gland, becomes activated in response to stress [2]. This activation is primarily instigated by corticotropin-releasing hormone (CRH) and vasopressin (VAP) secreted by the hypothalamus. In response, the pituitary gland releases adrenocorticotropic hormone (ACTH), which then stimulates the adrenal cortex to generate glucocorticoids [3]. These glucocorticoids serve as crucial stress hormones, overseeing a diverse array of physiological processes including metabolism, immune function, skeletal growth, cardiovascular function, reproduction, and cognition [4]. The regulation of the HPA axis is a promising avenue for the development of therapeutic agents for stress-related mental dysfunction or major depressive disorders [5,6].

Under normal conditions, blood glucocorticoid levels are tightly regulated via negative feedback on the HPA axis. This negative feedback mechanism involves the inhibition of CRH and ACTH production, facilitated by the action of glucocorticoids on their receptors within the hypothalamus and pituitary gland. However, when individuals experience chronic stress, their blood glucocorticoid levels may escalate, causing dysfunction in the HPA axis and potentially exacerbating depression [2,4,7]. It has been reported that the hyperfunction of the HPA axis leads to a sustained increase in the level of corticosterone, a type of glucocorticoid produced by the adrenal cortex [8]. Furthermore, excessive corticosterone production has been observed to reduce the release of serotonin (5-hydroxytryptamine, 5-HT), a neurotransmitter that plays crucial roles in mood regulation [9]. The cellular mechanisms underlying corticosterone-induced neurotoxicity involve oxidative stress characterized by the overproduction of reactive oxygen species (ROS) [10] as well as inflammatory responses [11]. Corticosterone has been reported to induce neuronal cell death and diminish neurogenesis in the hippocampal region [12,13]. These processes contribute to neuronal apoptosis and neurological alterations, potentially linking to the pathological changes observed in the brain.

Multiple molecular signaling mechanisms of anti-oxidation and cell survival are suggested to be involved in neuroprotective effects, including the mitogen-activated protein kinase (MAPK) pathways such as extracellular signal-regulated kinases (ERK), c-Jun amino-terminal kinases (JNK), and p38/MAPK, as well as the phosphoinositide 3-kinase (PI3K)/Akt pathway [14–16]. Previous studies demonstrated that the MAPK/ERK pathway plays a critical role in a variety of neuronal cell responses [17]. Upon the activation of ERK, it regulates the activity of various transcription factors, including the cAMP-response element binding protein (CREB), through phosphorylation [18]. When CREB is activated, it enhances the transcriptional activity of the cAMP response element, subsequently promoting the expression of genes are responsible for neuronal cell survival, growth, proliferation, plasticity, and differentiation [19–21]. The PI3K/Akt pathway plays a significant role in governing diverse neurological processes, including neuronal cell survival, differentiation, and synaptogenesis [22,23]. The modulation of the PI3K/Akt-forkhead box O3 (FOXO3a) pathway holds crucial importance in regulating neuronal cell survival. FOXO3a, a member of the forkhead box protein O (FOXO) family, functions as a critical transcription factor that regulates cell proliferation, apoptosis, oxidative stress, and DNA damage [24]. Upon the activation of the PI3K/Akt kinase signaling, there is a subsequent promotion of FOXO3a phosphorylation, leading to the translocation of the FOXO3a protein from the nucleus to the cytoplasm for degradation [25,26]. The down-regulation or inactivation of FOXO3a, achieved by reducing its nuclear retention within cells, hinders apoptosis and promotes cell survival.

The investigation of effective agents with antioxidant or anti-inflammatory activities from natural phytochemicals is emerging as a prominent strategy for the development of neuroprotective agents [27,28]. Fisetin (3,3′,4′,7-tetrahydroxyflavone), a dietary flavonoid, is widely present in fruits and vegetables, including strawberries, persimmons, grapes, onions, cucumbers, and nuts [29]. Fisetin has been reported to exhibit antioxidative, anti-inflammatory, anti-tumorigenic, anti-diabetic, and cardioprotective properties [29–32]. It has demonstrated neuroprotective effects by suppressing oxidative stress and neuronal cell death, promoting neuronal survival, and enhancing learning and memory in various neurodegenerative diseases [33–36]. Additionally, fisetin has been found to alleviate depressive-like behaviors induced by lipopolysaccharide (LPS) through the inhibition of inflammatory responses [37] and the reduction in ROS generation in animal models [38]. Moreover, fisetin has shown to increase the levels of phosphorylated tropomyosin receptor kinase B (TrkB) and exhibit an antidepressant effect by activating the TrkB signaling pathway in mice [39].

The aim of this study was to explore the potential neuroprotective attributes of fisetin against corticosterone-induced neurotoxicity in a rat pheochromocytoma PC12 cell model. Additionally, our objective was to uncover the underlying molecular mechanism involved.

## 2. Materials and Methods

### 2.1. Chemicals

Fisetin (3,3′,4′,7-tetrahydroxyflavone), dimethyl sulfoxide (DMSO), poly-L-lysine, thiazoly blue tetrazolium bromide (MTT), nonessential amino acids (NEAAs), RPMI-1640 medium, and other chemicals were purchased from Sigma–Aldrich (St. Louis, MO, USA) unless otherwise indicated. Corticosterone and SP600125 were purchased from Enzo Life Sciences (Farmingdale, New York, NY, USA). Horse serum (HS) and fetal bovine serum (FBS) were purchased from Thermo Fisher Scientific (Rockford, IL, USA). U0126, SB203580 and LY294002 were purchased from Promega (Madison, WI, USA).

### 2.2. Cell Culture

The PC12 cells were obtained from the Bioresource Collection and Research Center (BCRC, Hsinchu, Taiwan). Cells were maintained in RPMI-1640 medium composed of 10% heat-inactivated HS, 5% FBS, and 1% NEAA. The cells were cultured in a 5% $CO_2$ incubator at 37 °C.

### 2.3. Compounds Treatment

For the treatment of compounds, PC12 cells were seeded at a density of $2 \times 10^5$ cells/mL in 24-well plates using a low-serum RPMI medium (1% HS, 0.5% FBS, and 1% NEAA). The cells were then treated with the specified compounds for subsequent analysis. In the case of corticosterone treatment, the cells were exposed to either the vehicle or corticosterone (100–800 μM) for 24 h. In experiments involving the co-treatment of corticosterone and fisetin, the cells were subjected to corticosterone treatment (300 μM) in the absence or presence of fisetin, which was dissolved in a vehicle containing 0.1% DMSO, at the indicated concentrations. For the treatment involving kinase inhibitors, the cells were pre-treated with various inhibitors: MAPK/ERK kinase1/2 (MEK1/2) inhibitor U0126 (10 μM), JNK inhibitor SP600125 (10 μM), p38 MAPK inhibitor SB203580 (10 μM), and PI3K inhibitor LY294002 (40 μM) for 30 min. Subsequently, the cells were treated with corticosterone (300 μM) in the absence or presence of fisetin (40 μM) for an additional 24 h.

### 2.4. Analysis of Cell Viability

The MTT assay was employed to assess cell viability as previously described [40]. Following the compounds treatment, the cells were exposed to an MTT reagent (1 mg/mL) and incubated for 3 h at 37 °C. After the incubation, the cells were collected through centrifugation at 13,000 rpm for 5 min. The supernatant was then discarded, and the resulting purple crystals were dissolved using DMSO. Cell viability was assessed by measuring the absorbance at 550 nm.

### 2.5. Analysis of ROS Generation

The ROS production was measured by flow cytometry as previously described [36]. PC12 cells were seeded onto poly-L-lysine-coated 6-well plates and cultured in normal serum medium for 24 h. The medium was subsequently removed and replaced with low-serum RPMI medium. Then, 5 μM of 2′, 7′-dichlorodihydrofluorescein diacetate ($H_2$DCFDA) (Thermo Fisher Scientific) was introduced to the cells and incubated at 37 °C for 30 min in a dark environment. After the incubation, the cells were subjected to treatment with corticosterone (300 μM), either in the absence or presence of fisetin, for a 24 h duration. Subsequently, the cells were treated with 1× trypsin–EDTA for 2 min to detach adherent cells, facilitating the analysis of ROS. The level of intracellular ROS was assessed by detecting the fluorescence. Three independent samples, each containing 10,000 cells, were analyzed using the Beckman CytoFLEX Flow Cytometer equipped with a FITC emission filter (Beckman Coulter, Inc., Brea, CA, USA). The resulting data were expressed as the relative percentage of the geometric mean fluorescence intensity.

### 2.6. Analysis of Cell Apoptosis by Fluorescent Microscopy

PC12 cells were seeded onto poly-L-lysine coated coverslips and cultured in normal serum medium for 24 h. On the following day, the culture medium was replaced with low-serum RPMI medium, and the cells were treated with corticosterone (300 μM) in the absence or presence of fisetin (40 μM) for 24 h. After incubation, the coverslips were carefully lifted and placed onto slides. Subsequently, they were mounted using 20 μL of VECTASHIELD® Vibrance™ Antifade Mounting Medium with 4′,6-diamidino-2-phenylindole (DAPI) (Vector Laboratories, Burlingame, CA, USA). The images were then observed and captured using a Nikon Ti2-E inverted fluorescent microscope along with the NIS-Element imaging software Version 5.5 (Nikon Instruments, Tokyo, Japan). The resulting images were subjected to analysis using a tally counter. The apoptotic cells were analyzed across 35 randomly chosen microscopic fields, with each group containing over 600 cells.

### 2.7. Analysis of Cell Apoptosis by Flow Cytometry

The cell apoptosis was measured by flow cytometry as previously described [40]. To determine the proportion of apoptotic cells, cells were treated with corticosterone (300 μM) in the absence or presence of fisetin (40 μM) for 24 h. After incubation, cells were washed in ice-cold PBS and the apoptotic cells were analyzed by an Annexin V-FITC Apoptosis Detection Kit (Strong Biotech, Taipei, Taiwan) according to the manufacturer's instructions. The population of apoptotic cells was detected using a Beckman CytoFLEX Flow Cytometer (Beckman Coulter).

### 2.8. Proteins Preparation and Western Blot Analysis

For the preparation of total cellular proteins, the cell lysates were isolated using RIPA buffer (Thermo Fisher Scientific) containing 1× Protease Inhibitor Single-Use Cocktail (Thermo Fisher Scientific) and PhosphataseArrest™ II (G-Biosciences, Saint Louis, MO, USA). For the preparation of nuclear proteins, the protein lysates were extracted by Nuclear Extract Kit (Active Motif, Carlsbad, CA, USA) according to the manufacturer's instruction. The protein concentration was determined by Bio-Rad Bradford protein assay reagents. Equal amounts of protein lysates were separated by 8%, 10%, or 12% SDS-PAGE and transferred onto a PolyScreen PVDF Transfer Membrane (Cytiva, Buckinghamshire, UK). The membranes were incubated with specific primary antibodies for rat proteins: Phospho-p44/42 MAPK kinase (Thr202/Tyr204), p44/42 MAPK kinase (Thr202/Tyr204), Phospho-p38 MAPK (Thr180/Tyr182), p38 MAPK, Phospho-Akt (Ser473), Akt, Phospho-CREB (Ser133), CREB, Phospho-FOXO3a, and FOXO3a (Cell Signaling Technology, Danvers, MA, USA); α-Tubulin and HDAC2 (GeneTex, Irvine, CA, USA); and actin (Thermo Fisher Scientific). The membranes were then incubated with horseradish peroxidase (HRP)-conjugated goat anti-mouse (Cell Signaling Technology) or anti-rabbit (GeneTex) IgG secondary antibodies. The blots were then rinsed with Amersham ECL™ Prime Western blotting detection reagent, and the chemiluminescent protein signals were detected on Amersham Hyperfilm™ ECL (Cytiva).

### 2.9. Transfection of FOXO3a-GFP Expression Plasmid

PC12 cells were seeded on poly-L-lysine coated coverslip and cultured in a normal serum medium for 24 h. For plasmid transfection, cells were transfected with the pCMV6-

AC-FOXO3-GFP plasmid (OriGene Technologies, Rockville, MD, USA) using Lipofectamine 2000 Transfection Reagent (Thermo Fisher Scientific). Following a 4 h transfection period, the medium supplemented with 20% HS, 10% FBS, and 1% NEAA was added and incubated for 24 h. Subsequently, the medium was replaced with low-serum RPMI medium, and the cells were treated with vehicle or fisetin (40 μM) for an additional 24 h. After the treatments, the coverslips containing the cells were carefully collected and positioned on the slides. VECTASHIELD® Vibrance™ Antifade Mounting Medium was applied to the samples for mounting. Imaging was conducted using a Nikon Ti2-E inverted fluorescent microscopy. The images were captured and documented using the NIS-Element imaging software.

### 2.10. Statistical Analysis

The experiments were conducted independently on a minimum of three occasions, and within each experiment, the procedures were replicated three times. The data are presented as mean ± standard deviation (SD). Graph Pad Prism (version 9) (Graph Pad Software Inc., San Diego, CA, USA) was used to analyze the statistical significance. For comparisons involving two groups, Student's *t*-test was utilized. When multiple groups were being compared, a one-way ANOVA followed by Tukey's post hoc test was employed. A *p*-value of less than 0.05 was considered statistically significant.

## 3. Results

### 3.1. Effects of Fisetin on Cell Viability in Corticosterone-Treated PC12 Cells

To confirm the cytotoxic impact of corticosterone (depicted in Figure 1a), PC12 cells were subjected to treatment with either the vehicle (0.1% DMSO) or varying concentrations of corticosterone (ranging from 100 to 800 μM) for a duration of 24 h. As shown in Figure 1b, the treatment with corticosterone led to a significant reduction in cell viability: 94.73 ± 4.20% ($p < 0.05$), 82.56 ± 5.99% ($p < 0.01$), 50.76 ± 7.17% ($p < 0.01$), and 27.12 ± 3.03% ($p < 0.01$), respectively, compared to the group treated solely with the vehicle (100.00 ± 5.40%). These data indicated that corticosterone is capable of inducing cell death in PC12 cells in a dose-dependent manner. To investigate the impact of fisetin (depicted in Figure 1c) on the viability of PC12 cells, the cells were treated with either the vehicle or various concentrations of fisetin (ranging from 2.5 to 60 μM). The results, as shown in Figure 1d, revealed that fisetin treatment did not induce any cytotoxic effects on PC12 cells. To further explore the protective effects of fisetin against corticosterone-induced cell death, PC12 cells were subjected to a 24 h exposure to corticosterone (300 μM), either in the presence or absence of fisetin. As shown in Figure 1e, the treatment with corticosterone resulted in a significant reduction in the cell viability of PC12 cells to 62.40 ± 3.50%. Conversely, the co-administration of fisetin (at concentrations of 2.5, 5, 10, 20, 40, and 60 μM) markedly elevated the cell viability to 68.75 ± 4.37%, 70.98 ± 4.62%, 73.97 ± 3.68%, 80.93 ± 3.39%, 95.83 ± 8.72%, and 104.21 ± 11.33%, respectively, in comparison to cells only treated with the vehicle (100.00 ± 2.17%). These findings reveal the significant protective effect of fisetin in guarding PC12 cells against cell death induced by corticosterone.

### 3.2. Effects of Fisetin on ROS Production in Corticosterone-Treated PC12 Cells

Oxidative stress contributes to numerous neurodegenerative disorders, and the generation of ROS is a critical indicator of this process. Previous studies have reported that corticosterone can induce cytotoxicity by elevating cellular ROS levels [41]. Hence, we proceeded to investigate whether fisetin possesses the ability to mitigate corticosterone-induced cell death in PC12 cells through the reduction in ROS production. As shown in Figure 2a,b, corticosterone increased cellular ROS levels by 154.03 ± 12.92% ($p < 0.01$) compared to the vehicle group (100.00 ± 7.19%). Co-treatment with fisetin significantly reduced measured ROS levels in corticosterone-treated cells to 112.83 ± 10.51% ($p < 0.01$). These findings suggest that fisetin exhibits the function to mitigate corticosterone-induced ROS production, thereby offering protection to neuronal cells against cell death.

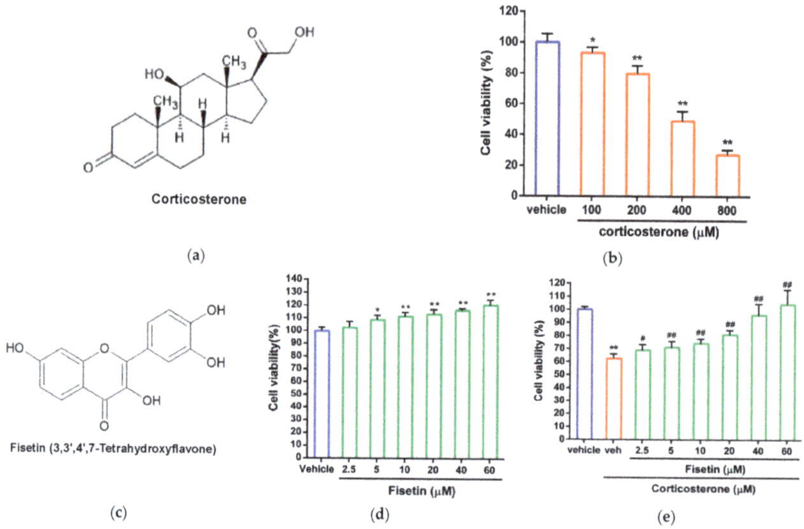


**Figure 1.** The effect of fisetin on cell viability in corticosterone-treated PC12 cells. (**a**) The chemical structure of corticosterone. (**b**) PC12 cells were treated with vehicle (0.1% DMSO) or corticosterone (100, 200, 400, and 800 μM) for 24 h. (**c**) The chemical structure of fisetin (3,3′,4′,7-tetrahydroxyflavone). (**d**) PC12 cells were treated with vehicle or fisetin (2.5, 5, 10, 20, 40, and 60 μM) for 24 h. (**e**) PC12 cells were treated with vehicle or corticosterone (300 μM) in the absence (veh) or presence of fisetin (2.5, 5, 10, 20, 40, and 60 μM) for 24 h. Cell viability was measured by MTT assay. The data represent the mean ± SD of three independent experiments. * $p < 0.05$ and ** $p < 0.01$ represent significant differences compared to the vehicle-treated group. # $p < 0.05$ and ## $p < 0.01$ represent significant differences compared to the corticosterone alone (veh)-treated group.

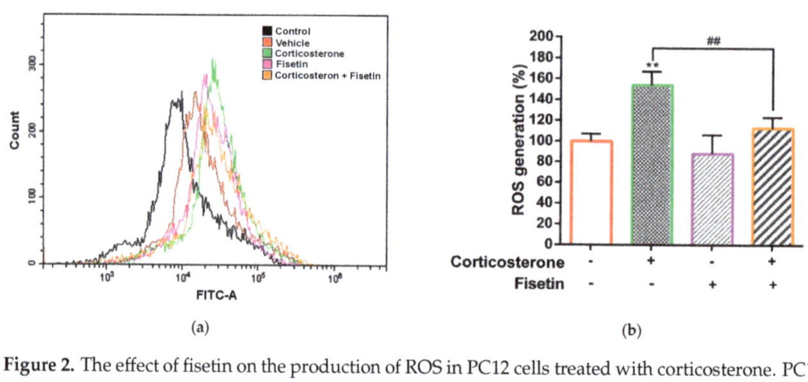


**Figure 2.** The effect of fisetin on the production of ROS in PC12 cells treated with corticosterone. PC12 cells were cultured in a poly-L-lysine-coated 6-well plate and pre-incubated with $H_2$DCFDA (5 μM) at 37 °C for 30 min in the dark. Following incubation, the cells were treated with corticosterone (300 μM), either in the absence or presence of fisetin (40 μM), for a duration of 24 h. The level of ROS was quantified by flow cytometry analysis. (**a**) A representative histogram featuring the control group (the black line), cells subjected to vehicle treatment (the red line), cells treated with corticosterone (the green line), cells treated with fisetin (the pink line), and cells co-treated with both corticosterone and fisetin (the orange line). (**b**) Quantification of ROS levels. The data represent the mean ± SD of three independent experiments. ** $p < 0.01$ represents significant differences compared to vehicle-treated group. ## $p < 0.01$ represent significant differences compared to corticosterone alone-treated group.

### 3.3. Effects of Fisetin on Cell Apoptosis in Corticosterone-Treated PC12 Cells

Previous studies have indicated that corticosterone can induce apoptosis in neuronal cells [42]. As indicated in the previous section, fisetin was able to mitigate corticosterone-induced ROS production. This function may contribute to the inhibition of cellular apoptosis. Therefore, we conducted further investigations to explore the protective effect of fisetin against corticosterone-induced apoptosis in PC12 cells. As shown in Figure 3a,b, the DAPI fluorescent staining of cells indeed revealed that corticosterone increased the percentage of apoptotic cells to 13.55 ± 1.65% ($p < 0.01$), as compared to the vehicle-treated group (5.87 ± 0.78%). On the other hand, co-treatment with fisetin significantly reduced the occurrence of cell apoptosis to 5.83 ± 1.40% ($p < 0.01$). The flow cytometric analysis of Annexin V/PI-stained cells confirmed that fisetin treatment significantly reduced the proportion of cells undergoing corticosterone-induced apoptosis, decreasing from approximately 16% (corticosterone alone treatment) to around 6% ($p < 0.01$) (Figure 3c,d). These findings provide further support for the ability of fisetin to inhibit corticosterone-induced neuronal cell apoptosis.

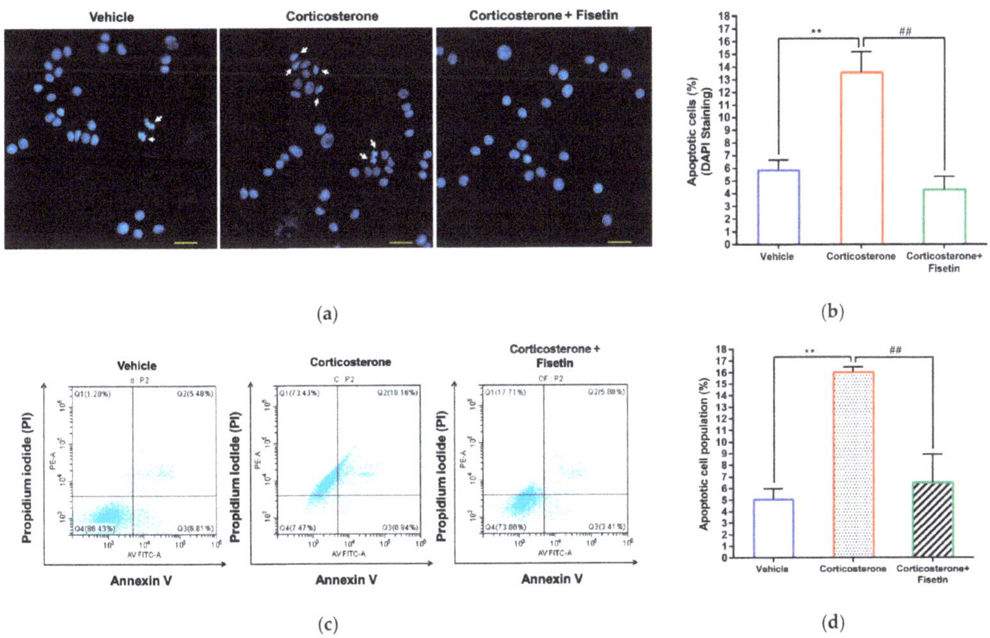


**Figure 3.** The effect of fisetin on cell apoptosis in corticosterone-treated PC12 cells. PC12 cells were treated with corticosterone (300 μM) in the absence or presence of fisetin (40 μM) for 24 h. After incubation, the cells were further incubated with DAPI. The stained cells were observed and photographed using fluorescence microscopy. (**a**) The nuclei were stained with DAPI (blue). A representative image is presented, with an arrow indicating apoptotic cells, and a scale bar measuring 20 μm. (**b**) Quantification of apoptotic cells. Apoptotic cells were quantified across 35 randomly selected fields from three independent replicates. The cell counts for vehicle-treated cells ($n$ = 805), corticosterone-treated cells ($n$ = 760), and cells co-treated with corticosterone and fisetin ($n$ = 658) were calculated. (**c**) The cell apoptosis was assessed using flow cytometric analysis. A representative histogram illustrating cell apoptosis is shown. (**d**) The percentage of the apoptotic cell population was quantified. The data represent the mean ± SD of three independent experiments. ** $p < 0.01$ indicates significant differences compared to the vehicle-treated group. ## $p < 0.01$ indicates significant differences compared to the corticosterone alone-treated group.

### 3.4. Involvement of MAPK/ERK, p38 MAPK and PI3/Akt Pathways in the Cytoprotective Effect of Fisetin against Corticosterone-Induced Cell Death

To further elucidate the signaling pathways underlying the neuroprotective effect of fisetin against corticosterone-induced cell death, we employed specific kinase inhibitors, including U0126, SP600125, SB203580, and LY294002, for investigations. As shown in Figure 4, the restoration of cell viability in PC12 cells treated with corticosterone, mediated by fisetin, was observed to be less effective when specific inhibitors U0126, SB203580, and LY294002 were introduced ($p < 0.01$), strongly suggesting the potential involvement of the ERK, p38 MAPK, and PI3/Akt pathways in the cytoprotective effect of fisetin against corticosterone-induced cell death.

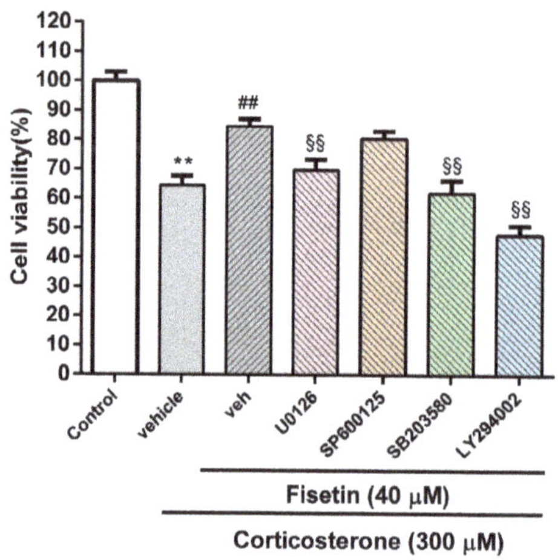


**Figure 4.** The kinase signaling pathways involved in fisetin-mediated neuroprotection in corticosterone-treated PC12 cells. Cells were pretreated with inhibitors, including U0126 (10 μM), SP600125 (10 μM), SB203580 (10 μM), and LY294002 (40 μM), respectively, for 30 min. Subsequently, the cells were exposed to corticosterone (300 μM) in the absence or presence of fisetin (40 μM) for 24 h, and cell viability was assessed using the MTT assay. The presented data represent the mean ± SD of three independent experiments. ** $p < 0.01$ indicates significant differences compared to the control group. ## $p < 0.01$ indicates significant differences compared to the corticosterone alone-treated (vehicle) group. §§ $p < 0.01$ indicates significant differences compared to the inhibitor-untreated (veh) group.

Additionally, we evaluated the activation status of key signaling molecules, namely ERK, p38 MAPK, and Akt, following treatment with fisetin using Western blot analysis. The results demonstrated that upon treatment with fisetin (40 μM) for 0, 0.5, 1, 2, and 4 h, the protein ratio of phospho-ERK1/2 (p-ERK) to ERK1/2 increased from 1.00 ± 0.42-fold to 4.51 ± 0.51-fold ($p < 0.01$), 10.26 ± 0.19-fold ($p < 0.01$), 9.98 ± 0.20-fold ($p < 0.01$), and 13.41 ± 0.59-fold ($p < 0.01$), respectively, compared to the 0 h group. (Figure 5a,b). Similarly, the protein ratio of phospho-p38 (p-p38) to p38 demonstrated significant increases due to fisetin treatment at 0, 0.5, 1, 2, and 4 h, increasing from 1.00 ± 0.19-fold to 1.37 ± 0.12-fold ($p < 0.05$), 1.98 ± 0.30-fold ($p < 0.01$), 2.61 ± 0.12-fold ($p < 0.01$), and 2.74 ± 0.23-fold ($p < 0.01$), respectively (Figure 5c,d). Notably, fisetin also significantly increased the protein ratio of phospho-Akt (p-Akt) to Akt from 1.00 ± 0.11-fold to 2.87 ± 0.56-fold ($p < 0.01$), 3.41 ± 0.32-fold ($p < 0.01$), 4.15 ± 0.74-fold ($p < 0.01$), and 5.92 ± 0.70-fold ($p < 0.01$), respectively, over the 0.5~4 h interval when compared to the 0 h group (Figure 5e,f). Collectively, these findings strongly indicate that fisetin might augment cell survival by triggering the activation of the ERK, p38, and PI3K/Akt pathways, thereby acting as a defense mechanism against corticosterone-induced cell death in PC12 cells.

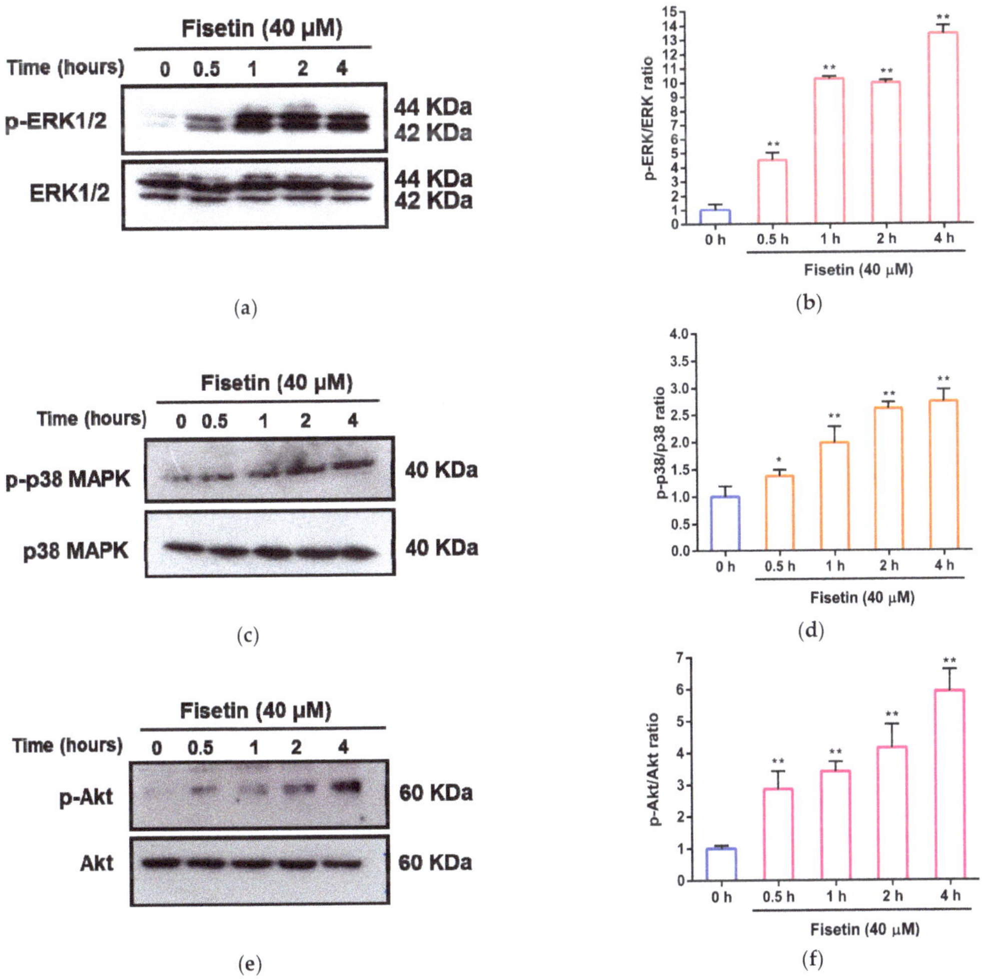


**Figure 5.** Fisetin enhanced the phosphorylation of ERK, p38, and Akt in PC12 cells. PC12 cells were treated with fisetin (40 μM) for 0.5, 1, 2, and 4 h. (**a**) p-ERK and ERK; (**c**) p-p38 and p38; (**e**) p-Akt and Akt proteins were measured by Western blot analysis, and a representative blot is shown. Quantitative analysis of normalized intensity of (**b**) p-ERK versus ERK, (**d**) p-p38 versus p38, (**f**) p-Akt versus Akt. The data represent the mean ± SD of three independent experiments. * $p < 0.05$ and ** $p < 0.01$ indicate significant differences compared to 0 h group.

### *3.5. Effects of Fisetin on CREB Activation in Corticosterone-Treated PC12 Cells*

Previous studies have highlighted the importance of CREB as a pivotal mediator in signal transduction pathways associating with cellular survival and mechanisms [43,44]. To investigate the possible involvement of CREB in the protective effects of fisetin in corticosterone-treated cells, we employed Western blot analysis to determine whether fisetin has the capacity to elevate the level of phospho-CREB proteins. As shown in Figure 6a,b, fisetin (40 μM) demonstrated a significant increase in the ratio of phospho-CREB (p-CREB) to CREB, rising from 1.00 ± 0.08-fold to 1.18 ± 0.11-fold ($p < 0.05$), 1.44 ± 0.14-fold ($p < 0.01$), 1.54 ± 0.14-fold ($p < 0.01$), and 1.64 ± 0.35-fold ($p < 0.01$), respectively, when compared to the 0 h group. Furthermore, to substantiate the effects of fisetin and corticosterone on phospho-CREB levels, our findings revealed that the group treated with corticosterone alone showed a reduction in phospho-CREB levels by approximately 0.28-fold ($p < 0.01$) when compared to the vehicle-treated group. In contrast, cells treated solely with fisetin demonstrated an increase in phospho-CREB levels by approximately 1.65-fold ($p < 0.01$). In the co-treatment scenario involving both corticosterone and fisetin, fisetin effectively counteracted the corticosterone-induced reduction in phospho-CREB levels from approximately 0.28-fold to 1.16-fold ($p < 0.01$) in PC12 cells (Figure 6c,d). The cumulative findings suggest that fisetin could potentially reverse the CREB inactivation induced by corticosterone in PC12 cells.

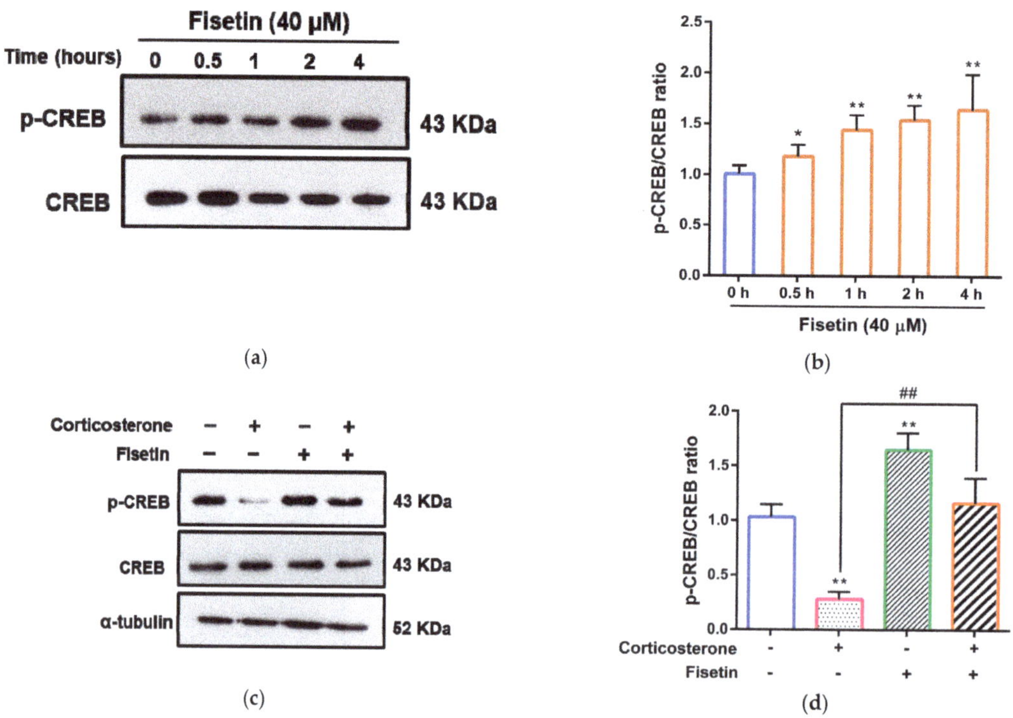


**Figure 6.** The effect of fisetin on the corticosterone-mediated inactivation of CREB. PC12 cells were treated with fisetin (40 μM) for 0, 0.5, 1, 2, and 4 h. (**a**) p-CREB and CREB proteins were measured by Western blot analysis, and a representative blot is shown. (**b**) Quantitative analysis of normalized intensity of p-CREB versus CREB. The data represent the mean ± SD of three independent experiments. * $p < 0.05$ and ** $p < 0.01$ indicate significant differences compared to the 0 h group. PC12 cells were treated with corticosterone (300 μM) in the absence or presence of fisetin (40 μM) for 24 h. (**c**) p-CREB, CREB, and α-tubulin proteins were measured by Western blot analysis, and a representative blot is shown. (**d**) Quantitative analysis of normalized intensity of p-CREB versus CREB. The data represent the mean ± SD of three independent experiments. ** $p < 0.01$ indicates significant differences compared to the vehicle-treated group. ## $p < 0.01$ indicates significant differences compared to the corticosterone alone-treated group.

We further investigated the potential crosstalk between the ERK pathway and CREB activation. Corticosterone prominently resulted in a reduction in phospho-ERK levels by approximately 0.08-fold, while fisetin alone significantly elevated phospho-ERK levels by approximately 1.50-fold compared to the vehicle-treated group in PC12 cells (Figure 7a,b). Importantly, in the co-treatment scenario involving both corticosterone and fisetin, fisetin effectively reversed the corticosterone-induced decrease in phospho-ERK levels from approximately 0.08-fold to 1.21-fold, compared to cells treated solely with corticosterone ($p < 0.01$). Furthermore, we employed the inhibitor U0126 to explore whether fisetin-induced CREB activation occurs through an ERK-dependent pathway. The data show that fisetin increased phospho-CREB levels by approximately 1.52-fold, and this effect remained unaltered by U0126 treatment in PC12 cells, where the levels increased by approximately 1.48-fold (Figure 7c,d). These results collectively indicate that fisetin induces CREB phosphorylation through a pathway that is independent of ERK activation.

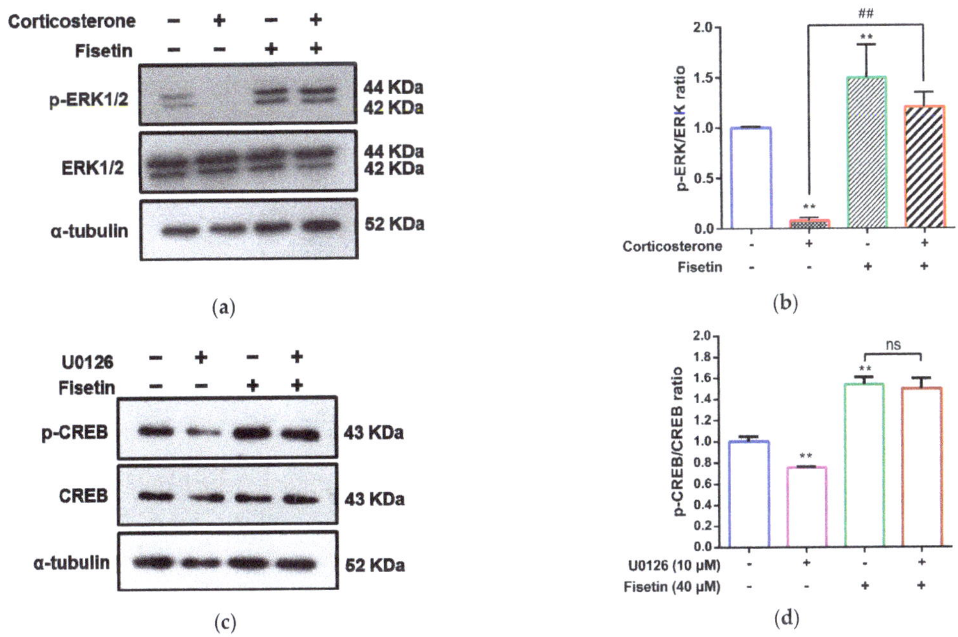


**Figure 7.** The impact of fisetin on crosstalk between the ERK pathway and CREB activation in corticosterone-treated PC12 cells. PC12 cells were treated with corticosterone (300 μM) in the absence or presence of fisetin (40 μM) for 24 h. (**a**) p-ERK, ERK, and α-tubulin proteins were measured by Western blot analysis, and a representative blot is shown. (**b**) Quantitative analysis of normalized intensity of p-ERK versus ERK. The data represent the mean ± SD of three independent experiments. ** $p < 0.01$ indicates significant differences compared to the vehicle-treated group. ## $p < 0.01$ indicates significant differences compared to the corticosterone alone-treated group. PC12 cells were pretreated with U0126 (10 μM) for 30 min, then treated fisetin (40 μM) for 24 h. (**c**) p-CREB, CREB and α-tubulin proteins were measured by Western blot analysis, and a representative blot is shown. (**d**) Quantitative analysis of the normalized intensity of p-CREB versus CREB. The data represent the mean ± SD of three independent experiments. ** $p < 0.01$ indicates statistically significant differences compared to the vehicle-treated group. "ns" indicates no statistically significant differences.

### *3.6. Fisetin Increases FOXO3a Phosphorylation and Promotes Cytosolic Localization of FOXO3a in PC12 Cells*

FOXO3a functions as a transcription factor, and its phosphorylation triggers translocation from the nucleus to the cytosol, subsequently downregulating the expression of genes linked to cell death and promoting cell survival [41]. As a result, we investigated the impact of fisetin on FOXO3a phosphorylation in PC12 cells using Western blot analysis. As depicted in Figure 8a,b, the protein ratio of phospho-FOXO3a (p-FOXO3a) to FOXO3a showed a significant increase in cells treated with fisetin. Furthermore, we utilized PC12 cells transfected with the plasmid expressing FOXO3-GFP proteins to confirm the effect of fisetin on the modulation of FOXO3a's subcellular localization. As shown in Figure 8c, administering fisetin increased the cytosolic localization of FOXO3a-GFP proteins in PC12 cells compared to the vehicle-treated cells. These findings suggest that fisetin may promote

FOXO3a phosphorylation, subsequently causing its translocation from the nucleus to the cytoplasm in neuronal cells.

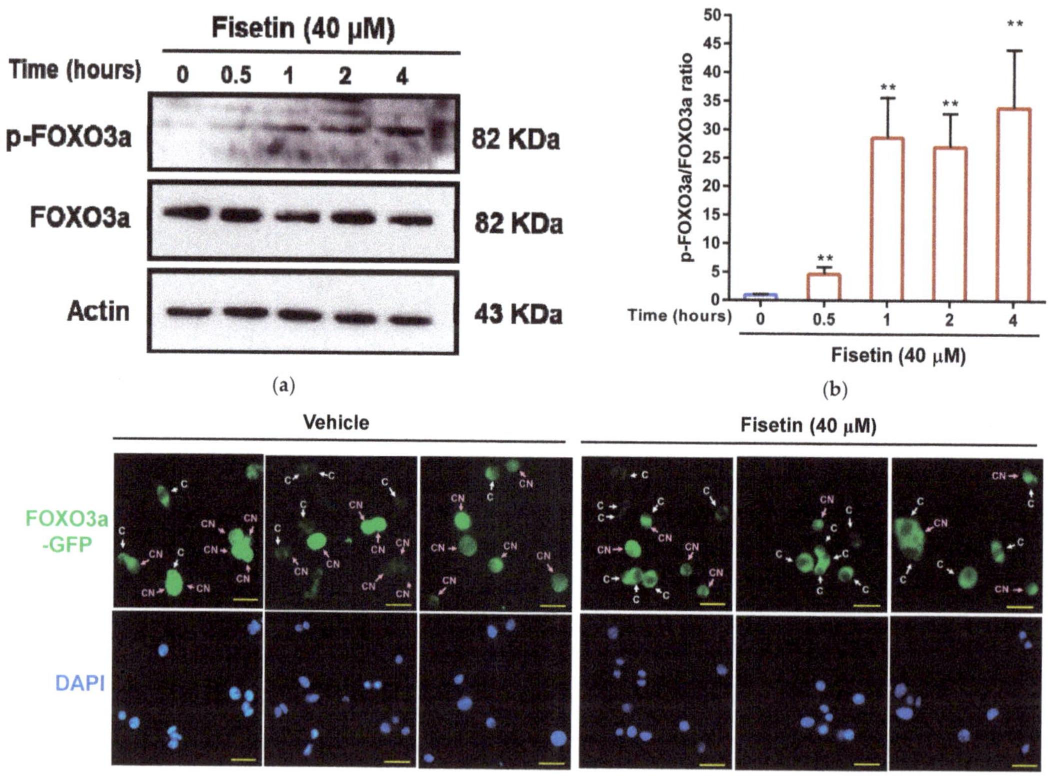


(c)

**Figure 8.** The effect of fisetin on the corticosterone-mediated retention of nuclear FOXO3a protein in PC12 cells. PC12 cells were treated with fisetin (40 μM) for 0, 0.5, 1, 2, and 4 h. (**a**) p-FOXO3a, FOXO3a, and actin proteins were measured by Western blot analysis, and a representative blot is shown. (**b**) Quantitative analysis of normalized intensity of p-FOXO3a versus FOXO3a. The data represent the mean ± SD of three independent experiments. ** $p < 0.01$ indicates significant differences compared to the 0 h group. (**c**) PC12 cells were cultured on a poly-L-lysine-coated coverslip, then transfected with pCMV6-AC-FOXO3 (GFP-tagged) plasmid for 24 h. The plasmid-transfected cells were treated with vehicle or fisetin (40 μM) for 24 h. After incubation, the cells were stained with DAPI. The FOXO3a-GFP protein (green) expressed in cells and the nuclei stained with DAPI (blue) were observed and photographed using fluorescence microscopy. The arrows showed the FOXO3a-GFP proteins located in the cytoplasm (C) and both in the cytoplasm and nucleus (CN). The representative images are shown, and the scale bar is 20 μm.

### *3.7. Fisetin Counteracts the Corticosterone-Induced Nuclear Accumulation of FOXO3a through the PI3K/Akt Pathway*

To investigate the modulation of FOXO3a's subcellular localization in corticosterone-treated cells, we performed a Western blot analysis to quantify nuclear FOXO3a protein levels in PC12 cells subjected to various treatments involving corticosterone, fisetin, or their combination. As illustrated in Figure 9a,b, corticosterone treatment significantly increased nuclear FOXO3a protein levels by approximately 1.58-fold compared to the vehicle-treated cells ($p < 0.01$). Conversely, fisetin treatment alone markedly reduced nuclear FOXO3a accumulation by approximately 0.33-fold ($p < 0.01$) in PC12 cells. Notably, in the co-treatment scenario involving corticosterone and fisetin, fisetin effectively attenuated the nuclear FOXO3a levels from around 1.58-fold to 1.21-fold ($p < 0.01$) when compared to cells exclusively treated with corticosterone. These results indicated that fisetin treatment promoted a pronounced translocation of FOXO3a proteins to the cytoplasm, contrasting with cells treated solely with corticosterone. Collectively, these findings reveal that fisetin significantly counteracts the corticosterone-induced nuclear retention of FOXO3a protein in PC12 cells.

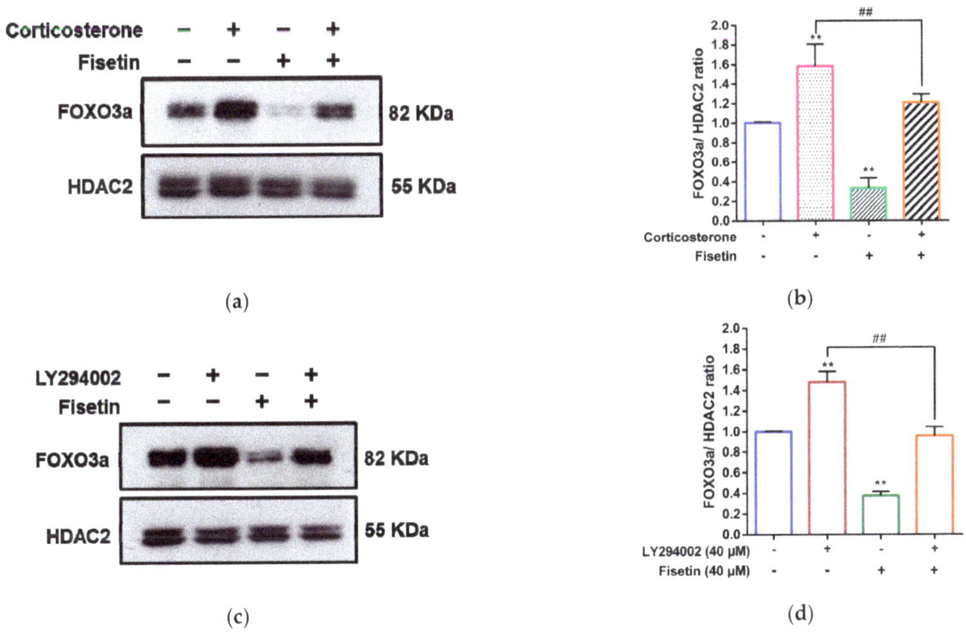


**Figure 9.** Fisetin counteracts the corticosterone-mediated retention of nuclear FOXO3a protein via the PI3K/Akt pathway. PC12 cells were treated with corticosterone (300 µM) in the absence or presence of fisetin (40 µM) for 24 h. (**a**) FOXO3a and HDAC2 proteins in the nucleus were measured by Western blot analysis, and a representative blot is shown. (**b**) A quantitative analysis of normalized intensity of FOXO3a versus HDAC2. The data represent the mean ± SD of three independent experiments. ** $p < 0.01$ indicates significant differences compared to the vehicle-treated group. ## $p < 0.01$ indicates significant differences compared to the corticosterone alone-treated group. PC12 cells were pre-treated with LY294002 (40 µM) for 30 min, then treated with fisetin (40 µM) for 24 h. (**c**) FOXO3a and HDAC2 proteins were measured by Western blot analysis, and a representative blot is shown. (**d**) Quantitative analysis of normalized intensity of FOXO3a versus HDAC2. The data represent the mean ± SD of three independent experiments. ** $p < 0.01$ indicates statistically significant differences compared to the vehicle-treated group. ## $p < 0.01$ indicates significant differences compared to the LY294002 alone-treated group.

Previous research has indicated that the activation of the PI3/Akt signaling pathway leads to the phosphorylation of FOXO3a, thus preventing the nuclear retention of FOXO3a [45]. We further investigated the effect of fisetin on the regulation of the PI3/Akt-FOXO3a pathway in corticosterone-treated PC12 cells. To explore this aspect, the experiments were initiated by pre-treating PC12 cells with LY294002, followed by their exposure to corticosterone and fisetin. The results, as shown in Figure 9c,d, demonstrated that the application of the inhibitor LY294002 led to an increase in nuclear FOXO3a levels within fisetin-treated cells. This observation suggests that fisetin possesses the capability to induce the cytoplasmic translocation of FOXO3a by activating the PI3/Akt pathway in corticosterone-treated cells. These results suggest fisetin's potential in promoting cell survival against corticosterone-induced cell death, achieved through the orchestrated modulation of the PI3/Akt-FOXO3a pathway in PC12 cells.

## 4. Discussion

In the context of this study, the flavonoid fisetin demonstrates its neuroprotective potential against corticosterone-induced cell death in PC12 cells. Fisetin exhibits the ability to potentially attenuate corticosterone-triggered ROS production and apoptosis in cells. Through the activation of ERK, p38, and CREB phosphorylation, fisetin enhances cell survival and safeguards against neuronal cell death caused by corticosterone. Furthermore, fisetin activates the PI3K/Akt pathway, leading to enhanced FOXO3a phosphorylation and cytoplasmic translocation, thereby decreasing the nuclear FOXO3a accumulation and preventing cell apoptosis prompted by corticosterone in PC12 cells (Figure 10). The PC12 cell line is recognized for displaying various typical neuronal characteristics and is extensively utilized in studies related to neurotoxicity and neuroprotection. Moreover, it is known for its ability to express a high level of glucocorticoid receptors. Due to these attributes, the PC12 cell line is widely employed as a model cell line in the development of neuroprotective agents [46,47]. Our current findings suggest that the neuroprotective efficacy of fisetin against corticosterone-induced neurotoxicity, combined with its ability to traverse the blood–brain barrier (BBB), holds the potential for it to emerge as a novel neuroprotective agent.

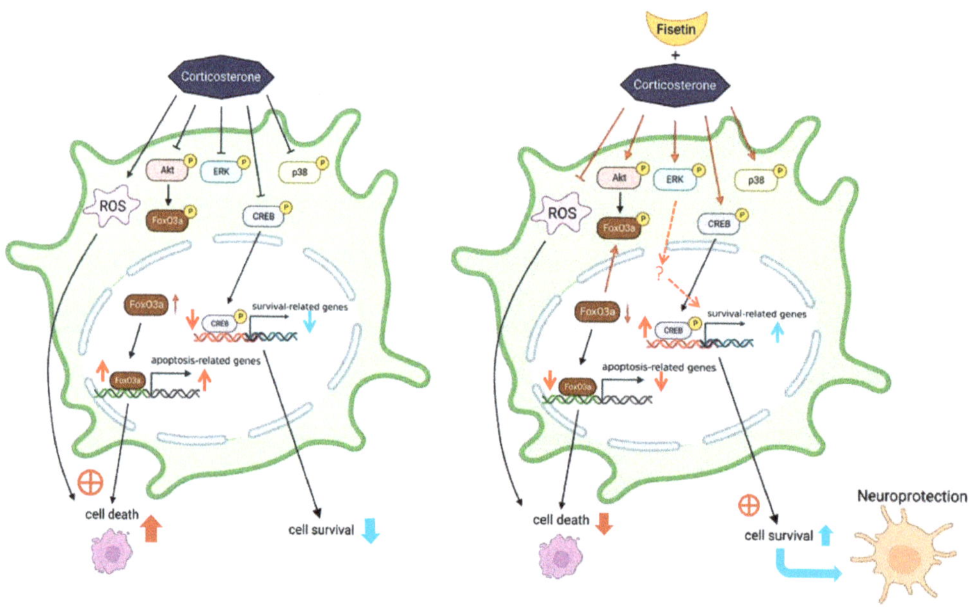


**Figure 10.** Hypothetical mechanism of the neuroprotective effect of fisetin against corticosterone-induced neuronal cell death. Fisetin has been shown to mitigate the corticosterone-mediated production of ROS and has the potential to inhibit apoptosis induced by corticosterone. Furthermore, fisetin demonstrates the capability to enhance cell survival and provide protection against corticosterone-induced cell death by modulating various survival pathways, including MAPK/ERK, p38 MAPK, CREB, and PI3K/Akt/FOXO3a, in PC12 cells. Thus, fisetin exerts a neuroprotective effect against corticosterone-induced neuronal cell death. (Created with BioRender.com).

It is established that corticosterone can trigger cellular damage in PC12 cells. Moreover, corticosterone has been associated with the induction of depression-like behaviors [48], alterations in brain anatomy, including the inhibition of structural plasticity in hippocampal neurons in animal models [49]. In this present study, we substantiated that corticosterone significantly induced cytotoxic effects on the neuronal PC12 cell line in a dose-dependent manner. Phytochemicals encompass bioactive chemical compounds derived from plants, such as polyphenols and flavonoids, which are abundant in fruits, vegetables, herbs, and Chinese medicines [50,51]. Numerous studies underscore the multifaceted beneficial effects of phytochemicals, including antioxidant, anti-inflammatory, anti-cancer, and neuroprotective activities [51,52]. Research has indicated that certain dietary phytochemicals, like polyphenols and flavonoids, can effectively traverse the BBB, enhancing their potential as agents for the prevention or treatment of neurological disorders [53,54]. In prior investigations conducted by our research group, it was demonstrated that various phytochemicals, including fisetin [36], 5-demethylnobiletin [55], and luteolin [56], exhibited the capacity to stimulate neuritogenesis and provide neuroprotection in PC12 cells. In this study, we present evidence showcasing the significant ability of fisetin to counteract corticosterone-induced ROS generation and cell apoptosis, thereby promoting cell survival within PC12 cell models. Previous reports have demonstrated that fisetin prevents dexamethasone-induced cytotoxicity by attenuating ROS levels and apoptosis in hippocampal HT22 cell lines [57]. These findings suggest that fisetin could protect neuronal cells against corticosteroid-induced oxidative damage. It has been reported that high stress-induced releases of glucocorticoids can affect synaptic plasticity and the neuronal network in the hippocampus, impairing cognitive functions [58,59]. Gite et al. reported that fisetin could elevate brain-derived neurotrophic factor (BDNF) expression in corticosterone-treated human neuroblastoma cell lines [60]. BDNF is a neurotrophin in the brain that regulates neuronal development, synaptic plasticity, and neuroprotection, playing essential roles in cognition and memory [61]. These findings suggest that fisetin may modulate neuronal cell plasticity in the brain under corticosteroid exposure. Our current findings support the notion that dietary fisetin has the potential to act as a neuroprotective agent, preserving neuronal cell survival and regulating neuronal plasticity against corticosteroid-induced neurotoxicity in the brain.

It has been reported that intracellular kinase signaling cascades are considered responsible for promoting neuronal survival [62]. In this study, we demonstrated that fisetin significantly enhances the phosphorylation of ERK, p38, and Akt, leading to increased cell survival in corticosterone-treated PC12 cells. The MAPK/ERK pathway is known to play a vital role in regulating cell survival and proliferation [63]. In this study, fisetin was able to enhance the phosphorylation level of ERK1/2, in concordance with the results from our previous study [36,64]. Additionally, the p38 MAPK pathway plays a significant role in cell differentiation and regulating cell survival against oxidative stress [65]. Furthermore, it may contribute to inflammation, apoptosis, and cell cycle regulation [66]. Akt phosphorylation has been reported to regulate neuronal toxicity through various substrates, including FOXOs and GSK3β, among others [67]. These findings support the notion that fisetin modulates the MAPK and PI3K/Akt pathways, thereby offering protection to neuronal cells against corticosterone-induced cell death, which could potentially contribute to the prevention or treatment of corticosteroids-induced neurotoxicity.

CREB is a transcription factor that is involved in numerous functions within neuronal cells, including the promotion of neuronal cell survival, neuronal differentiation, migration, and synaptogenesis [68]. In this study, our results revealed that fisetin significantly elevates the phosphorylation level of CREB, thereby effectively reversing the corticosterone-induced inactivation of CREB. This restoration of CREB activity is associated with enhanced neuronal cell survival. In our results, fisetin not only reversed the corticosterone-mediated inactivation of CREB, but also countered the corticosterone-induced inactivation of ERK. Nevertheless, our results demonstrated that the elevation of phosphorylated CREB levels induced by fisetin remained unaffected by the MAPK/ERK inhibitor U0126. This suggests that CREB might not be directly modulated by fisetin-mediated ERK activation in PC12 cells. As previously mentioned, the phosphorylation of the CREB protein can also be facilitated by the PI3K/Akt kinase. Our results demonstrated that fisetin could indeed enhance the phosphorylation of Akt. However, whether fisetin-induced CREB activation occurs through a PI3K/Akt-dependent pathway requires further investigation.

It has been reported that the activity of FOXO3a can be regulated through post-translational modifications, including phosphorylation, methylation, acetylation, and ubiquitination. The post-translational phosphorylation of the FOXO3a protein influences its localization within various cellular compartments, ultimately determining whether it becomes inactivated (in the cytoplasm) or activated (in the nucleus) [45,69]. In this study, fisetin treatment increased the levels of FOXO3a phosphorylation and induced the cytoplasmic localization of FOXO3a through a PI3K/Akt-dependent pathway in PC12 cells. Brunet et al. reported that PI3K/Akt activation enhances cell survival by phosphorylating and inhibiting the activity of the FOXO3a protein [70]. Previous studies have shown that activated FOXO3a can induce the expression of downstream death-related genes, such as the p53-upregulated modulator of apoptosis (Puma) and Bcl-2-interacting mediator of cell death (Bim) [41]. In our study, we observed an increased amount of FOXO3a in the nucleus, alterations in cell morphology, and cell death due to corticosterone treatment. Our current findings support the notion that fisetin-mediated neuroprotection against corticosterone-induced cell death may be correlated with the phosphorylation of FOXO3a through the PI3K/Akt pathway in neuronal cells. However, the specific signaling molecules involved in fisetin-mediated FOXO3a translocation remain unclear and require further investigation.

In conclusion, our findings suggest that fisetin exerts a neuroprotective effect against corticosterone-induced cell death through multiple mechanisms, including ROS reduction, apoptosis inhibition, and the promotion of cell survival. These effects are achieved through the modulation of various pathways such as MAPK/ERK-, p38-, and the PI3/Akt/FOXO3a-dependent pathways. However, it is important to note that the PC12 cell lines used in this study may not perfectly reflect the complexities of neuronal cells. Given these promising results, further investigation is warranted to fully understand the neuroprotective effects of fisetin in primary neurons. This should include animal and human studies to better evaluate the potential preventive or therapeutic benefits of fisetin in corticosteroid-induced neuronal or psychiatric disorders.

**Author Contributions:** Conceptualization, J.-W.L. and J.-H.Y.; Data curation, P.-R.C., J.-W.L. and P.-Y.C.; Funding acquisition, J.-H.Y.; Investigation, P.-R.C. and P.-Y.C.; Methodology, P.-R.C., W.-Y.G., C.-L.W. and M.-J.W.; Project administration, J.-H.Y.; Supervision, J.-H.Y.; Visualization, J.-H.Y.; Writing—original draft, P.-R.C. and J.-H.Y.; Writing—review and editing, J.-W.L. and J.-H.Y. All authors have read and agreed to the published version of the manuscript.

**Funding:** This research was supported by the grant MOST 111-2320-B-320-004 (to Dr. J-H Yen) from the National Science and Technology Council, Taiwan.

**Institutional Review Board Statement:** Not applicable.

**Informed Consent Statement:** Not applicable.

**Data Availability Statement:** The data can be shared up on request.

**Acknowledgments:** We are grateful for the support from the Core Research Laboratory, Tzu Chi University, for the assistance in operating the Beckman CytoFLEX Flow Cytometer and Nikon Ti2-E inverted fluorescent microscope.

**Conflicts of Interest:** The authors declare no conflict of interest.

## References

1. Vyas, S.; Rodrigues, A.J.; Silva, J.M.; Tronche, F.; Almeida, O.F.; Sousa, N.; Sotiropoulos, I. Chronic Stress and Glucocorticoids: From Neuronal Plasticity to Neurodegeneration. *Neural Plast.* **2016**, *2016*, 6391686. [CrossRef]
2. Herman, J.P.; McKlveen, J.M.; Ghosal, S.; Kopp, B.; Wulsin, A.; Makinson, R.; Scheimann, J.; Myers, B. Regulation of the Hypothalamic-Pituitary-Adrenocortical Stress Response. *Compr. Physiol.* **2016**, *6*, 603–621.

3. Peng, G.J.; Tian, J.S.; Gao, X.X.; Zhou, Y.Z.; Qin, X.M. Research on the Pathological Mechanism and Drug Treatment Mechanism of Depression. *Curr. Neuropharmacol.* **2015**, *13*, 514–523. [CrossRef] [PubMed]
4. Ramamoorthy, S.; Cidlowski, J.A. Corticosteroids: Mechanisms of Action in Health and Disease. *Rheum. Dis. Clin. N. Am.* **2016**, *42*, 15–31. [CrossRef] [PubMed]
5. Ising, M.; Horstmann, S.; Kloiber, S.; Lucae, S.; Binder, E.B.; Kern, N.; Kunzel, H.E.; Pfennig, A.; Uhr, M.; Holsboer, F. Combined dexamethasone/corticotropin releasing hormone test predicts treatment response in major depression—A potential biomarker? *Biol. Psychiatry* **2007**, *62*, 47–54. [CrossRef] [PubMed]
6. Almeida, F.B.; Pinna, G.; Barros, H.M.T. The Role of HPA Axis and Allopregnanolone on the Neurobiology of Major Depressive Disorders and PTSD. *Int. J. Mol. Sci.* **2021**, *22*, 5495. [CrossRef] [PubMed]
7. Pearson Murphy, B.E. Antiglucocorticoid therapies in major depression: A review. *Psychoneuroendocrinology* **1997**, *22*, S125–S132. [CrossRef]
8. Joseph, J.J.; Golden, S.H. Cortisol dysregulation: The bidirectional link between stress, depression, and type 2 diabetes mellitus. *Ann. N. Y. Acad. Sci.* **2017**, *1391*, 20–34. [CrossRef]
9. Pretorius, E. Corticosteroids, depression and the role of serotonin. *Rev. Neurosci.* **2004**, *15*, 109–116. [CrossRef]
10. Spiers, J.G.; Chen, H.J.; Sernia, C.; Lavidis, N.A. Activation of the hypothalamic-pituitary-adrenal stress axis induces cellular oxidative stress. *Front. Neurosci.* **2014**, *8*, 456. [CrossRef]
11. Picard, K.; St-Pierre, M.K.; Vecchiarelli, H.A.; Bordeleau, M.; Tremblay, M.E. Neuroendocrine, neuroinflammatory and pathological outcomes of chronic stress: A story of microglial remodeling. *Neurochem. Int.* **2021**, *145*, 104987. [CrossRef]
12. Hill, A.S.; Sahay, A.; Hen, R. Increasing Adult Hippocampal Neurogenesis is Sufficient to Reduce Anxiety and Depression-Like Behaviors. *Neuropsychopharmacology* **2015**, *40*, 2368–2378. [CrossRef]
13. Podgorny, O.V.; Gulyaeva, N.V. Glucocorticoid-mediated mechanisms of hippocampal damage: Contribution of subgranular neurogenesis. *J. Neurochem.* **2021**, *157*, 370–392. [CrossRef]
14. Kwon, M.S.; Kim, M.H.; Kim, S.H.; Park, K.D.; Yoo, S.H.; Oh, I.U.; Pak, S.; Seo, Y.J. Erythropoietin exerts cell protective effect by activating PI3K/Akt and MAPK pathways in C6 Cells. *Neurol. Res.* **2014**, *36*, 215–223. [CrossRef]
15. Owuor, E.D.; Kong, A.N. Antioxidants and oxidants regulated signal transduction pathways. *Biochem. Pharmacol.* **2002**, *64*, 765–770. [CrossRef] [PubMed]
16. Singh, E.; Devasahayam, G. Neurodegeneration by oxidative stress: A review on prospective use of small molecules for neuroprotection. *Mol. Biol. Rep.* **2020**, *47*, 3133–3140. [CrossRef] [PubMed]
17. Albert-Gascó, H.; Ros-Bernal, F.; Castillo-Gómez, E.; Olucha-Bordonau, F.E. MAP/ERK Signaling in Developing Cognitive and Emotional Function and Its Effect on Pathological and Neurodegenerative Processes. *Int. J. Mol. Sci.* **2020**, *21*, 4471. [CrossRef] [PubMed]
18. Chen, H.; Teng, Y.; Chen, X.; Liu, Z.; Geng, F.; Liu, Y.; Jiang, H.; Wang, Z.; Yang, L. Platelet-derived growth factor (PDGF)-BB protects dopaminergic neurons via activation of Akt/ERK/CREB pathways to upregulate tyrosine hydroxylase. *CNS Neurosci. Ther.* **2021**, *27*, 1300–1312. [CrossRef] [PubMed]
19. Cohen-Cory, S.; Kidane, A.H.; Shirkey, N.J.; Marshak, S. Brain-derived neurotrophic factor and the development of structural neuronal connectivity. *Dev. Neurobiol.* **2010**, *70*, 271–288. [CrossRef]
20. Wang, H.; Xu, J.; Lazarovici, P.; Quirion, R.; Zheng, W. cAMP Response Element-Binding Protein (CREB): A Possible Signaling Molecule Link in the Pathophysiology of Schizophrenia. *Front. Mol. Neurosci.* **2018**, *11*, 255. [CrossRef]
21. Monje, P.; Hernández-Losa, J.; Lyons, R.J.; Castellone, M.D.; Gutkind, J.S. Regulation of the transcriptional activity of c-Fos by ERK. A novel role for the prolyl isomerase PIN1. *J. Biol. Chem.* **2005**, *280*, 35081–35084. [CrossRef] [PubMed]
22. Gilbert, J.; Man, H.Y. Fundamental Elements in Autism: From Neurogenesis and Neurite Growth to Synaptic Plasticity. *Front. Cell Neurosci.* **2017**, *11*, 359. [CrossRef]
23. Sharma, A.; Mehan, S. Targeting PI3K-AKT/mTOR signaling in the prevention of autism. *Neurochem. Int.* **2021**, *147*, 105067. [CrossRef] [PubMed]
24. Nho, R.S.; Hergert, P. FoxO3a and disease progression. *World J. Biol. Chem.* **2014**, *5*, 346–354. [CrossRef]
25. Xie, Y.; Shi, X.; Sheng, K.; Han, G.; Li, W.; Zhao, Q.; Jiang, B.; Feng, J.; Li, J.; Gu, Y. PI3K/Akt signaling transduction pathway, erythropoiesis and glycolysis in hypoxia (Review). *Mol. Med. Rep.* **2019**, *19*, 783–791. [CrossRef]
26. Zhang, X.; Tang, N.; Hadden, T.J.; Rishi, A.K. Akt, FoxO and regulation of apoptosis. *Biochim. Biophys. Acta* **2011**, *1813*, 1978–1986. [CrossRef]
27. Martins, J.; Brijesh, S. Phytochemistry and pharmacology of anti-depressant medicinal plants: A review. *Biomed. Pharmacother.* **2018**, *104*, 343–365. [CrossRef]
28. Khan, A.; Jahan, S.; Imtiyaz, Z.; Alshahrani, S.; Antar Makeen, H.; Mohammed Alshehri, B.; Kumar, A.; Arafah, A.; Rehman, M.U. Neuroprotection: Targeting Multiple Pathways by Naturally Occurring Phytochemicals. *Biomedicines* **2020**, *8*, 284. [CrossRef]
29. Pal, H.C.; Pearlman, R.L.; Afaq, F. Fisetin and Its Role in Chronic Diseases. *Adv. Exp. Med. Biol.* **2016**, *928*, 213–244.
30. Kubina, R.; Krzykawski, K.; Kabala-Dzik, A.; Wojtyczka, R.D.; Chodurek, E.; Dziedzic, A. Fisetin, a Potent Anticancer Flavonol Exhibiting Cytotoxic Activity against Neoplastic Malignant Cells and Cancerous Conditions: A Scoping, Comprehensive Review. *Nutrients* **2022**, *14*, 2604. [CrossRef] [PubMed]
31. Rahmani, A.H.; Almatroudi, A.; Allemailem, K.S.; Khan, A.A.; Almatroodi, S.A. The Potential Role of Fisetin, a Flavonoid in Cancer Prevention and Treatment. *Molecules* **2022**, *27*, 9009. [CrossRef]

32. Prem, P.N.; Sivakumar, B.; Boovarahan, S.R.; Kurian, G.A. Recent advances in potential of Fisetin in the management of myocardial ischemia-reperfusion injury-A systematic review. *Phytomedicine* **2022**, *101*, 154123. [CrossRef]
33. Ravula, A.R.; Teegala, S.B.; Kalakotla, S.; Pasangulapati, J.P.; Perumal, V.; Boyina, H.K. Fisetin, potential flavonoid with multifarious targets for treating neurological disorders: An updated review. *Eur. J. Pharmacol.* **2021**, *910*, 174492. [CrossRef]
34. Hassan, S.S.U.; Samanta, S.; Dash, R.; Karpinski, T.M.; Habibi, E.; Sadiq, A.; Ahmadi, A.; Bunagu, S. The neuroprotective effects of fisetin, a natural flavonoid in neurodegenerative diseases: Focus on the role of oxidative stress. *Front. Pharmacol.* **2022**, *13*, 1015835. [CrossRef]
35. Jiang, Y.; Tang, X.; Deng, P.; Jiang, C.; He, Y.; Hao, D.; Yang, H. The Neuroprotective Role of Fisetin in Different Neurological Diseases: A Systematic Review. *Mol. Neurobiol.* **2023**. [CrossRef] [PubMed]
36. Chen, P.Y.; Ho, Y.R.; Wu, M.J.; Huang, S.P.; Chen, P.K.; Tai, M.H.; Ho, C.T.; Yen, J.H. Cytoprotective effects of fisetin against hypoxia-induced cell death in PC12 cells. *Food Funct.* **2015**, *6*, 287–296. [CrossRef] [PubMed]
37. Yu, X.; Jiang, X.; Zhang, X.; Chen, Z.; Xu, L.; Chen, L.; Wang, G.; Pan, J. The effects of fisetin on lipopolysaccharide-induced depressive-like behavior in mice. *Metab. Brain Dis.* **2016**, *31*, 1011–1021. [CrossRef] [PubMed]
38. Yao, X.; Li, L.; Kandhare, A.D.; Mukherjee-Kandhare, A.A.; Bodhankar, S.L. Attenuation of reserpine-induced fibromyalgia via ROS and serotonergic pathway modulation by fisetin, a plant flavonoid polyphenol. *Exp. Ther. Med.* **2020**, *19*, 1343–1355. [CrossRef]
39. Wang, Y.; Wang, B.; Lu, J.; Shi, H.; Gong, S.; Wang, Y.; Hamdy, R.C.; Chua, B.H.L.; Yang, L.; Xu, X. Fisetin provides antidepressant effects by activating the tropomyosin receptor kinase B signal pathway in mice. *J. Neurochem.* **2017**, *143*, 561–568. [CrossRef]
40. Chen, P.Y.; Wang, C.Y.; Tsao, E.C.; Chen, Y.T.; Wu, M.J.; Ho, C.T.; Yen, J.H. 5-Demethylnobiletin Inhibits Cell Proliferation, Downregulates ID1 Expression, Modulates the NF-kappaB/TNF-alpha Pathway and Exerts Antileukemic Effects in AML Cells. *Int. J. Mol. Sci.* **2022**, *23*, 7392. [CrossRef]
41. Lin, R.; Liu, L.; Silva, M.; Fang, J.; Zhou, Z.; Wang, H.; Xu, J.; Li, T.; Zheng, W. Hederagenin Protects PC12 Cells Against Corticosterone-Induced Injury by the Activation of the PI3K/AKT Pathway. *Front. Pharmacol.* **2021**, *12*, 712876. [CrossRef]
42. Shi, X.; Zhou, N.; Cheng, J.; Shi, X.; Huang, H.; Zhou, M.; Zhu, H. Chlorogenic acid protects PC12 cells against corticosterone-induced neurotoxicity related to inhibition of autophagy and apoptosis. *BMC Pharmacol. Toxicol.* **2019**, *20*, 56. [CrossRef]
43. Guo, Y.J.; Pan, W.W.; Liu, S.B.; Shen, Z.F.; Xu, Y.; Hu, L.L. ERK/MAPK signalling pathway and tumorigenesis. *Exp. Ther. Med.* **2020**, *19*, 1997–2007. [CrossRef]
44. Qi, X.; Lin, W.; Li, J.; Li, H.; Wang, W.; Wang, D.; Sun, M. Fluoxetine increases the activity of the ERK-CREB signal system and alleviates the depressive-like behavior in rats exposed to chronic forced swim stress. *Neurobiol. Dis.* **2008**, *31*, 278–285. [CrossRef]
45. Fasano, C.; Disciglio, V.; Bertora, S.; Lepore Signorile, M.; Simone, C. FOXO3a from the Nucleus to the Mitochondria: A Round Trip in Cellular Stress Response. *Cells* **2019**, *8*, 1110. [CrossRef]
46. Tian, J.S.; Liu, S.B.; He, X.Y.; Xiang, H.; Chen, J.L.; Gao, Y.; Zhou, Y.Z.; Qin, X.M. Metabolomics studies on corticosterone-induced PC12 cells: A strategy for evaluating an in vitro depression model and revealing the metabolic regulation mechanism. *Neurotoxicol. Teratol.* **2018**, *69*, 27–38. [CrossRef] [PubMed]
47. Oprea, D.; Sanz, C.G.; Barsan, M.M.; Enache, T.A. PC-12 Cell Line as a Neuronal Cell Model for Biosensing Applications. *Biosensors* **2022**, *12*, 500. [CrossRef] [PubMed]
48. Zhao, Y.; Ma, R.; Shen, J.; Su, H.; Xing, D.; Du, L. A mouse model of depression induced by repeated corticosterone injections. *Eur. J. Pharmacol.* **2008**, *581*, 113–120. [CrossRef]
49. Zhang, H.Y.; Zhao, Y.N.; Wang, Z.L.; Huang, Y.F. Chronic corticosterone exposure reduces hippocampal glycogen level and induces depression-like behavior in mice. *J. Zhejiang Univ. Sci. B* **2015**, *16*, 62–69. [CrossRef] [PubMed]
50. Fantini, M.; Benvenuto, M.; Masuelli, L.; Frajese, G.V.; Tresoldi, I.; Modesti, A.; Bei, R. In vitro and in vivo antitumoral effects of combinations of polyphenols, or polyphenols and anticancer drugs: Perspectives on cancer treatment. *Int. J. Mol. Sci.* **2015**, *16*, 9236–9282. [CrossRef]
51. Calis, Z.; Mogulkoc, R.; Baltaci, A.K. The Roles of Flavonols/Flavonoids in Neurodegeneration and Neuroinflammation. *Mini-Rev. Med. Chem.* **2020**, *20*, 1475–1488. [CrossRef] [PubMed]
52. Balakrishnan, R.; Azam, S.; Cho, D.Y.; Su-Kim, I.; Choi, D.K. Natural Phytochemicals as Novel Therapeutic Strategies to Prevent and Treat Parkinson's Disease: Current Knowledge and Future Perspectives. *Oxid. Med. Cell. Longev.* **2021**, *2021*, 6680935. [CrossRef] [PubMed]
53. Ullah, H.; Khan, H. Anti-Parkinson Potential of Silymarin: Mechanistic Insight and Therapeutic Standing. *Front. Pharmacol.* **2018**, *9*, 422. [CrossRef]
54. Khan, H.; Ullah, H.; Aschner, M.; Cheang, W.S.; Akkol, E.K. Neuroprotective Effects of Quercetin in Alzheimer's Disease. *Biomolecules* **2019**, *10*, 59. [CrossRef] [PubMed]
55. Chiu, S.P.; Wu, M.J.; Chen, P.Y.; Ho, Y.R.; Tai, M.H.; Ho, C.T.; Yen, J.H. Neurotrophic action of 5-hydroxylated polymethoxyflavones: 5-demethylnobiletin and gardenin A stimulate neuritogenesis in PC12 cells. *J. Agric. Food Chem.* **2013**, *61*, 9453–9463. [CrossRef]
56. Lin, L.F.; Chiu, S.P.; Wu, M.J.; Chen, P.Y.; Yen, J.H. Luteolin induces microRNA-132 expression and modulates neurite outgrowth in PC12 cells. *PLoS ONE* **2012**, *7*, e43304. [CrossRef]
57. Inkielewicz-Stepniak, I.; Radomski, M.W.; Wozniak, M. Fisetin prevents fluoride- and dexamethasone-induced oxidative damage in osteoblast and hippocampal cells. *Food Chem. Toxicol.* **2012**, *50*, 583–589. [CrossRef]

58. Madalena, K.M.; Lerch, J.K. The Effect of Glucocorticoid and Glucocorticoid Receptor Interactions on Brain, Spinal Cord, and Glial Cell Plasticity. *Neural Plast.* **2017**, *2017*, 8640970. [CrossRef]
59. McEwen, B.S. Corticosteroids and hippocampal plasticity. *Ann. N. Y. Acad. Sci.* **1994**, *746*, 134–142; discussion 142-4, 178–179. [CrossRef]
60. Gite, S.; Ross, R.P.; Kirke, D.; Guiheneuf, F.; Aussant, J.; Stengel, D.B.; Dinan, T.G.; Cryan, J.F.; Stanton, C. Nutraceuticals to promote neuronal plasticity in response to corticosterone-induced stress in human neuroblastoma cells. *Nutr. Neurosci.* **2019**, *22*, 551–568. [CrossRef]
61. Kowianski, P.; Lietzau, G.; Czuba, E.; Waskow, M.; Steliga, A.; Morys, J. BDNF: A Key Factor with Multipotent Impact on Brain Signaling and Synaptic Plasticity. *Cell. Mol. Neurobiol.* **2018**, *38*, 579–593. [CrossRef]
62. Frebel, K.; Wiese, S. Signalling molecules essential for neuronal survival and differentiation. *Biochem. Soc. Trans.* **2006**, *34 Pt 6*, 1287–1290. [CrossRef]
63. Hossain, M.S.; Ifuku, M.; Take, S.; Kawamura, J.; Miake, K.; Katafuchi, T. Plasmalogens rescue neuronal cell death through an activation of AKT and ERK survival signaling. *PLoS ONE* **2013**, *8*, e83508. [CrossRef] [PubMed]
64. Yen, J.H.; Wu, P.S.; Chen, S.F.; Wu, M.J. Fisetin Protects PC12 Cells from Tunicamycin-Mediated Cell Death via Reactive Oxygen Species Scavenging and Modulation of Nrf2-Driven Gene Expression, SIRT1 and MAPK Signaling in PC12 Cells. *Int. J. Mol. Sci.* **2017**, *18*, 852. [CrossRef]
65. Cuadrado, A.; Nebreda, A.R. Mechanisms and functions of p38 MAPK signalling. *Biochem. J.* **2010**, *429*, 403–417. [CrossRef]
66. Cargnello, M.; Roux, P.P. Activation and function of the MAPKs and their substrates, the MAPK-activated protein kinases. *Microbiol. Mol. Biol. Rev.* **2011**, *75*, 50–83. [CrossRef] [PubMed]
67. Manning, B.D.; Toker, A. AKT/PKB Signaling: Navigating the Network. *Cell* **2017**, *169*, 381–405. [CrossRef] [PubMed]
68. Esvald, E.E.; Tuvikene, J.; Sirp, A.; Patil, S.; Bramham, C.R.; Timmusk, T. CREB Family Transcription Factors Are Major Mediators of BDNF Transcriptional Autoregulation in Cortical Neurons. *J. Neurosci.* **2020**, *40*, 1405–1426. [CrossRef]
69. Wang, Z.; Yu, T.; Huang, P. Post-translational modifications of FOXO family proteins (Review). *Mol. Med. Rep.* **2016**, *14*, 4931–4941. [CrossRef]
70. Brunet, A.; Bonni, A.; Zigmond, M.J.; Lin, M.Z.; Juo, P.; Hu, L.S.; Anderson, M.J.; Arden, K.C.; Blenis, J.; Greenberg, M.E. Akt promotes cell survival by phosphorylating and inhibiting a Forkhead transcription factor. *Cell* **1999**, *96*, 857–868. [CrossRef]

*pharmaceutics*

MDPI

*Article*

# (-)-5-Demethoxygrandisin B a New Lignan from *Virola surinamensis (Rol.) Warb.* Leaves: Evaluation of the Leishmanicidal Activity by In Vitro and In Silico Approaches

**Steven Souza Paes [1,†], João Victor Silva-Silva [2,3,†], Paulo Wender Portal Gomes [4], Luely Oliveira da Silva [5], Ana Paula Lima da Costa [1], Manoel Leão Lopes Júnior [1], Daiana de Jesus Hardoim [2], Carla J. Moragas-Tellis [6], Noemi Nosomi Taniwaki [7], Alvaro Luiz Bertho [8,9], Fábio Alberto de Molfetta [10], Fernando Almeida-Souza [2,11,*], Lourivaldo Silva Santos [1,*] and Kátia da Silva Calabrese [2]**

1 Institute of Exact and Natural Sciences, Federal University of Pará, Belém 66075-110, PA, Brazil
2 Laboratory of Protozoology, Oswaldo Cruz Institute, Oswaldo Cruz Foundation, Rio de Janeiro 21041-250, RJ, Brazil; calabrese@ioc.fiocruz.br (K.d.S.C.)
3 Laboratory of Medicinal and Computational Chemistry, Institute of Physics of São Carlos, University of São Paulo, São Carlos 13418-900, SP, Brazil
4 Collaborative Mass Spectrometry Innovation Center, Skaggs School of Pharmacy and Pharmaceutical Sciences, University of California, San Diego, CA 92123, USA
5 Department of Natural Sciences, Pará State University, Belém 66095-015, PA, Brazil
6 Laboratory of Natural Products for Public Health, Pharmaceutical Technology Institute, Farmanguinhos, Oswaldo Cruz Foundation, Rio de Janeiro 21040-900, RJ, Brazil
7 Electron Microscopy Nucleus, Adolfo Lutz Institute, São Paulo 01246-000, SP, Brazil
8 Laboratory of Immunoparasitology, Oswaldo Cruz Institute, Oswaldo Cruz Foundation, Rio de Janeiro 21040-900, RJ, Brazil; alvaro.bertho@ioc.fiocruz.br
9 Flow Cytometry Core Facility, Oswaldo Cruz Institute, Oswaldo Cruz Foundation, Rio de Janeiro 21040-900, RJ, Brazil
10 Laboratory of Molecular Modeling, Institute of Exact and Natural Sciences, Federal University of Pará, Belém 66075-110, PA, Brazil; fabioam@ufpa.br
11 Postgraduate Program in Animal Science, State University of Maranhão, Sao Luis 65055-310, MA, Brazil
* Correspondence: fernandoalsouza@gmail.com (F.A.-S.); lss@ufpa.br (L.S.S.)
† These authors contributed equally to this work.

**Abstract:** Leishmaniasis is a complex disease caused by infection with different *Leishmania* parasites. The number of medications used for its treatment is still limited and the discovery of new drugs is a valuable approach. In this context, here we describe the in vitro leishmanicidal activity and the in silico interaction between trypanothione reductase (TryR) and (-)-5-demethoxygrandisin B from the leaves of *Virola surinamensis* (Rol.) Warb. The compound (-)-5-demethoxygrandisin B was isolated from *V. surinamensis* leaves, a plant found in the Brazilian Amazon, and it was characterized as (7R,8S,7′R,8′S)-3,4,5,3′,4′-pentamethoxy-7,7′-epoxylignan. In vitro antileishmanial activity was examined against *Leishmania amazonensis*, covering both promastigote and intracellular amastigote phases. Cytotoxicity and nitrite production were gauged using BALB/c peritoneal macrophages. Moreover, transmission electron microscopy was applied to probe ultrastructural alterations, and flow cytometry assessed the shifts in the mitochondrial membrane potential. In silico methods such as molecular docking and molecular dynamics assessed the interaction between the most stable configuration of (-)-5-demethoxygrandisin B and TryR from *L. infantum* (PDB ID 2JK6). As a result, the (-)-5-demethoxygrandisin B was active against promastigote ($IC_{50}$ 7.0 μM) and intracellular amastigote ($IC_{50}$ 26.04 μM) forms of *L. amazonensis*, with acceptable selectivity indexes. (-)-5-demethoxygrandisin B caused ultrastructural changes in promastigotes, including mitochondrial swelling, altered kDNA patterns, vacuoles, vesicular structures, autophagosomes, and enlarged flagellar pockets. It reduced the mitochondria membrane potential and formed bonds with important residues in the TryR enzyme. The molecular dynamics simulations showed stability and favorable interaction with TryR. The compound targets *L. amazonensis* mitochondria via TryR enzyme inhibition.

**Keywords:** *Leishmania amazonensis*; mitochondrial membrane potential; flow cytometry; trypanothione reductase; computational studies

## 1. Introduction

Herbal preparations are plant-based and are often used for the treatment of many diseases. For instance, species from the Myristicaceae family are used by indigenous people of the Amazon region as a source of powerfully hallucinogenic narcotics, triggering chemical and pharmacological interest in the species of this family during the 1950s [1]. Belonging to the Myristicaceae family, *Virola surinamensis* is among the best-known *Virola* species and is popularly known as ucuúba, ucuúba-branca, ucuúba-cheirosa, ucuúba-de-igapó, ucuúba-da-várzea, ucuúba-verdadeira, bicuíba, bicuíba-branca, tree-de-tallow, and virola [2].

Lignans and neolignans have been successfully extracted from the leaves of *V. pavoni*, *V. surinamensis*, *V. michelli*, and *V. sebifera* [3–5]. Among these, (-)-5-demethoxygrandisin B arises as a tetrahydrofuran lignan, constituting its first occurrence of isolation within the *Virola* genus. Its structure bears similarity to (-)-5-demethoxygrandisin, another tetrahydrofuran lignan previously isolated from the leaves and tree barks of *V. surinamensis*. The key distinction between (-)-5-demethoxygrandisin and (-)-5-demethoxygrandisin B lies in the configuration of the furan ring. *V. surinamensis* is widely found in Central and South America, including Brazil, and its constituents have been reported with interesting pharmacological properties, such as anti-chagasic, anti-malarial, and leishmanicidal activities [6–11]. Additionally, the leaves and seeds of *V. surinamensis* contain a high content of lignans [12], well-known as a promising class of compounds against *Leishmania* spp. [13].

Leishmaniases is a complex of diseases caused by infection with different *Leishmania* parasites from the Trypanosomatidae family. It is transmitted through the bite of infected female sandflies, with diverse clinical manifestations such as a cutaneous, mucocutaneous, and visceral diseases in the Old and New World [14]. Leishmaniasis is considered a neglected disease, and it is estimated that more than 1 billion people are at risk of infection because they live in areas endemic to leishmaniasis [15].

Leishmaniasis predominantly impacts communities already facing vulnerability due to factors such as poverty, limited healthcare access, and various adversities. Finding new medicines represents a humanitarian endeavor aimed at enhancing the well-being of these individuals and guaranteeing their access to treatments that are both impactful and accessible. Moreover, its treatment remains a huge challenge due to serious limitations, such as few effective drugs, resistance, toxicity, high cost, and a lack of adherence to the therapeutic protocol [16]. Thus, leishmaniasis remains a major public health problem worldwide.

Among the molecular targets used to search for new drugs against *Leishmania* and *Trypanosoma*, trypanothione reductase (TryR) has been identified as unique to parasites. Also, TryR has been described as an effective target against trypanosomatids. Thus, that enzyme plays a crucial role in the control of redox homeostasis by the parasite, which is vital for its survival in a hostile environment generated by the host as a response to the infection. The TryR has been an interesting binding site for many drug candidates that exhibit inhibitory activities against parasites [17,18].

Therefore, in order to discover new drugs against *Leishmania*, here we describe (-)-5-demethoxygrandisin B as a potential candidate and its behavior to the TryR enzyme.

## 2. Materials and Methods

### 2.1. Botanical Material, Extraction, and Isolation

The species under study was collected by a parataxonomist from Embrapa Amazônia Oriental (1°26′27.5″ S 48°26′22.8″ W), where an exsiccate from 180,980 was cataloged in the Herbarium. The leaves were dried in an oven at a temperature of 40 °C for a period of two days. After drying, the leaves were crushed and ground in a knife mill, resulting in 6.2 kg of material. The dried and ground leaves of *V. surinamensis* (4.0 Kg) were sequentially extracted by maceration with hexane (Tedia, Fairfield, CT, USA), ethyl acetate (Tedia, Fairfield, CT, USA), and methanol (Tedia, Fairfield, CT, USA) for a period of seven days for each solvent. The obtained masses of the hexane, EtOAc, and methanol extracts were 61.7 g, 120 g, and 244 g, respectively. EtOAc was selected to exhibit a superior chromatographic profile in thin-layer chromatography. A 6.0 g portion of the EtOAc extract was dissolved in 1 L of MeOH/$H_2O$ solution (7:3), followed by successive partitioning with 700 mL of hexane (×4) and 700mL of ethyl acetate (×4). After drying with anhydrous sodium sulfate, both yielded 0.2 g and 2.8 g, respectively. The ethyl acetate phase (2.8 g) was submitted to a fractionation on column chromatography using Silica gel 60 (70–230 mesh) and 230–400 mesh (SILICYCLE, Québec, QC, Canada) as the stationary phase and polarity gradients of acetone in hexane and methanol in acetone, as elution systems. The subfractions were analyzed by thin-layer chromatography (TLC) (silica gel F254, Merck, Darmstadt, Germany) using Hexane/EtOAc (6:4) as the elution system, and ceric sulfate was used as a detection spray. A yellowish oil, obtained in 30% hexane/acetone, was isolated and characterized as the (-)-5-demethoxygrandisin B: (7R,8S,7′R,8′S)-3,4,5,3′,4′-pentamethoxy-7,7′-epoxylignan (40 mg). Obtaining the extracts and the substance (-)-5-demethoxygrandisin B is presented in Figure 1.

### *2.2. Spectroscopy of HRMS and NMR for Structural Characterization, and Optic Rotation Determination*

The assessment of (-)-5-demethoxygrandisin B at 100 ppm in methanol: $H_2O$ (80:20) was conducted using a Xevo® G2-S QTof coupled with an ACQUITY Ultra Performance LC™ system (Waters Corp., Milford, MA, USA). The ionization source was configured with a desolvation gas flow (N2) at 600 L/h and a desolvation temperature of 150 °C. The cone gas flow (N2) was set at 50 L/h, and the source temperature was maintained at 120 °C. Adjustments were made to the capillary and sampling cone voltages, setting them at 1.0 kV and 40 V, respectively. The data acquisition was executed utilizing MassLynx 4.1 software (Waters, Milford, USA) [19,20]. NMR 1D ($^1$H and $^{13}$C) and 2D homonuclear-correlated spectroscopy (COSY $^1$Hx$^1$H), HMBC, and HSQC analyses were acquired with a Bruker 400 spectrometer AscendTM (Rheinstetten, Germany) model at 400.15 MHz ($^1$H) and 100.62 MHz ($^{13}$C). The chemical shifts were determined relative to $CDCl_3$ at 0 ppm. A total of 20 mg of (-)-5-demethoxygrandisin B was solubilized in 600 µL of $CDCl_3$. TopSpin 3.6.0 software was used for data control and processing. Thereby, the spectra were manually evaluated. The optical rotation of the (-)-5-demethoxygrandisin was measured on a Perkin–Elmer 341 polarimeter (Perkin–Elmer Inc., Waltham, MA, USA).

### *2.3. Ethical Statements and Animals*

The implementation of animal procedures adhered to the guidelines established by the National Council for Control of Animal Experimentation (Conselho Nacional de Controle de Experimentação Animal—CONCEA) and received endorsement from the institutional Ethics Committee on Animal Care and Utilization (Comissão de Ética no Uso de Animais do Instituto Oswaldo Cruz—CEUA-IOC L53/2016-A3). Female BALB/c mice aged 4 to 6 weeks were purchased from the Institute of Science and Technology in Biomodels of the Oswaldo Cruz Foundation.

### *2.4. Peritoneal Macrophage Isolation and Parasite Cultures*

Thioglycollate-elicited peritoneal macrophages were obtained from BALB/c mice that were injected intraperitoneally 72 h previously with 3 mL of sterile thioglycollate broth as described by Silva-Silva et al. [21]. Then, cells were maintained in RPMI 1640 medium and cultured overnight at 37 °C, under a humidified atmosphere of 5% $CO_2$. Promastigotes of *L. amazonensis* H21 (MHOM/BR/76/MA-76) were cultivated at 26 °C in Schneider's Insect medium (Sigma, St Louis, MO, USA). All media were supplemented with 10% fetal bovine serum (Gibco, Gaithersburg, MD, USA), 100 IU/mL penicillin, and 100 μg/mL streptomycin as previously described [22].

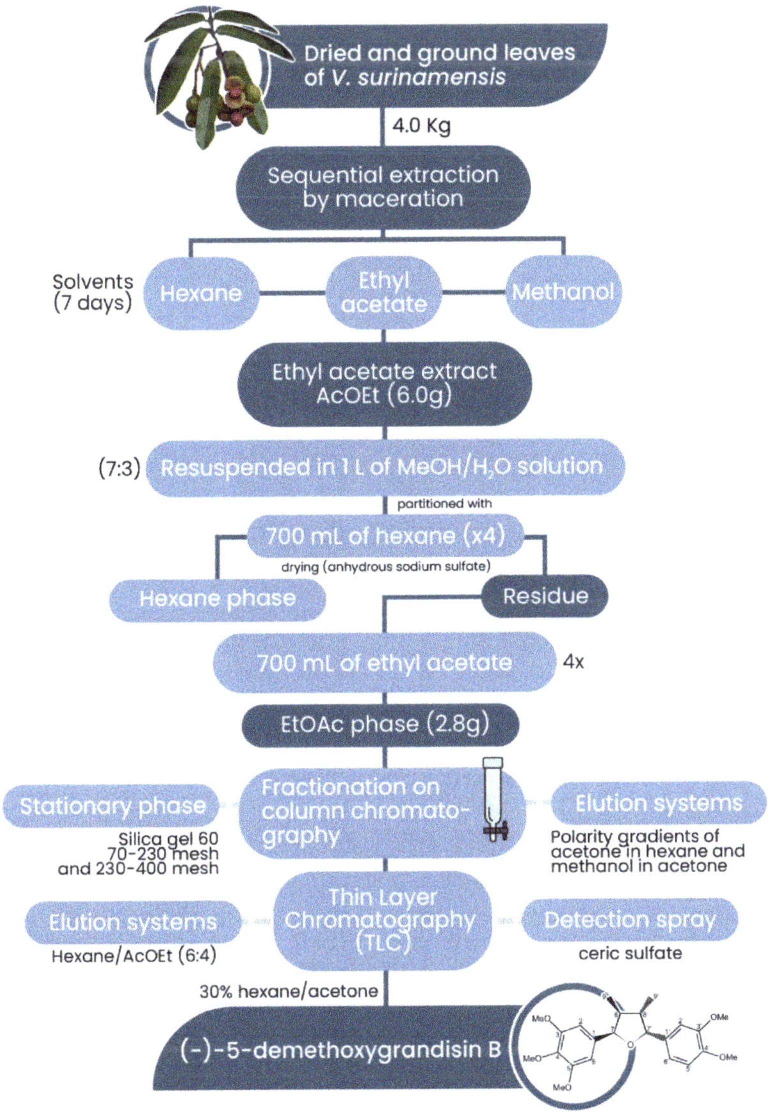


**Figure 1.** Flowchart of obtaining the extracts and (-)-5-demethoxygrandisin B.

## *2.5. Cytotoxicity Assay*

To determine the cytotoxic effects of (-)-5-demethoxygrandisin B, the peritoneal macrophages were cultured in 96-well plates ($5 \times 10^5$ cells/mL) with different concentrations of (-)-5-demethoxygrandisin B (31–994 μM) at least in triplicate up to a final volume of 100 μL per well in 5% $CO_2$ for 24 h at 37 °C. Wells without cells were used as blanks, and wells with cells and dimethyl sulfoxide (DMSO) 1% were used as controls. All assays were performed in triplicate at three different times using amphotericin B (0.20–27 μM) as the reference drug. Cell viability was measured using a colorimetric methyl thiazole tetrazolium (MTT) assay [23]. The results were used to calculate the 50% cell cytotoxicity ($CC_{50}$).

*2.6. Antileishmanial Activity Assay and Selectivity Index*

The promastigote forms of *L. amazonensis* were used at a $10^6$ parasites/mL concentration from a 3- to 5-day-old culture. The assay was performed in 96-well plates in the presence of different concentrations of (-)-5-demethoxygrandisin B at 16–497 μM, in a final volume of 100 μL per well, and was then incubated for 24 h at 26 °C. Wells with medium and without parasites were used as blanks, and wells with parasites incubated with DMSO 1% were used as controls. Amphotericin B at 0.0169–1 μM was used as the reference drug. The parasite viability post-treatment was ascertained by enumerating live promastigotes, factoring in flagellar motility, employing Neubauer's camera in conjunction with an optical light microscope [24]. This count was compared with the score of the nontreated promastigote growth. The results were expressed as a parasite growth inhibitory concentration of 50% ($IC_{50}$). The effect of (-)-5-demethoxygrandisin B on the intracellular amastigote form was performed in 24-well plates, with coverslips, with peritoneal macrophages ($5 \times 10^5$ cells per well) cultured and infected with promastigote forms of *L. amazonensis* using a ratio of 10:1 parasite per cell. After 6 h, the cells were washed three times with PBS to remove the free parasites. The infected cells were treated with (-)-5-demethoxygrandisin B at 8–124 μM, or amphotericin B at concentrations spanning 0.169 to 2.7 μM, over a 24 h period. Following this, the coverslips containing the treated and infected cells were fixed using Bouin solution and subsequently stained with Giemsa for subsequent light microscopy analysis. The determination of the IC50 values ensued from quantifying the intracellular amastigotes within 100 host cells, adopting the methodology outlined by Silva-Silva et al. [22]. These experiments were carried out in independent triplicate with each condition performed in triplicate. The selectivity index (SI) was obtained from the ratio of the peritoneal macrophages $CC_{50}$ and $IC_{50}$.

*2.7. Nitrite Quantification*

The impact of (-)-5-demethoxygrandisin B on nitric oxide (NO) release was assessed indirectly through the quantification of nitrite in the culture supernatants of macrophages ($5 \times 10^6$ cells/mL) using the Griess reaction. Subsequent to treatment with (-)-5-demethoxygrandisin B (62 μM) and/or stimulation with *L. amazonensis* ($3 \times 10^7$ parasites/mL) or LPS (10 μg/mL), approximately 50 μL of the supernatants was collected and introduced into designated wells of 96-well plates. To these supernatants, 50 μL of Griess reagent (comprising 25 μL of 1% sulfanilamide in a 2.5% $H_3PO_4$ solution and 25 μL of 0.1% N-(1-naphthyl) ethylenediamine solution) was added. Following a 10 min incubation, the plates were subjected to spectrophotometric analysis at 570 nm, and the nitrite levels were deduced through reference to the standard sodium nitrite curve (ranging from 1.5 to 100 μM) [21].

*2.8. Transmission Electron Microscopy*

*L. amazonensis* promastigotes treated with $IC_{50}$ for (-)-5-demethoxygrandisin B and incubated for 24 h at 26 °C were collected by centrifugation at 5000 rpm for 5 min. The parasites underwent fixation for an overnight duration within a solution comprising 2.5% glutaraldehyde (Sigma, St Louis, MO, USA) immersed in 0.1 M sodium cacodylate buffer (pH 7.2). After this, a sequence of three washes utilizing 0.1 M sodium cacodylate buffer followed. Subsequently, a post-fixation procedure was carried out through immersion in a solution containing 1% osmium tetroxide, 0.8% potassium ferrocyanide, and 5 mM calcium chloride. Following further washing steps in 0.1 M sodium cacodylate buffer, the parasites underwent dehydration through a graduated acetone series before being embedded within EPON 812 resin (Sigma, St Louis, MO, USA). Ultrathin sections, spanning 100 nm, were then attained via Sorvall MT 2-B (Porter Blum) ultramicrotome (Sorvall, Newtown, CT, USA). These sections were subsequently stained with a 5% uranyl acetate aqueous solution and lead citrate (comprising 1.33% lead nitrate and 1.76% sodium citrate) prior to observation under a transmission electron microscope JEM-1011 (JEOL, Tokyo, Japan) operating at 80 kV [25].

### *2.9. Determination of Mitochondrial Membrane Potential (MMP) (ΔΨm)*

For the determination of the mitochondrial membrane potential in the promastigote forms of *L. amazonensis*, we used tetramethylrhodamine ethyl ester (TMRE) (Molecular Probes, Carlsbad, CA, USA) and flow cytometry. For this method, $2 \times 10^6$ parasites/mL were treated with (-)-5-demethoxygrandisin B $IC_{50}$ for 24 h at 26 °C. Nontreated parasites were used as a negative control, and heat-killed parasites (60 °C bath for 30 min) were used as a positive control. Post-incubation, the parasites were subjected to centrifugation at 1500× *g* for 5 min at room temperature. Afterward, they were washed with PBS and incubated in a solution of 300 µL TMRE (50 nM) in the absence of light for 15 min at room temperature. Subsequently, a flow cytometry analysis was conducted using a CytoFlex flow cytometer (Beckman Coulter Life Sciences, Inc., Brea, CA, USA). TMRE excitation was achieved using a 488 nm blue laser, and the emitted fluorescence was captured with a 585/42 nm bandpass filter. The flow cytometry data were analyzed using CytExpert software version 2.1 (Beckman Coulter Life Sciences, Inc., Brea, CA, USA) [26].

### *2.10. Molecular Docking*

The enzyme trypanothione reductase (TryR) is a promising target for the design of new drugs with antileishmanial activity, because it plays a fundamental role in the intracellular redox balance necessary for the survival of parasites of the *Leishmania* genus and is not present in the host system, where it is replaced by its human homologous enzyme (glutathione reductase) [17]. In this sense, the docking calculations were performed using the GOLD program (Genetic Optimization for Ligand Docking) version 5.5 [27] with the crystallographic structure of the TryR enzyme from *L. infantum*, retrieved from the Protein Data Bank (PDB ID 2JK6) with a resolution of 2.95 Å and complexed by the FAD cofactor [28]. GOLD is a docking program that uses the genetic algorithm to explore the conformational flexibility of the ligand in a complex with a partially flexible macromolecule [29]. As the TryR enzyme is a homodimer [30] composed of two identical monomers, the A chain present in the enzyme structure was selected for the docking step, while the B chain was removed in the Chimera program [31]. To perform the docking, the two-dimensional structure of neolignan (-)-5-demethoxygrandisin B was previously constructed in the MarvinSketch program (https://chemaxon.com/presentation/marvin-the-nextgeneration-of-chemical-drawing, accessed on 10 July 2020) and later optimized using the molecular mechanic's force field MMFF94 [32] in the Avogadro program [33] for the in silico study of the compound. Subsequently, the GOLD program removed all the water molecules, and hydrogen atoms were added to the TryR enzyme. Then, the binding site was defined based on the location of the residues that participate in the catalytic mechanism (Cys52, Cys57, His461′, and Glu466′) [34]. Finally, the composite was docked by applying the GoldScore and ChemScore scoring functions with a search efficiency of 100% in GOLD. The GoldScore and ChemScore functions provide rapid results that predict binding affinity with the macromolecule [35]. Because interactions with the amino acid residues of the active site are essential to promote the inhibition of enzyme activity, a visual inspection of the interactions of the more thermodynamically stable conformation of (-)-5-demethoxygrandisin B was performed on the online server PoseView [36,37].

### *2.11. Molecular Dynamics*

Initially, the calculation of the electrostatic potential charge for the structure obtained during docking was performed by the Gaussian 03 program [38], applying the restricted electrostatic potential (RESP) [39] combined with the Hartree–Fock method together with the basis (6–31 G(d,p)) [40]. The protonation states of all the amino acid residues present in the enzyme were determined under pH 7.0 by the PROPKA server [41]. The molecular dynamics (MD) simulations were performed by the AMBER 18 program, implemented by the *pmemd.CUDA* module [42]. The general AMBER force field (GAFF) [43] was used to describe the (-)-5-demethoxygrandisin B ligand, while the MMFF99SB force field was used for the amino acid residues of the enzyme [44]. Using the *tleap* module of the AmberTools18 package, the enzyme–ligand complex was solvated with the TIP3P explicit solvent model in a cubic box with an edge equal to 12 Å, followed by the addition of Cl- counterions to neutralize the resulting charges [45]. The system went through 4 energy minimization steps by the SANDER module [46], applying 10,000 steps in the first step with the steepest descent method and 15,000 with the conjugate gradient method, with the following 3 steps being carried out in 10,000 steps evenly split for both methods. Subsequently, the system was gradually heated from 0K to 298K in the NVT set, using the Langevin algorithm [47] with the restriction of the atoms by a force constant equal to 25 kcal/mol·Å$^2$. The periodic boundary conditions employed by the Particle Mesh Ewald (PME) method [48] were used to identify long-range electrostatic interactions. The limit condition of 1 atm and the temperature of 298 K were used to maintain constant temperature and pressure set NPT. The SHAKE algorithm [49] was employed to constrain all hydrogen bonds. The equations of motion were integrated every 2 femtoseconds (2 fs) using the Verlet algorithm [50] based on the principle of finite differences. Thus, the enzyme–ligand complex was simulated at a time of 50 ns by the NPT set at a temperature of 298 K. In addition, the CPPTRAJ module [51] available in AMBERTOOLS 18 was applied for a stability analysis, in terms of the root mean square deviation (RMSD) calculation and structural flexibility through the B-factor.

Binding Free-Energy Calculations

The molecular mechanics Poisson–Boltzmann surface area (MMPBSA) [52] and molecular mechanics generalized Born surface area (MMGBSA) [53] methods were used to calculate the binding free energy ($\Delta G_{bind}$) from the complex to the last 10 ns of the MD trajectory, extracted using the CPPTRAJ and MMPBSA.py modules available in AmberTools 18 [54]. Thus, the binding free energy was calculated according to the following equations:

$$\Delta G_{bind} = \Delta E_{MM} + \Delta G_{bind,solv} - T.\Delta S \quad (1)$$

$$\Delta E_{MM} = \Delta E_{int} + \Delta E_{ele} + \Delta E_{vdW} \quad (2)$$

$$\Delta Eint = \Delta E_{bond} + \Delta E_{angle} + \Delta E_{torsion} \quad (3)$$

$$\Delta G_{bind,solv} = \Delta G_{GB/PB} + \Delta G_{n\text{-}polar} \quad (4)$$

$$\Delta G_{n\text{-}polar} = \gamma \times SASA \quad (5)$$

$$\Delta G_{bind} = \Delta G_{vdW} + \Delta G_{ele} + \Delta G_{GB} + \Delta G_{n\text{-}polar} \quad (6)$$

The term $\Delta E_{MM}$ in Equation (1) represents the molecular mechanical energy; $\Delta G_{bind}$, solve is the free energy of solvation; and the term T.ΔS represents the product of the absolute temperature and conformational entropy. In Equation (2), the individual contributions to the mechanical energy are presented, composed of the internal ($\Delta E_{int}$), electrostatic ($\Delta E_{ele}$), and van der Waals ($\Delta E_{vdW}$) energy. The free energy of solvation ($\Delta Gb_{ind,solv}$) results from the sum of polar ($\Delta G_{PB/GB}$) and non-polar ($\Delta G_{n\text{-}polar}$) free energy, as shown in Equation (3). The polar electrostatic contribution was calculated by the Poisson–Boltzmann (PB) and by the generalized Born approximation (GB) methods, while the non-polar energy was estimated by the product between the empirical parameters of the surface tension (γ) and solvent accessible area (SASA), presented in Equation (4). Finally, the interaction free energy obtained in the last 10 ns of the simulation by the MMGBSA method was decomposed according to Equation (6), for the analysis of the individual contribution of the amino acid residues. This method has been widely applied in drug design studies due to its satisfactory approach that successfully reproduces experimental results. The main amino acid residues that contribute to the total energy MMGBSA were visualized using the CHEWD plugin, to generate a color-coding model that expresses the magnitude of the contribution by residues [55].

### 2.12. Statistical Analysis

The numerical results were expressed as the mean ± standard deviation, and the statistical analyses were conducted through the statistical software GraphPad Prism® version 7 (GraphPad Software Inc., San Diego, CA, USA). Differences were considered significant when $p < 0.05$.

## 3. Results

### 3.1. Lignan Structure Elucidation

The (-)-5-demethoxygrandisin B (40 mg) was isolated from the ethyl acetate extract and showed yellowish oil characteristics. High-resolution mass spectrometry (HRMS) data provided a molecular formula $C_{23}H_{30}O_6$ with a calculated and measured precursor monoisotopic mass of 403.2126 Da, corresponding to an error of 0.00 ppm (Supplementary Material Figure S1). After that, it was confirmed by spectroscopic experiments of $^1$H and $^{13}$C nuclear magnetic resonance (NMR) spectra. The $^1$H, $^{13}$C, heteronuclear single-quantum correlation (HSQC), and heteronuclear multiple-bond coherence (HMBC) NMR data are available in Table 1 and in Supplementary Material Figures S2–S6.

**Table 1.** $^1$H, $^{13}$C, HSQC, and HMBC NMR data from the (-) 5-demethoxygrandisin B.

| No. | $\delta^1{}_H$ in ppm (CDCl$_3$) (400 MHz) | $\delta_C$ in ppm (100 MHz) | HMBC |
|---|---|---|---|
| 1 | - | 137.46 | - |
| 2 | 6.63 (*s*) | 103.14 | 88.47 (3*J*) |
| 3 | - | 153.27 | - |
| 4 | - | 138.12 | - |
| 5 | - | 153.27 | - |
| 6 | 6.63 (*s*) | 103.14 | 153.27 (2*J*)<br>138.12 (3*J*)<br>137.87 (3*J*) |
| 7 | 4.65 (*d*, 4.5 Hz) | 88.35 | - |
| 8 | 1.80–1.78 (*m*) | 51.06 | 88.35 (2*J*) |
| 9 | 1.09 (*d*, 6.0 Hz) | 14.07/13.83 | - |
| 1′ | - | 134.86 | - |
| 2′ | 6.96 (*d*, 1.8 Hz) | 109.27 | 148.55 (2*J*) |
| 3′ | - | 148.60 | - |
| 4′ | - | 149.14 | - |
| 5′ | 6.85 (*d*, 8.1 Hz) | 110.97 | 134.86 (3*J*) |
| 6′ | 6.92 *dd* (1.8 and 8.1 Hz) | 118.61 | 109.27 (3*J*) |
| 7′ | 4.63 (*d*, 4.5 Hz) | 88.47 | - |
| 8′ | 1.80–1.78 (*m*) | 50.97 | 88.35 (2*J*) |
| 9′ | 1.06 (*d*, 6.0 Hz) | 14.07/13.83 | 88.47 (3*J*) |
| OMe-3/5 | 3.88 (*s*) | 56.18 | 153.27 (2*J*) |
| OMe-4 | 3.83 (*s*) | 60.83 | 138.12 (2*J*) |
| OMe-3′ | 3.87 (*s*) | 55.95 | 148.60 (2*J*) |
| OMe-4′ | 3.91 (*s*) | 55.95 | 149.14 (2*J*) |

Multiplicities and coupling constants (*J*) in Hertz are shown in parentheses. HSQC: Heteronuclear Single-Quantum Correlation; HMBC: Heteronuclear Multiple-Bond Coherence; NMR: Nuclear Magnetic Resonance.

The (-)-5-demethoxygrandisin B showed oxybenzyl proton absorptions H-7 and H-7′ ($\delta$ 4.65, *d*, $J$ = 4.5 Hz, $^1$H and 4.63, *d*, $J$ = 4.5 Hz, $^1$H) and methyl protons H-9 ($\delta$ 1.09, *d*, $J$ = 6 Hz, 3H) and H-9′ ($\delta$ 1.06 (*d*, $J$ = 6 Hz, 3H) related to tetrahydrofuran. The presence of oxybenzyl proton was confirmed by an analysis of the literature on the classes of substances isolated from *V. surinamensis* [11,56]. Based on the $^{1}$H NMR spectrum, (-)-5-demethoxygrandisin B showed common signals to tetrahydrofuran lignan. This hypothesis was confirmed once signals were observed in the region of aromatic hydrogens at $\delta_H$ 6.96 (*d*, $J$ = 1.8 Hz, $^{1}$H), 6.92 (*dd*, $J$ = 1.8 and 8.1 Hz, $^{1}$H), and 6.85 (*d*, $J$ = 8.1 Hz, $^{1}$H) in an AMX coupling system, assigned to the hydrogens of a 1,3,4-trisubstituted aromatic ring, with the COSY correlation map showing the existing correlations between these signals (Supplementary Material Figures S2–S6).

An analysis of the aromatic hydrogen region of the $^{1}$H NMR spectrum of the (-)-5-demethoxygrandisin B leads to the conclusion that the two aromatic rings were 3′,4′-dimethoxyphenyl and 3,4,5-trimethoxyphenyl. Interpretations of the $^{1}$H and $^{13}$C NMR, HSQC, and HMBC spectra for (-)-5-demethoxygrandisin B are described in Table 1, where the carbon signals at $\delta_C$ 88.35 and 88.47, as well as those at $\delta_C$ 51.06 and 50.97, were typical, respectively, of carbons C-7, C-7′, C-8, and C-8′ of tetrahydrofuran lignans [12,57]. Thus, it was possible to infer that the profile of the $^{1}$H NMR spectrum of (-)-5-demethoxygrandisin B is common to tetrahydrofuran lignan, which has two phenylpropane units linked by an oxygen atom as illustrated in Figure 2. The stereochemistry of the furan ring was identified by analyzing the coupling constants ($J$) with data well-established in the literature presented in the discussion of this paper [9,28,56,58–60].

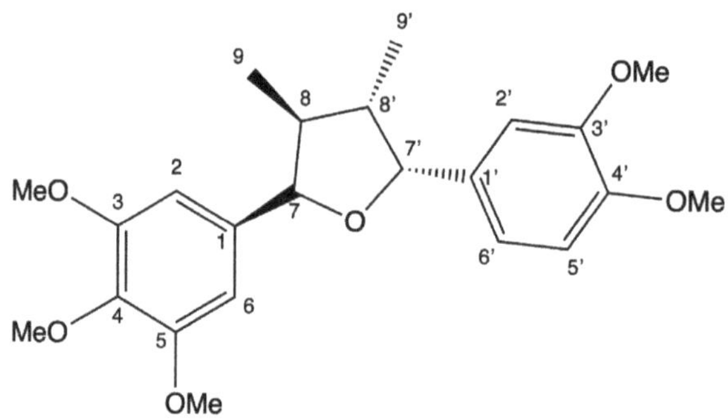


**Figure 2.** Chemical structure of the (-)-5-demethoxygrandisin B.

### 3.2. *Antileishmanial Activity and Cytotoxicity*

In vitro antileishmanial assays were performed to evaluate the effect of the (-)-5-demethoxygrandisin B in the promastigote and intracellular amastigote forms of *L. amazonensis*, as well as its cytotoxic effect against mammal cells. The (-)-5-demethoxygrandisin B exhibited inhibitory activity against both forms in a concentration-dependent manner (Figure 3). (-)-5-demethoxygrandisin B was able to inhibit the promastigote forms of *L. amazonensis* with a 50% inhibitory concentration ($IC_{50}$) of 7.0 μM. Furthermore, (-)-5-demethoxygrandisin B was active against the intracellular amastigote forms, displaying $IC_{50}$ values of 26.04 μM. These results suggest that promastigotes were more sensitive to (-)-5-demethoxygrandisin B treatment. The cytotoxic effect on peritoneal macrophages showed a $CC_{50}$ value of 193.37 μM, lower than the values observed for the standard drug amphotericin B. The selectivity index (SI) for promastigotes and intracellular amastigotes showed that (-)-5-demethoxygrandisin B was 26.6 and 7.4 times less toxic for macrophages than parasites, respectively. The reference drug amphotericin B showed antileishmanial activity and cytotoxicity as expected, after 24h of treatment (Table 2).

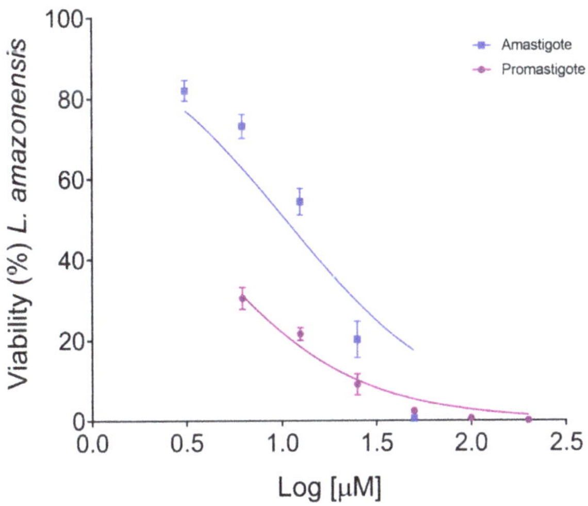


**Figure 3.** Concentration–response curve of (-)-5-demethoxygrandisin B effects on the viability of *Leishmania amazonensis* promastigote and intracellular amastigote forms. Data represent the mean ± standard error of two independent experiments carried out in triplicate.

**Table 2.** $IC_{50}$ and $CC_{50}$ values (μM) for (-)-5-demethoxygrandisin B and its selectivity index (SI) against promastigote and intramacrophage amastigote forms of *Leishmania amazonensis*.

| Compounds | Cytotoxicity | *L. amazonensis* | | | |
|---|---|---|---|---|---|
| | $CC_{50}$ (μM) | Promastigote | | Intracellular Amastigote | |
| | | $IC_{50}$ (μM) | $SI_{pro}$ | $IC_{50}$ (μM) | $SI_{ama}$ |
| (-)-5-demethoxygrandisin B | 193.37 | 7.0 | 26.6 | 26.04 | 7.4 |
| Amphotericin B | 8.82 μM | 0.02226 | 396.1 | 0.1898 | 46.5 |

Data represent mean ± SD of at least two experiments realized in triplicate; $CC_{50}$: cytotoxic concentration for 50% of peritoneal macrophages; $IC_{50}$: inhibitory concentration for 50% of parasites; SI: selectivity index about cytotoxicity for the BALB/c peritoneal macrophages.

### *3.3. Nitrite Quantification in L. amazonensis-Infected Peritoneal Macrophages Treated with (-)-5-Demethoxygrandisin B*

As shown in Figure 4, the NO levels were indirectly estimated by nitrite quantification in the supernatant of BALB/c peritoneal macrophages. An increase in the nitrite levels in the supernatant of cells treated with (-)-5-demethoxygrandisin B (5.75 ± 0.24 μM $NaNO_2$, $p = 0.1587$) when compared to untreated cells (5.21 ± 0.21 μM $NaNO_2$) was observed, although this increase was not statistically significant. This effect remained even when macrophages were infected with *L. amazonensis* and treated with (-)-5-demethoxygrandisin B (6.67 + 1.53 μM $NaNO_2$, $p = 0.8286$) when compared to stimulated and untreated cells (5.69 ± 0.20 μM $NaNO_2$). Macrophages stimulated with lipopolysaccharide (LPS) showed the expected higher levels of nitrite compared to cultures not stimulated with LPS.

### *3.4. Ultrastructural Changes*

A transmission electron microscopy analysis revealed untreated parasites displaying organelles with no alterations (Figure 5A), while the main alteration induced by (-)-5-demethoxygrandisin B treatment was kinetoplast swelling (Figure 5B), with alteration in a kDNA compacting pattern (yellow arrow). In addition, vacuoles (asterisks) near the flagellar pocket, vesicular electron-dense structures with an outer bilayer membrane (thin arrows), a structure that looks like an autophagosome (thick arrow), and enlargement of the flagellar pocket with electron-dense filaments inside were also observed in the (-)-5-demethoxygrandisin B-treated parasite.

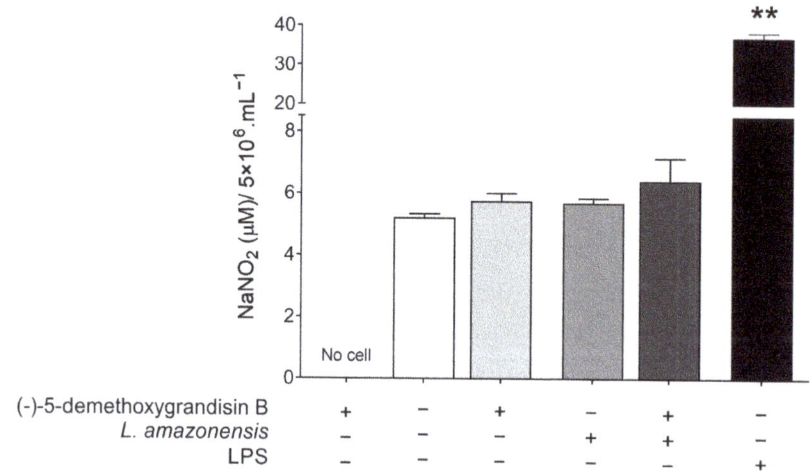


**Figure 4.** Nitrite quantification in the supernatant of BALB/c peritoneal macrophages treated with compound (-)-5-demethoxygrandisin B at 62 µM, stimulated or not with *Leishmania amazonensis*. LPS (lipopolysaccharide from *Escherichia coli*; 10 µg/mL) was used as the positive control. ** $p < 0.01$ when compared with the untreated group by Mann–Whitney test.

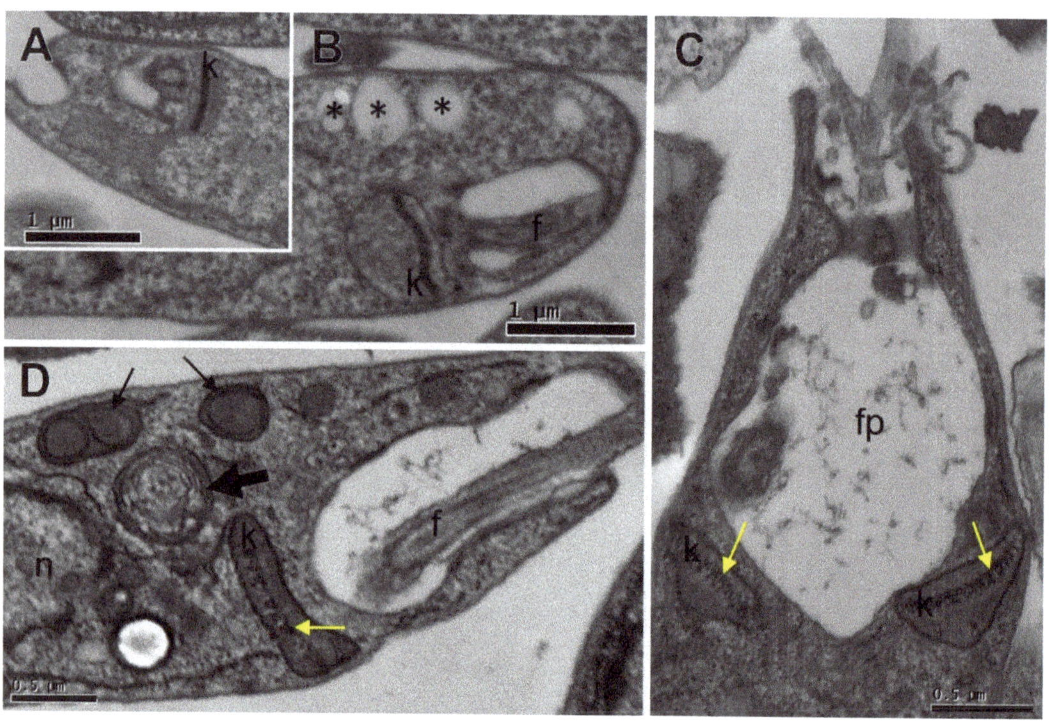


**Figure 5.** Transmission electron microscopy analysis of *Leishmania amazonensis* promastigote forms treated with (-)-5-demethoxygrandisin B at 25 ug/mL for 24 h. (**A**) Untreated parasites display organelles

with no alterations. (**B–D**) Parasites treated with (-)-5-demethoxygrandisin B compound exhibit kinetoplast swelling (**B**), with altered kDNA compacting pattern (yellow arrow), vacuoles (asterisks) near flagellar pocket, vesicular electron-dense structures with outer bilayer membrane (thin arrows), a structure that looks like an autophagosome (thick arrow), and enlargement of flagellar pocket. k: kinetoplast; f: flagellum; fp: flagellar pocket; n: nucleus.

### 3.5. Mitochondrial Membrane Potential (Δψm) by Flow Cytometry

To investigate the influence of treatment with (-)-5-demethoxygrandisin in the mitochondrial membrane potential of *L. amazonensis* by flow cytometry, $2 \times 10^6$ promastigotes/mL were incubated with 50 nM tetramethylrhodamine ethyl ester (TMRE), a cell-permeant fluorescent dye that is readily sequestered by active mitochondria. For this, in Figure 6, we represent the flow cytometric protocol used to assess the frequency of *Leishmania* promastigotes with mitochondrial activity (TMRE+), where a dot plot of morphological patterns of size and granularity (forward scatter (FSC) versus side scatter (SSC)) was created to define the population of *L. amazonensis* promastigotes (Region P4) (Figure 6A). From this region, we created a TMRE-PE-fluorescence histogram to quantify the TMRE+ promastigotes; Figure 5B,C represent an unstained promastigotes sample and (-)-5-demethoxygrandisin B-untreated-*L. amazonensis* promastigotes, which showed 83.52 ± 4.55% of the TMRE+ promastigotes, respectively. The inhibition of the mitochondrial membrane potential of the heat-treated (60 °C) parasites showed a significant decrease in the Δψm to 49.42 ± 1.60% ($p$ = 0.0286) (Figure 6D). Statistically significant changes in the mitochondrial membrane potential were also observed after treatment with 7.0 μM of (-)-5-demethoxygrandisin B (38.76 ± 0.67%, $p$ = 0.0286) (Figure 6E,F).

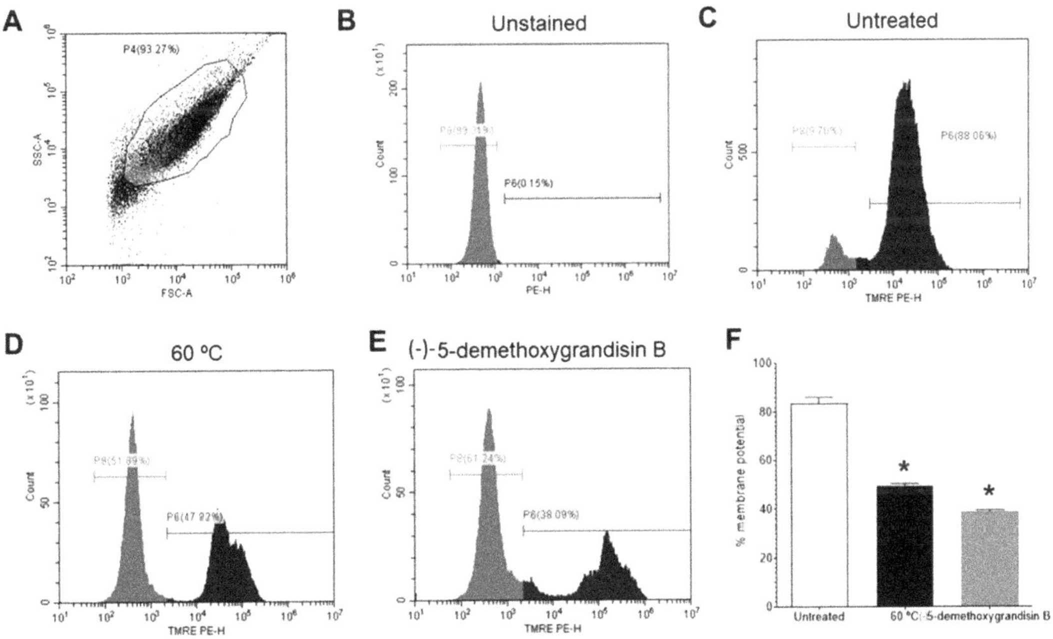


**Figure 6.** Flow cytometry of mitochondrial membrane potential (ΔΨm) in promastigote forms of *Leishmania amazonensis* treated with (-)-5-demethoxygrandisin B. (**A**) FSC versus SSC dot plot to define *L. amazonensis*-promastigotes population. (**B**) TMRE-staining histogram of control samples (unstained and untreated parasites) gated on "promastigotes." (**C**) TMRE-staining histogram of (-)-5-demethoxygrandisin B-untreated parasites. (**D**) *L. amazonensis* promastigotes treated by heat

(60 °C). (**E**) Representative histogram of *L. amazonensis* promastigotes treated with $IC_{50}$ of (-)-5-demethoxygrandisin B. (**F**) Statistically significant differences were observed between the percentages of cells marked with TMRE in the untreated group and the groups treated with (-)-5-demethoxygrandisin B, at the $IC_{50}$ concentration (7.0 µM). * $p < 0.05$, when compared with the untreated group by Mann–Whitney test. Images are representative of two independent experiments carried out at least in triplicate.

### 3.6. Molecular Docking

To assess whether (-)-5-demethoxygrandisin B has potential antileishmanial activity, the enzyme trypanothione reductase (TryR) from *Leishmania infantum* was obtained from the Protein Data Bank (PDB) database under code 2JK6 [28] and selected as a molecular target for the docking simulation. The GoldScore and ChemScore scoring functions were used to predict the binding affinity of the more stable configuration of (-)-5-demethoxygrandisin B in the redox catalytic site (Cys52, Cys57, His461′, and Glu466′) and generated the respective scores of 52.94 and 40.76. The predicted binding modes on the docking for (-)-5-demethoxygrandisin B in the *L. infantum* TryR enzyme cavity were analyzed in the PoseView online server. As shown in Figure 7, it can be seen that the conformation of the molecule interacts with the enzyme through hydrophobic interactions with the Thr51, Cys57, and Leu334 amino acids residues and hydrogen bonds with the Lys60 and Asp327 residues.

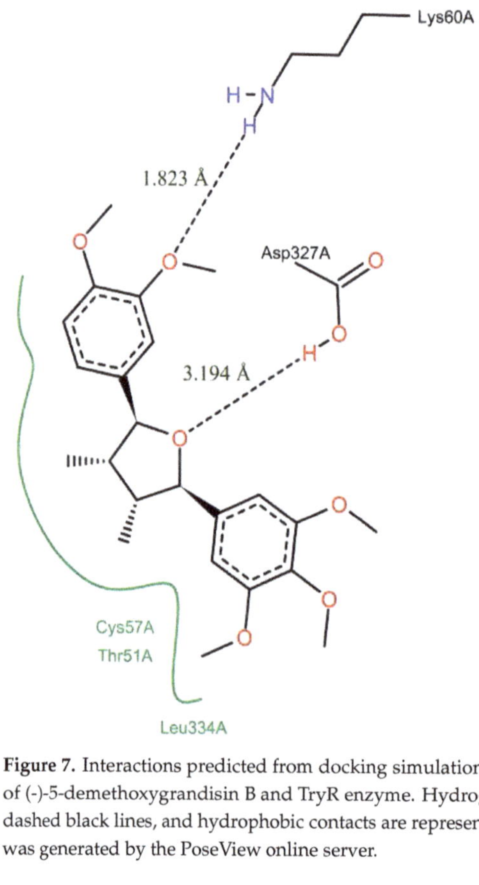


**Figure 7.** Interactions predicted from docking simulations between the more stable configuration of (-)-5-demethoxygrandisin B and TryR enzyme. Hydrogen bond interactions are represented by dashed black lines, and hydrophobic contacts are represented by continuous green lines. The figure was generated by the PoseView online server.

### 3.7. Molecular Dynamics

From the coordinates obtained in the molecular docking procedures, the more stable configuration of (-)-5-demethoxygrandisin B was submitted to 50 ns of molecular dynamics (MD) simulations. The root mean square deviation (RMSD) values were calculated to evaluate the structural stability of the enzyme–ligand complex. The plot of the RMSD (Å) versus simulation time (ns) (Figure 8) demonstrates that the TryR-ligand complex stabilized after 20 ns, with a mean RMSD value of 2.63 Å and standard deviation ± 0.41. Thus, it is possible to infer that the compound remained in the enzyme cavity, showing few conformational changes in the complex's structure.

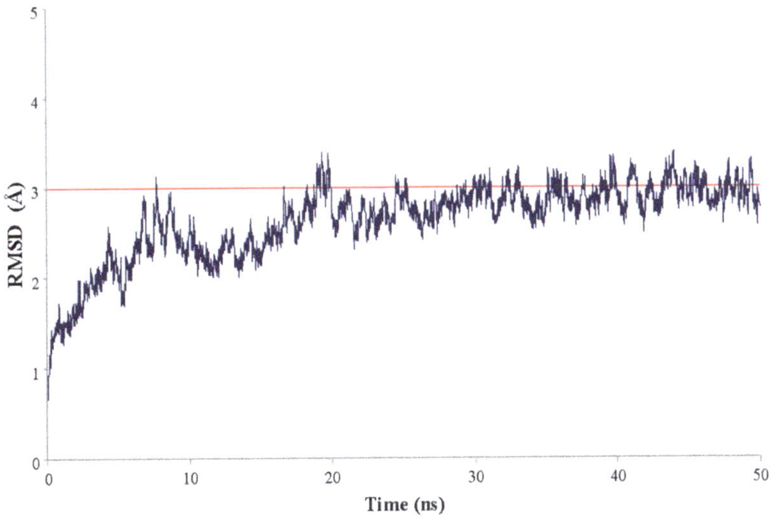


**Figure 8.** Graphical representation of the root mean square deviation (RMSD) values versus simulation time for the studied complex.

To evaluate the flexibility of the protein regions concerning (-)-5-demethoxygrandisin B, the B-factor analysis was carried out. From Figure 9, we can note the complex generally presented low values of flexibility, except the region formed to the Ser76-Asn91 amino acid sequence (highlighted in red in Figure 9), which is the most flexible region of the enzyme.

### 3.8. Binding Free-Energy Calculations

The results in Table 3 show that the binding free energy by the molecular mechanics generalized Born surface area (MMGBSA) method for the TryR-(-)-5-demethoxygrandisin B complex was $-30.41 \pm 2.79$ kcal/mol, while the $\Delta G_{bind}$ value by the molecular mechanics Poisson–Boltzmann surface area (MMPBSA) was calculated to be $-27.66 \pm 3.55$ kcal/mol. Furthermore, by comparing the binding free-energy values of these two methods, it is possible to observe that both are favorable and stable for the formation of the complex. Note that the van der Waals energy ($\Delta E_{vdW}$) presents the greatest contribution to the total free energy, indicating that it acts as a determining factor for the affinity of (-)-5-demethoxygrandisin B for the active site of the TryR enzyme. Furthermore, electrostatic energy ($\Delta E_{ele}$) and non-polar solvation energy ($\Delta E_{n\text{-}polar}$, GB/$\Delta E_{n\text{-}polar, PB}$) also showed favorable contributions, while polar solvation energy ($\Delta G_{solv, GB}/\Delta G_{solv, PB}$ $\Delta E_{PB}$) and the solvation free-energy total ($\Delta G_{solv, GB}/\Delta G_{solv, PB}$) showed unfavorable contributions.

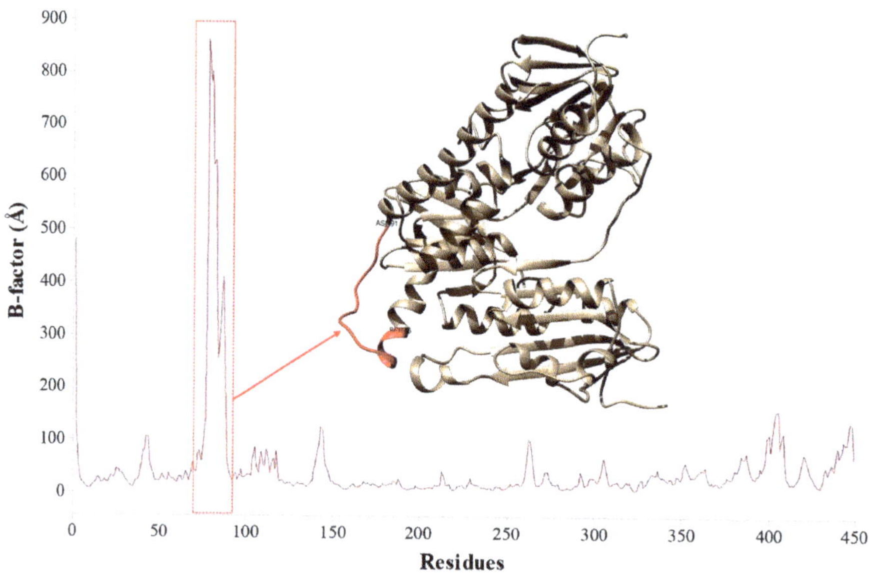


**Figure 9.** Graphical representation of the structural fluctuations taking into account the enzyme's residues from the molecular dynamics simulations.

**Table 3.** Binding free energy ($\Delta G_{bind}$) of the TryR-(-)-5-demethoxygrandisin B more stable configuration complex and the respective energy components obtained with the MMGBSA/PBSA methods.

| MMGBSA | | MMPBSA | |
|---|---|---|---|
| **Contribution** | **TryR-(-)-5-Demethoxygrandisin B (kcal/mol)** | **Contribution** | **TryR-(-)-5-Demethoxygrandisin B (kcal/mol)** |
| $\Delta E_{vdw}$ | −45.57 (2.33) | $\Delta E_{vdw}$ | −45.57 (2.33) |
| $\Delta E_{ele}$ | −5.10 (2.70) | $\Delta E_{ele}$ | −5.10 (2.70) |
| $\Delta E_{GB}$ | 26.00 (3.17) | $\Delta E_{PB}$ | 29.95 (3.87) |
| $\Delta E_{n\text{-polar, GB}}$ | −5.74 (0.25) | $\Delta E_{n\text{-polar, PB}}$ | −6.94 (0.24) |
| $\Delta G_{solv, GB}$ | 20.26 (3.08) | $\Delta G_{solv, PB}$ | 23.01 (3.83) |
| $\Delta G_{bind, GB}$ | −30.41 (2.79) | $\Delta G_{bind, PB}$ | −27.66 (3.55) |

The standard deviations are given in parentheses. MMPBSA: molecular mechanics Poisson–Boltzmann surface area; MMGBSA: molecular mechanics generalized Born surface area; $\Delta E_{vdW}$: van der Waals energy; $\Delta E_{ele}$: electrostatic energy; $\Delta E_{GB}$: polar solvation energy by the generalized Born method; $\Delta E_{n\text{-polar, GB/PB}}$: non-polar solvation energy; $\Delta G_{solv, GB/PB}$: total solvation free energy; $\Delta G_{bind, GB/PB}$: total free energy.

### *3.9. Per-Residue Energy Decomposition*

Energy decomposition by residue was employed to identify the TryR enzyme residues involved in the interactions with (-)-5-demethoxygrandisin B, based on the energy contribution calculated by the MMGBSA method for the last 10 ns of the MD simulations. The binding free-energy values below -1 kcal/mol were defined as a criterion for the selection of amino acid residues that performed the most favorable interactions, that is, that contributed to the stabilization of the ligand in the complex. An analysis of the graph (Figure 10) indicates that Cys52 (−1.17 kcal/mol), Gly326 (−1.18 kcal/mol), Asp327 (−1.86 kcal/mol), Leu334 (−2.20 kcal/mol), and Thr335 (−1.59 kcal/mol) contributed most significantly to the total free energy of the complex.

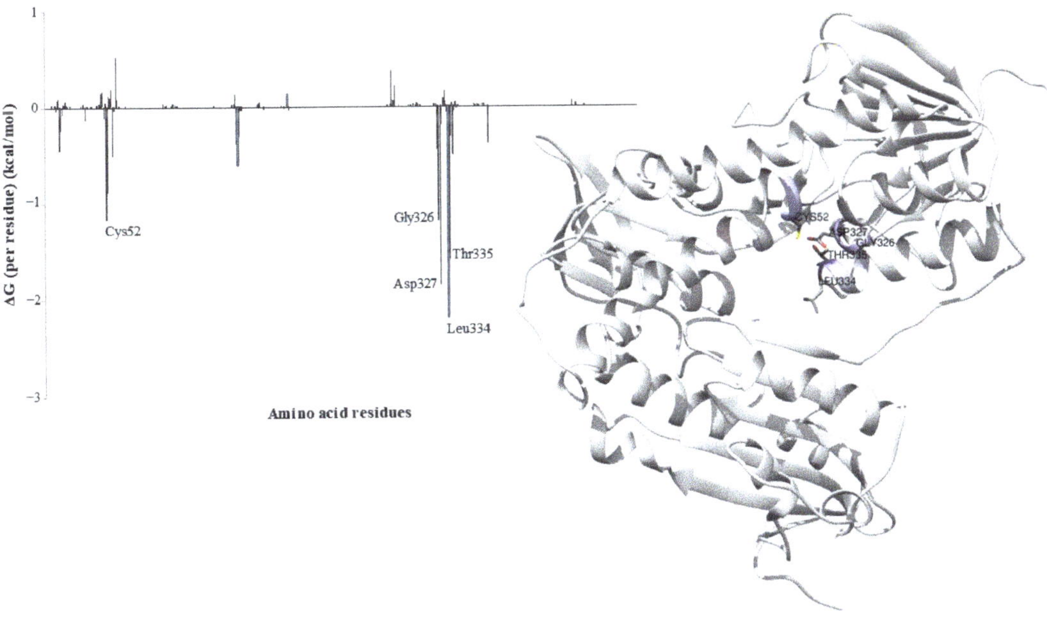


**Figure 10.** Interaction plot by residues obtained by the MMBGSA method. On the right, the residues that most contributed to the binding free energy, visualized by the CHEWD plugin in the Chimera program.

## 4. Discussion

(-)-5-demethoxygrandisin B was detected as a protonated molecule $[M + H]^+$ of m/z 403.2126 and error = 1.24 ppm. These mass spectrometry data presented high precision and accuracy in the high-resolution mass. Additionally, $^1H$ and $^{13}C$ NMR spectra to (-)-5-demethoxygrandisin B showed signals attributed to a tetrahydrofuran lignan. Thus, the molecular formula to (-)-5-demethoxygrandisin B was correctly attributed as $C_{23}H_{30}O_6$. The stereochemistry was established based on the literature data [56], i.e., the methyl groups have a *cis* relationship with the aromatic rings, because when methyl groups present a *cis* relation, the shift expected is approximately or greater than at $\delta_H$ 1.0, close to the signals revealed to (-)-5-demethoxygrandisin B ($\delta_H$ 1.06 and 1.09).

Given these considerations, it is possible to conclude that the signals in $\delta_H$ 4.67–4.63 are two doublets, one in $\delta_H$ 4.65 (*d*, $J$ = 4.5 Hz, $^1$H) for H-7 and the other in $\delta_H$ 4.63 (*d*, $J$ = 4.5 Hz, $^1$H) for H-7′. The coupling constant ($J$ = 4.5 Hz) is by the Karplus vicinal correlation [58], where these hydrogens are expected to have a smaller dihedral angle with their respective vicinal hydrogens H-8 and H-8′, in the gauche position. Lopes et. al. [12], when isolating 3,4,5,3′,4′-pentamethoxy-5-demethoxygrandisin, assigned to carbons C-7 and C-7′ the signals at $\delta_C$ 88.36 and 88.49, respectively. Therefore, the carbon signals at $\delta_C$ 88.35 and 88.47 of the $^{13}C$ NMR spectrum of (-)-5-demethoxygrandisin B were attributed to carbons C-7 and C-7′, respectively. To determine the conformation of the tetrahydrofuran ring of (-)-5-demethoxygrandisin B, the four chiral centers of the substance (C-7, C-7′, C-8, C-8′) were considered, so there are several possible isomers; however, when performing projections and taking into account the coupling constant of H-7 and H-7′ ($J$ = 4.5 Hz), where vicinal hydrogens are expected to be in the *cis* position, only two possible pairs of enantiomers remain.

$^{13}C$ NMR studies carried out by Fonseca [57] concluded that tetrahydrofuran lignans with the ferryl group in the equatorial position present a C-1 or C-1′ carbon signal shift at $\delta_C$ 134.6, while for (-)-5-demethoxygrandisin B it is observed that the C-1′ carbon showed a signal at $\delta_C$ 134.86, suggesting that the aryl group attached to C-1′ is in the pseudoequatorial position. Fonseca [57] also carried out $^{13}$C NMR studies for the methyl groups of the aforementioned lignans, where the methyl groups presented the following displacements for carbon: in structure I at $\delta_C$ 13.7; in structure II at $\delta_C$ 12.9; and in structure III at $\delta_C$ 11.6. Therefore, it is observed that when the methyl groups are in the *trans* position there is a greater chemical shift for the respective carbon signals. For (-)-5-demethoxygrandisin B, it was observed in the $^{13}$C NMR spectrum that the methyl groups presented signals at $\delta_C$ 13.83 and 14.07, thus concluding that they are in the *trans* position. Thus, there are the following possibilities for determining the absolute configuration of (-)-5-demethoxygrandisin B: 7S,8R,7′S,8′R or 7R,8S,7′R,8′S. As the substance presented an optical rotation [α]D = −11.15° (MeOH) and according to the studies byKubanek et al. [59], Kubanek et al. [60], Hwang et al. [61], and Biftu et al. [62], tetrahydrofuran lignans with absolute configuration 7S,8R,7′S,8′R present optical rotation (+), while with absolute configuration 7R,8S,7′R,8′S present optical rotation (−); therefore, for (-)-5-demethoxygrandisin B which showed optical rotation (−), the absolute configuration of said substance is 7R,8S,7′R,8′S. This is where the meeting of the cited analyses allowed us to conclude that (-)-5-demethoxygrandisin B is the lignan (7R,8S,7′R,8′S)-3,4,5,3′,4′-pentamethoxy-7,7′-epoxylignan (Figure 2).

Excitedly, the tetrahydrofuran lignan (-)-5-demethoxygrandisin B isolated in *V. surinamensis* leaves demonstrated leishmanicidal activity. Grandisin, another tetrahydrofuran neolignan found in *V. surinamensis* and *Piper solmsianum*, exhibits potent trypanocidal activity against trypomastigote forms of *Trypanosoma cruzi* and shows promising potential for applications in medicine as an antileishmanial agent [63]. The literature data report that lignans have excellent potential as growth inhibitors for the promastigote and intracellular amastigote stage of *Leishmania* [64,65]. The present data showed that the selectivity index for (-)-5-demethoxygrandisin B was more prominent for *Leishmania* than for macrophages. Ten lignans isolated from the hexane-ethyl acetate extract of *Phyllanthus amarus* leaves showed that the $CC_{50}$ for all the samples was >100 μg/mL, thus revealing low cytotoxicity against macrophages and selectivity against the parasite [66].

(-)-5-demethoxygrandisin B showed greater selectivity against the promastigote stage of *L. amazonensis* compared to the intracellular amastigote, demonstrating the direct action of this compound on this parasite. This can also be attributed to the lower compound bioavailability within the macrophage, which may have resulted in reduced activity against the intracellular amastigote. An indirect mechanism involved with the inhibition of intracellular amastigotes is related to macrophage activation, especially the induction of nitric oxide (NO) [22]. However, there were no significant changes in this assessment parameter when we analyzed the nitrite quantification of *L. amazonensis*-stimulated peritoneal macrophages treated with (-)-5-demethoxygrandisin B.

Therefore, to understand the leishmanicidal activity of the (-)-5-demethoxygrandisin B against promastigote forms, observations of ultrastructural and morphological alterations were performed as a direct way to determine parasite inhibition. Our observations revealed the presence of numerous large vacuoles, vesicular structures with electron-dense content and outer bilayer membranes, autophagosome-like structures, and an enlargement of the flagellar pocket with electron-dense filaments inside it. However, the most significant alteration was observed in the mitochondrion/kinetoplast, which displayed swelling and an altered kDNA compacting pattern. *Leishmania* parasites, like other species of trypanosomatids, have a single mitochondrion with specific energetic and antioxidant enzymes that regulate oxidative stress and bioenergetics and a unique arrangement of mitochondrial DNA (kinetoplast DNA). The kDNA is visible as a prominent disk-shaped spot located in the paraflagellar region of the single elongated mitochondrion, and its gene encoding information is far from that set needed for mitochondrial processes, revealing its importance in the regulation of pro- and antioxidant processes not only in the mitochondria [67,68]. The mitochondrial differences between mammals and trypanosomatids make this organelle a natural drug target, and the observed ultrastructural alteration led us to evaluate the mitochondrial membrane potential of parasites treated with (-)-5-demethoxygrandisin B.

Our results show that (-)-5-demethoxygrandisin B induced the depolarization of the mitochondrial membrane of the promastigote parasite, suggesting inhibition of the metabolic activity of parasites. *L. amazonensis* promastigotes treated with compounds like lignans change the mitochondrial membrane potential and lower ATP production [66]. Impairment of the membrane potential allows compounds to cross the mitochondrial membrane which may lead to a loss of the impermeability to intracellular electrolytes and consequently leads to inhibition of ATP production, resulting in parasite death [69–71]. This decrease in mitochondrial membrane potential can occur due to the oxidative imbalance caused by competitively inhibiting TryR activity, a pathway responsible for the leishmanicidal effect [72].

The TryR enzyme plays a crucial role in the antioxidant defense of trypanosomatids and has been widely used in molecular modeling studies in the literature [73,74]. Tetrahydrofuran lignans isolated from *V. surinamensis* showed potent trypanocidal activity against the trypomastigote form of *T. cruzi* [11]. A study by Oliveira et al. [75] suggests that the inhibition of *T. cruzi* TryR is related to the trypanocidal action of these lignans. Thus, to try to understand the inhibition mechanism of the compound against *Leishmania*, the enzyme TryR from *L. infantum* (PDB ID 2JK6) was selected as a molecular target for the docking simulation.

In this sense, the links with the residues Thr51, Lys60, Asp327, and Leu334 observed in the docking are considered important for biological activity against the TryR enzyme, as they participate in the binding of the FAD cofactor at its respective site. It is worth noting that the catalytic mechanism of the TryR occurs due to the transfer of electrons from NADPH via FAD to the two catalytic cysteines (Cys52 and Cys57), and the possible interference in this process prevents the reduction in the disulfide bridge [34]. Furthermore, the interaction with Cys57 is crucial for the potential inhibition of the enzyme as it makes the inactivation of the catalytic cysteine possible and thus stops the nucleophilic attack on Cys52 necessary for the release of the reduced substrate (T[SH]2) [17]. Therefore, it is observed that the more stable configuration obtained by the molecular docking of (-)-5-demethoxygrandisin B has an affinity for the amino acid residues of the active site and performs interactions with a catalytic cysteine considered important for the enzymatic activity.

In evaluating the MD simulations, it was found that the complex exhibited a high B-factor value due to its location in the N-terminal portion of an α-helix loop [76]. The studies carried out by Kuldeep et al. [74] report that the structure of the TryR enzyme of *L. infantum* shows characteristic fluctuations in the N- and C-terminal regions, mainly caused by the presence of unstable loops in these positions.

The binding affinity between the studied compound and the macromolecule was determined by calculating the binding free energy, which is an important tool for designing potent enzymatic inhibitors [77]. The amino acids Cys52, Gly326, Asp327, Leu334, and Thr335 contributed to the stabilization of the (-)-5-demethoxygrandisin B ligand in the complex. These key residues played an important role in stabilizing the ligand in the active site of the TryR enzyme, where they are located. Furthermore, the MD results corroborate those obtained in the molecular docking, which also identified interactions with Gly326, Asp327, and Leu334. Residues Gly326, Asp327, and Leu334 are part of the FAD cofactor binding site, whose interactions are also considered relevant for molecular recognition and the binding affinity of inhibitors [78]. The contribution of Cys52 suggests potential bioactivity of the more stable configuration of (-)-5-demethoxygrandisin B against the TryR enzyme, as this binding prevents the reaction of cysteine with the T[S]2 substrates for the production of mixed disulfide and, consequently, inhibits enzymatic activity [79]. A potent inhibitor studied by Baiocco et al. [28] presents a tetrahedral geometry at the redox site, whose active conformation is oriented by binding to catalytic residues and Thr335, demonstrating the significant role of this amino acid for the antileishmanial activity of drug candidates. A study carried out by Feitosa et al. [80] showed from the interaction energy analysis per residue that the contributions of Gly326 and Thr335 are responsible for the permanence of the ligand in the TryR site. Therefore, the results corroborate previous studies investigating the trypanothione catalytic mechanism and provide a theoretical basis for the design of lignan (-)-5-demethoxygrandisin B as a potential inhibitor of the TryR enzyme of *L. infantum*.

## 5. Conclusions

The (7R,8S,7′R,8′S)-3,4,5,3′,4′-pentamethoxy-7,7′-epoxylignan, namely (-)-5-demethoxygrandisin B, isolated from the leaves of *V. surinamensis* showed to be selective for the parasites, inhibiting the promastigote and intracellular amastigote forms of *L. amazonensis*. The antiparasitic effect of this compound was confirmed by the ultrastructural changes with decreased mitochondrial membrane potential in *L. amazonensis* promastigotes. The docking analyses and molecular dynamics simulation showed a possible mechanism of inhibition promoted by a more thermodynamically stable conformation of (-)-5-demethoxygrandisin B to the TryR enzyme.

**Supplementary Materials:** The following supporting information can be downloaded at: https://www.mdpi.com/article/10.3390/pharmaceutics15092292/s1, Figure S1: Mass spectrum of the precursor ion (-)-5-demethoxygrandisin B) and its isotopes; Figure S2: $^1$H NMR for the (-)-5-demethoxygrandisin B in $CDCl_3$ (400 MHz); Figure S3: $^{13}$C NMR for the (-)-5-demethoxygrandisin B in $CDCl_3$ (100 MHz); Figure S4: COSY for the (-)-5-demethoxygrandisin B in $CDCl_3$; Figure S5: HSQC for the (-)-5-demethoxygrandisin B in $CDCl_3$; Figure S6: HMBC for the (-)-5-demethoxygrandisin B in $CDCl_3$.

**Author Contributions:** Conceptualization, S.S.P. and J.V.S.-S.; data curation, L.S.S., F.A.d.M., A.L.B., and F.A.-S.; formal analysis, L.S.S., F.A.d.M., A.L.B., K.d.S.C. and F.A.-S.; funding acquisition, L.S.S., K.d.S.C. and F.A.-S.; investigation, S.S.P., J.V.S.-S., L.O.d.S., A.P.L.d.C., M.L.L.J., P.W.P.G., D.d.J.H., C.J.M.-T. and N.N.T.; methodology, S.S.P., J.V.S.-S., L.O.d.S. and F.A.d.M.; project administration, L.S.S., F.A.d.M., K.d.S.C. and F.A.-S.; resources, N.N.T., A.L.B., L.S.S., F.A.d.M., K.d.S.C. and F.A.-S.; software, S.S.P., A.P.L.d.C. and F.A.d.M.; supervision, L.S.S., F.A.d.M., K.d.S.C. and F.A.-S.; validation, L.S.S., F.A.d.M., A.L.B., N.N.T., K.d.S.C. and F.A.-S.; visualization, S.S.P., P.W.P.G. and J.V.S.-S.; writing—original draft preparation, S.S.P., P.W.P.G. and J.V.S.-S.; writing—review and editing, all authors. All authors have read and agreed to the published version of the manuscript.

**Funding:** Pró-reitoria de Pesquisa—Universidade Federal do Pará (UFPA), Oswaldo Cruz Institute, Coordination for the Improvement of Higher Education Personnel (Coordenação de Aperfeiçoamento de Pessoal de Nível Superior do Brazil; CAPES), grant number 8887.200524/2018-00, 88887.368507/2019-00, Finance Code 001, and Carlos Chagas Filho Foundation for Research Support of the State of Rio de Janeiro (Fundação Carlos Chagas Filho de Amparo à Pesquisa do Estado do Rio de Janeiro; FAPERJ), grant numbers E-26/210.344/2019, E-26/201.765/2019, and E-26/211.680/2021 (269680). The CytoFlex flow cytometer used at the Flow Cytometry Core Facility was acquired by FAPERJ, grant number E-26/110332/2014. The APC (Fund for Conjoint Research Project) was funded by the Oswaldo Cruz Institute/FIOCRUZ. F.A-S. is a postdoctoral research fellow and scholarship holder of CAPES, grant number 88887.363006/2019-00. K.S.C. (CNPq 315225/2021-1) is a senior researcher.

**Institutional Review Board Statement:** Not applicable.

**Informed Consent Statement:** Not applicable.

**Data Availability Statement:** All supporting data used in this study are available from the authors.

**Acknowledgments:** The authors would like to express their deepest gratitude to Pró-Reitoria de Pesquisa (Propesp) of the Federal University of Para (UFPA) and Instituto Oswaldo Cruz of the Oswaldo Cruz Foundation (IOC/FICORUZ) and Vanessa Costa and Raquel Paredes of the Flow Cytometry Core Facility for acquisition of the CytoFlex flow cytometer.

**Conflicts of Interest:** The authors declare no conflict of interest.

## References


1. Schultes, R.E. A New Narcotic Snuff from the Northwest Amazon. *Bot. Mus. Lealf. Harv. Univ.* **1954**, *16*, 241–260. [CrossRef]
2. Rodrigues, W.A. Revisão Taxonômica Das Espécies de *Virola aublet* (Myrısticaceae) Do Brasil. *Acta. Amaz.* **1980**, *10*, 3 127. [CrossRef]
3. Ferri, P.H.; Barata, L.E.S. Neolignans and a Phenylpropanoid from Virola Pavonis Leaves. *Phytochemistry* **1992**, *31*, 1375–1377. [CrossRef]
4. Lopes, L.M.X.; Yoshida, M.; Gottlieb, O.R. 1,11-Diarylundecan-1-One and 4-Aryltetralone Neolignans from Virola Sebifera. *Phytochemistry* **1982**, *21*, 751–755. [CrossRef]
5. Lopes, N.P.; dos Santos, P.A.; Kato, M.J.; Yoshida, M. New Butenolides in Plantlets of Virola Surinamensis (Myristicaceae). *Chem. Pharm. Bull.* **2004**, *52*, 1255–1257. [CrossRef]
6. Rye, C.E.; Barker, D. Asymmetric Synthesis and Anti-Protozoal Activity of the 8,4′-Oxyneolignans Virolin, Surinamensin and Analogues. *Eur. J. Med. Chem.* **2013**, *60*, 240–248. [CrossRef] [PubMed]
7. Messiano, G.B.; Santos, R.A.d.S.; Ferreira, L.D.S.; Simões, R.A.; Jabor, V.A.P.; Kato, M.J.; Lopes, N.P.; Pupo, M.T.; de Oliveira, A.R.M. In Vitro Metabolism Study of the Promising Anticancer Agent the Lignan (−)-Grandisin. *J. Pharm. Biomed. Anal.* **2013**, *72*, 240–244. [CrossRef] [PubMed]
8. Carvalho, A.A.V.; Galdino, P.M.; Nascimento, M.V.M.; Kato, M.J.; Valadares, M.C.; Cunha, L.C.; Costa, E.A. Antinociceptive and Antiinflammatory Activities of Grandisin Extracted from *Virola surinamensis*. *Phytother. Res.* **2010**, *24*, 113–118. [CrossRef] [PubMed]
9. Lopes, N.P.; Kato, M.J.; de Andrade, E.H.A.; Maia, J.G.S.; Yoshida, M.; Planchart, A.R.; Katzin, A.M. Antimalarial Use of Volatile Oil from Leaves of Virola Surinamensis (Rol.) Warb. by Waiãpi Amazon Indians. *J. Ethnopharmacol.* **1999**, *67*, 313–319. [CrossRef] [PubMed]
10. Cabral, M.M.O.; Barbosa-Filho, J.M.; Maia, G.L.A.; Chaves, M.C.O.; Braga, M.V.; De Souza, W.; Soares, R.O.A. Neolignans from Plants in Northeastern Brazil (Lauraceae) with Activity against Trypanosoma Cruzi. *Exp. Parasitol.* **2010**, *124*, 319–324. [CrossRef]
11. Lopes, N.; Chicaro, P.; Kato, M.; Albuquerque, S.; Yoshida, M. Flavonoids and Lignans from *Virola surinamensis* Twigs and Their In Vitro Activity against *Trypanosoma cruzi*. *Planta Med.* **1998**, *64*, 667–669. [CrossRef]
12. Lopes, N.P.; de Almeida Blumenthal, E.E.; Cavalheiro, A.J.; Kato, M.J.; Yoshida, M. Lignans, γ-Lactones and Propiophenones of Virola Surinamensis. *Phytochemistry* **1996**, *43*, 1089–1092. [CrossRef]
13. dos Santos Maia, M.; Raimundo e Silva, J.P.; de Lima Nunes, T.A.; Saraiva de Sousa, J.M.; Soares Rodrigues, G.C.; Messias Monteiro, A.F.; Fechine Tavares, J.; da Franca Rodrigues, K.A.B.; Mendonça-Junior, F.J.; Scotti, L.; et al. Virtual Screening and the In Vitro Assessment of the Antileishmanial Activity of Lignans. *Molecules* **2020**, *25*, 2281. [CrossRef] [PubMed]
14. Mann, S.; Frasca, K.; Scherrer, S.; Henao-Martínez, A.F.; Newman, S.; Ramanan, P.; Suarez, J.A. A Review of Leishmaniasis: Current Knowledge and Future Directions. *Curr. Trop. Med. Rep.* **2021**, *8*, 121–132. [CrossRef] [PubMed]
15. WHO. Leishmaniasis. Available online: https://www.who.int/news-room/fact-sheets/detail/leishmaniasis (accessed on 15 March 2023).
16. Imran, M.; Khan, S.A.; Abida; Alshrari, A.S.; Eltahir Mudawi, M.M.; Alshammari, M.K.; Harshan, A.A.; Alshammari, N.A. Small Molecules as Kinetoplastid Specific Proteasome Inhibitors for Leishmaniasis: A Patent Review from 1998 to 2021. *Expert. Opin. Ther. Pat.* **2022**, *32*, 591–604. [CrossRef] [PubMed]
17. Battista, T.; Colotti, G.; Ilari, A.; Fiorillo, A. Targeting Trypanothione Reductase, a Key Enzyme in the Redox Trypanosomatid Metabolism, to Develop New Drugs against Leishmaniasis and Trypanosomiases. *Molecules* **2020**, *25*, 1924. [CrossRef] [PubMed]
18. Fiorillo, A.; Colotti, G.; Exertier, C.; Liuzzi, A.; Seghetti, F.; Salerno, A.; Caciolla, J.; Ilari, A. Innovative Approach for a Classic Target: Fragment Screening on Trypanothione Reductase Reveals New Opportunities for Drug Design. *Front. Mol. Biosci.* **2022**, *9*, 900882. [CrossRef]
19. Gomes, P.; Quirós-Guerrero, L.; Muribeca, A.; Reis, J.; Pamplona, S.; Lima, A.; Trindade, M.; Silva, C.; Souza, J.; Boutin, J.; et al. Constituents of *Chamaecrista diphylla* (L.) Greene Leaves with Potent Antioxidant Capacity: A Feature-Based Molecular Network Dereplication Approach. *Pharmaceutics* **2021**, *13*, 681. [CrossRef]
20. Gomes, P.; Quirós-Guerrero, L.; Silva, C.; Pamplona, S.; Boutin, J.A.; Eberlin, M.; Wolfender, J.-L.; Silva, M. Feature-Based Molecular Network-Guided Dereplication of Natural Bioactive Products from Leaves of *Stryphnodendron pulcherrimum* (Willd.) Hochr. *Metabolites* **2021**, *11*, 281. [CrossRef] [PubMed]
21. Silva-Silva, J.V.; Moragas-Tellis, C.J.; Chagas, M.D.S.d.S.; de Souza, P.V.R.; Souza, C.d.S.F.d.; Hardoim, D.d.J.; Taniwaki, N.N.; Moreira, D.d.L.; Behrens, M.D.; Calabrese, K.d.S.; et al. Antileishmanial Activity of Flavones-Rich Fraction From *Arrabidaea chica* Verlot (Bignoniaceae). *Front. Pharmacol.* **2021**, *12*, 703985. [CrossRef] [PubMed]
22. Silva-Silva, J.V.; Moragas-Tellis, C.J.; Chagas, M.S.S.; Souza, P.V.R.; Moreira, D.L.; de Souza, C.S.F.; Teixeira, K.F.; Cenci, A.R.; de Oliveira, A.S.; Almeida-Souza, F.; et al. Carajurin: A Anthocyanidin from *Arrabidaea chica* as a Potential Biological Marker of Antileishmanial Activity. *Biomed. Pharmacother.* **2021**, *141*, 111910. [CrossRef] [PubMed]
23. Ramos, G.d.C.; Silva-Silva, J.V.; Watanabe, L.A.; de Sousa Siqueira, J.E.; Almeida-Souza, F.; Calabrese, K.S.; do Rosario Marinho, A.M.; Marinho, P.S.B.; de Oliveira, A.S. Phomoxanthone A, Compound of Endophytic Fungi Paecilomyces Sp. and Its Potential Antimicrobial and Antiparasitic. *Antibiotics* **2022**, *11*, 1332. [CrossRef]

24. Pina, J.R.S.; Silva-Silva, J.V.; Carvalho, J.M.; Bitencourt, H.R.; Watanabe, L.A.; Fernandes, J.M.P.; Souza, G.E.d.; Aguiar, A.C.C.; Guido, R.V.C.; Almeida-Souza, F.; et al. Antiprotozoal and Antibacterial Activity of Ravenelin, a Xanthone Isolated from the Endophytic Fungus Exserohilum Rostratum. *Molecules* **2021**, *26*, 3339. [CrossRef] [PubMed]
25. Silva-Silva, J.V.; Moreira, R.F.; Watanabe, L.A.; de Souza, C.D.S.F.; Hardoim, D.J.; Taniwaki, N.N.; Bertho, A.L.; Teixeira, K.F.; Cenci, A.R.; Doring, T.H.; et al. Monomethylsulochrin Isolated from Biomass Extract of Aspergillus Sp. against Leishmania Amazonensis: In Vitro Biological Evaluation and Molecular Docking. *Front. Cell. Infect. Microbiol.* **2022**, *12*, 974910. [CrossRef]
26. Silva-Silva, J.V.; Moragas-Tellis, C.J.; Chagas, M.S.S.; Souza, P.V.R.; Moreira, D.L.; Hardoim, D.J.; Taniwaki, N.N.; Costa, V.F.A.; Bertho, A.L.; Brondani, D.; et al. Carajurin Induces Apoptosis in *Leishmania amazonensis* Promastigotes through Reactive Oxygen Species Production and Mitochondrial Dysfunction. *Pharmaceuticals* **2022**, *15*, 331. [CrossRef] [PubMed]
27. Jones, G.; Willett, P.; Glen, R.C.; Leach, A.R.; Taylor, R. Development and Validation of a Genetic Algorithm for Flexible Docking 1 1Edited by F. E. Cohen. *J. Mol. Biol.* **1997**, *267*, 727–748. [CrossRef]
28. Baiocco, P.; Colotti, G.; Franceschini, S.; Ilari, A. Molecular Basis of Antimony Treatment in Leishmaniasis. *J. Med. Chem.* **2009**, *52*, 2603–2612. [CrossRef]
29. Lodhi, S.S.; Farmer, R.; Singh, A.K.; Jaiswal, Y.K.; Wadhwa, G. 3D Structure Generation, Virtual Screening and Docking of Human Ras-Associated Binding (Rab3A) Protein Involved in Tumourigenesis. *Mol. Biol. Rep.* **2014**, *41*, 3951–3959. [CrossRef]
30. Ruiz-Santaquiteria, M.; de Castro, S.; Toro, M.A.; de Lucio, H.; Gutiérrez, K.J.; Sánchez-Murcia, P.A.; Jiménez, M.Á.; Gago, F.; Jiménez-Ruiz, A.; Camarasa, M.-J.; et al. Trypanothione Reductase Inhibition and Anti-Leishmanial Activity of All-Hydrocarbon Stapled α-Helical Peptides with Improved Proteolytic Stability. *Eur. J. Med. Chem.* **2018**, *149*, 238–247. [CrossRef]
31. Pettersen, E.F.; Goddard, T.D.; Huang, C.C.; Couch, G.S.; Greenblatt, D.M.; Meng, E.C.; Ferrin, T.E. UCSF Chimera—A Visualization System for Exploratory Research and Analysis. *J. Comput. Chem.* **2004**, *25*, 1605–1612. [CrossRef]
32. Halgren, T.A. Merck Molecular Force Field. I. Basis, Form, Scope, Parameterization, and Performance of MMFF94. *J. Comput. Chem.* **1996**, *17*, 490–519. [CrossRef]
33. Hanwell, M.D.; Curtis, D.E.; Lonie, D.C.; Vandermeersch, T.; Zurek, E.; Hutchison, G.R. Avogadro: An Advanced Semantic Chemical Editor, Visualization, and Analysis Platform. *J. Cheminform.* **2012**, *4*, 17. [CrossRef]
34. Ilari, A.; Fiorillo, A.; Genovese, I.; Colotti, G. Polyamine-Trypanothione Pathway: An Update. *Future Med. Chem.* **2017**, *9*, 61–77. [CrossRef]
35. Verdonk, M.L.; Cole, J.C.; Hartshorn, M.J.; Murray, C.W.; Taylor, R.D. Improved Protein-Ligand Docking Using GOLD. *Proteins* **2003**, *52*, 609–623. [CrossRef]
36. Stierand, K.; Maaß, P.C.; Rarey, M. Molecular Complexes at a Glance: Automated Generation of Two-Dimensional Complex Diagrams. *Bioinformatics* **2006**, *22*, 1710–1716. [CrossRef] [PubMed]
37. Stierand, K.; Rarey, M. From Modeling to Medicinal Chemistry: Automatic Generation of Two-Dimensional Complex Diagrams. *ChemMedChem* **2007**, *2*, 853–860. [CrossRef] [PubMed]
38. Frisch, M.J.; Frisch, M.; Trucks, G.; Schlegel, K.; Scuseria, G.; Robb, M.; Cheeseman, J.; Montgomery, J.; Vreven, T.; Kudin, K.-N.; et al. *Gaussian 03: Revision C.02*; Gaussian Inc.: Pittsburgh, PA, USA, 2003.
39. Bayly, C.I.; Cieplak, P.; Cornell, W.; Kollman, P.A. A Well-Behaved Electrostatic Potential Based Method Using Charge Restraints for Deriving Atomic Charges: The RESP Model. *J. Phys. Chem.* **1993**, *97*, 10269–10280. [CrossRef]
40. Hariharan, P.C.; Pople, J.A. The Influence of Polarization Functions on Molecular Orbital Hydrogenation Energies. *Theor. Chim. Acta* **1973**, *28*, 213–222. [CrossRef]
41. Li, H.; Robertson, A.D.; Jensen, J.H. Very Fast Empirical Prediction and Rationalization of Protein PKa Values. *Proteins* **2005**, *61*, 704–721. [CrossRef] [PubMed]
42. Salomon-Ferrer, R.; Götz, A.W.; Poole, D.; Le Grand, S.; Walker, R.C. Routine Microsecond Molecular Dynamics Simulations with AMBER on GPUs. 2. Explicit Solvent Particle Mesh Ewald. *J. Chem. Theory Comput.* **2013**, *9*, 3878–3888. [CrossRef]
43. Wang, J.; Wolf, R.M.; Caldwell, J.W.; Kollman, P.A.; Case, D.A. Development and Testing of a General Amber Force Field. *J. Comput. Chem.* **2004**, *25*, 1157–1174. [CrossRef] [PubMed]
44. Hornak, V.; Abel, R.; Okur, A.; Strockbine, B.; Roitberg, A.; Simmerling, C. Comparison of Multiple Amber Force Fields and Development of Improved Protein Backbone Parameters. *Proteins* **2006**, *65*, 712–725. [CrossRef]
45. Jorgensen, W.L.; Chandrasekhar, J.; Madura, J.D.; Impey, R.W.; Klein, M.L. Comparison of Simple Potential Functions for Simulating Liquid Water. *J. Chem. Phys.* **1983**, *79*, 926–935. [CrossRef]
46. Case, D.A.; Cheatham, T.E.; Darden, T.; Gohlke, H.; Luo, R.; Merz, K.M.; Onufriev, A.; Simmerling, C.; Wang, B.; Woods, R.J. The Amber Biomolecular Simulation Programs. *J. Comput. Chem.* **2005**, *26*, 1668–1688. [CrossRef] [PubMed]
47. Loncharich, R.J.; Brooks, B.R.; Pastor, R.W. Langevin Dynamics of Peptides: The Frictional Dependence of Isomerization Rates OfN-Acetylalanyl-N?-Methylamide. *Biopolymers* **1992**, *32*, 523–535. [CrossRef] [PubMed]
48. Darden, T.; York, D.; Pedersen, L. Particle Mesh Ewald: An $N\cdot\log(N)$ Method for Ewald Sums in Large Systems. *J. Chem. Phys.* **1993**, *98*, 10089–10092. [CrossRef]
49. Ryckaert, J.-P.; Ciccotti, G.; Berendsen, H.J.C. Numerical Integration of the Cartesian Equations of Motion of a System with Constraints: Molecular Dynamics of n-Alkanes. *J. Comput. Phys.* **1977**, *23*, 327–341. [CrossRef]
50. Verlet, L. Computer "Experiments" on Classical Fluids. II. Equilibrium Correlation Functions. *Phys. Rev.* **1968**, *165*, 201–214. [CrossRef]

51. Roe, D.R.; Cheatham, T.E. PTRAJ and CPPTRAJ: Software for Processing and Analysis of Molecular Dynamics Trajectory Data. *J. Chem. Theory Comput.* **2013**, *9*, 3084–3095. [CrossRef]
52. Kollman, P.A.; Massova, I.; Reyes, C.; Kuhn, B.; Huo, S.; Chong, L.; Lee, M.; Lee, T.; Duan, Y.; Wang, W.; et al. Calculating Structures and Free Energies of Complex Molecules: Combining Molecular Mechanics and Continuum Models. *ACC Chem. Res.* **2000**, *33*, 889–897. [CrossRef]
53. Massova, I.; Kollman, P.A. Combined Molecular Mechanical and Continuum Solvent Approach (MM-PBSA/GBSA) to Predict Ligand Binding. *Perspect. Drug Discov. Des.* **2000**, *18*, 113–135. [CrossRef]
54. Miller, B.R.; McGee, T.D.; Swails, J.M.; Homeyer, N.; Gohlke, H.; Roitberg, A.E. *MMPBSA.Py*: An Efficient Program for End-State Free Energy Calculations. *J. Chem. Theory Comput.* **2012**, *8*, 3314–3321. [CrossRef]
55. Raza, S.; Ranaghan, K.E.; van der Kamp, M.W.; Woods, C.J.; Mulholland, A.J.; Azam, S.S. Visualizing Protein–Ligand Binding with Chemical Energy-Wise Decomposition (CHEWD): Application to Ligand Binding in the Kallikrein-8 S1 Site. *J. Comput. Aided Mol. Des.* **2019**, *33*, 461–475. [CrossRef]
56. Barata, L.E.S.; Baker, P.M.; Gottlieb, O.R.; Rùveda, E.A. Neolignans of Virola Surinamensis. *Phytochemistry* **1978**, *17*, 783–786. [CrossRef]
57. Fonseca, S.F. 1945-RMN-13c de Lignanas Da Araucaria Angustifolia, de Neolignanas Ariltetralinicas e Tetraidrofuranicas e de Derivados Da Podofilotoxina. Tese (Doutorado), Universidade Estadual de Campinas, Campinas, Brazil, 1980. [CrossRef]
58. Karplus, M. Contact Electron-Spin Coupling of Nuclear Magnetic Moments. *J. Chem. Phys.* **1959**, *30*, 11–15. [CrossRef]
59. Kubanek, J.; Hay, M.E.; Brown, P.J.; Lindquist, N.; Fenical, W. Lignoid Chemical Defenses in the Freshwater Macrophyte Saururus Cernuus. *Chemoecology* **2001**, *11*, 1–8. [CrossRef]
60. Kubanek, J.; Fenical, W.; Hay, M.E.; Brown, P.J.; Lindquist, N. Two Antifeedant Lignans from the Freshwater Macrophyte Saururus Cernuus. *Phytochemistry* **2000**, *54*, 281–287. [CrossRef]
61. Hwang, B.Y.; Lee, J.-H.; Nam, J.B.; Hong, Y.-S.; Lee, J.J. Lignans from Saururus Chinensis Inhibiting the Transcription Factor NF-KB. *Phytochemistry* **2003**, *64*, 765–771. [CrossRef] [PubMed]
62. Biftu, T.; Hazra, B.G.; Stevenson, R.; Williams, J.R. Syntheses of Lignans from 2,3-Diarolybutanes. *J. Chem. Soc. Perkin. Trans. 1* **1978**, 1147–1150. [CrossRef]
63. Patel, D.K. Grandisin and Its Therapeutic Potential and Pharmacological Activities: A Review. *Pharmacol. Res. Mod. Chin. Med.* **2022**, *5*, 100176. [CrossRef]
64. Veiga, A.; Albuquerque, K.; Corrêa, M.E.; Brigido, H.; Silva e Silva, J.; Campos, M.; Silveira, F.; Santos, L.; Dolabela, M. Leishmania Amazonensis and Leishmania Chagasi: In Vitro Leishmanicide Activity of Virola Surinamensis (Rol.) Warb. *Exp. Parasitol.* **2017**, *175*, 68–73. [CrossRef] [PubMed]
65. Gervazoni, L.F.O.; Barcellos, G.B.; Ferreira-Paes, T.; Almeida-Amaral, E.E. Use of Natural Products in Leishmaniasis Chemotherapy: An Overview. *Front. Chem.* **2020**, *8*, 579891. [CrossRef]
66. Conrado, G.G.; Grazzia, N.; de Oliveira, A.d.S.S.; Franco, C.H.; Moraes, C.B.; Gadelha, F.R.; Miguel, D.C.; Garcia, V.L. Prospecting and Identifying Phyllanthus Amarus Lignans with Antileishmanial and Antitrypanosomal Activity. *Planta Med.* **2020**, *86*, 782–789. [CrossRef]
67. Lukeš, J.; Hashimi, H.; Zíková, A. Unexplained Complexity of the Mitochondrial Genome and Transcriptome in Kinetoplastid Flagellates. *Curr. Genet.* **2005**, *48*, 277–299. [CrossRef] [PubMed]
68. Menna-Barreto, R.F.S.; de Castro, S.L. The Double-Edged Sword in Pathogenic Trypanosomatids: The Pivotal Role of Mitochondria in Oxidative Stress and Bioenergetics. *Biomed. Res. Int.* **2014**, *2014*, 614014. [CrossRef] [PubMed]
69. Rottini, M.M.; Amaral, A.C.F.; Ferreira, J.L.P.; de Andrade Silva, J.R.; Taniwaki, N.N.; da Silva Freitas de Souza, C.; d'Escoffier, L.N.; Almeida-Souza, F.; de Jesus Hardoim, D.; da Costa, S.C.G.; et al. In Vitro Evaluation of (-)α-Bisabolol as a Promising Agent against Leishmania Amazonensis. *Exp. Parasitol.* **2015**, *148*, 66–72. [CrossRef] [PubMed]
70. de Medeiros, M.d.G.F.; da Silva, A.C.; Citó, A.M. das G.L.; Borges, A.R.; de Lima, S.G.; Lopes, J.A.D.; Figueiredo, R.C.B.Q. In Vitro Antileishmanial Activity and Cytotoxicity of Essential Oil from *Lippia sidoides* Cham. *Parasitol. Int.* **2011**, *60*, 237–241. [CrossRef]
71. Teixeira de Macedo Silva, S.; Visbal, G.; Lima Prado Godinho, J.; Urbina, J.A.; de Souza, W.; Cola Fernandes Rodrigues, J. In Vitro Antileishmanial Activity of Ravuconazole, a Triazole Antifungal Drug, as a Potential Treatment for Leishmaniasis. *J. Antimicrob. Chemother.* **2018**, *73*, 2360–2373. [CrossRef]
72. Inacio, J.D.F.; Fonseca, M.S.; Limaverde-Sousa, G.; Tomas, A.M.; Castro, H.; Almeida-Amaral, E.E. Epigallocathechin-O-3-Gallate Inhibits *Trypanothione reductase* of *Leishmania infantum*, Causing Alterations in Redox Balance and Leading to Parasite Death. *Front. Cell. Infect. Microbiol.* **2021**, *11*, 640561. [CrossRef] [PubMed]
73. Tonelli, M.; Sparatore, A.; Basilico, N.; Cavicchini, L.; Parapini, S.; Tasso, B.; Laurini, E.; Pricl, S.; Boido, V.; Sparatore, F. Quinolizidine-Derived Lucanthone and Amitriptyline Analogues Endowed with Potent Antileishmanial Activity. *Pharmaceuticals* **2020**, *13*, 339. [CrossRef]
74. Kuldeep, J.R.; Karthik, R.; Kaur, P.; Goyal, N.; Siddiqi, M.I. Identification of Potential Anti-Leishmanial Agents Using Computational Investigation and Biological Evaluation against *Trypanothione reductase*. *J. Biomol. Struct. Dyn.* **2021**, *39*, 960–969. [CrossRef]
75. De Oliveira, R.B.; Vaz, A.B.; Alves, R.O.; Liarte, D.B.; Donnici, C.L.; Romanha, A.J.; Zani, C.L. Arylfurans as Potential Trypanosoma Cruzi Trypanothione Reductase Inhibitors. *Mem. Inst. Oswaldo Cruz.* **2006**, *101*, 169–173. [CrossRef] [PubMed]

76. Colotti, G.; Baiocco, P.; Fiorillo, A.; Boffi, A.; Poser, E.; Di Chiaro, F.; Ilari, A. Structural Insights into the Enzymes of the Trypanothione Pathway: Targets for Antileishmaniasis Drugs. *Future Med. Chem.* **2013**, *5*, 1861–1875. [CrossRef] [PubMed]
77. Souza, A.; Cardoso, F.; Martins, L.; Alves, C.; Silva, J.; Molfetta, F. Molecular Modelling Study of Heteroarylamide/Sulfonamide Compounds with Antitrypanosomal Activity. *J. Braz. Chem. Soc* **2021**, *32*, 83–97. [CrossRef]
78. Vargas, J.A.R.; Lopez, A.G.; Piñol, M.C.; Froeyen, M. Molecular Docking Study on the Interaction between 2-Substituted-4,5-Difuryl Imidazoles with Different Protein Target for Antileishmanial Activity. *J. App. Pharm. Sci.* **2018**, *8*, 14–22. [CrossRef]
79. da Silva, A.D.; dos Santos, J.A.; Machado, P.A.; Alves, L.A.; Laque, L.C.; de Souza, V.C.; Coimbra, E.S.; Capriles, P.V.S.Z. Insights about Resveratrol Analogs against Trypanothione Reductase of *Leishmania braziliensis*: Molecular Modeling, Computational Docking and In Vitro Antileishmanial Studies. *J. Biomol. Struct. Dyn.* **2019**, *37*, 2960–2969. [CrossRef]
80. Feitosa, A.O.; Ferreira, F.J.N.; Brigido, H.P.C.; Bastos, M.L.C.; Carvalho, J.M.; Carneiro, A.S.; Dolabela, M.F.; Marinho, P.S.B.; Marinho, A.M.R. Study on Experimental Leishmanicidal Activity and In Silico of Cytochalasin B. *J. Braz. Chem. Soc.* **2019**, *30*, 592–596. [CrossRef]

*pharmaceutics*

MDPI

*Article*

# Coleon U, Isolated from *Plectranthus mutabilis* Codd., Decreases P-Glycoprotein Activity Due to Mitochondrial Inhibition

**Sofija Jovanović Stojanov [1,†], Epole N. Ntungwe [2,3,†], Jelena Dinić [1], Ana Podolski-Renić [1], Milica Pajović [1], Patrícia Rijo [2,*] and Milica Pešić [1,*]**

1 Institute for Biological Research "Siniša Stanković"—National Institute of the Republic of Serbia, University of Belgrade, 11060 Belgrade, Serbia; sofija.jovanovic@ibiss.bg.ac.rs (S.J.S.); jelena.dinic@ibiss.bg.ac.rs (J.D.); ana.podolski@ibiss.bg.ac.rs (A.P.-R.); milica.pajovic@ibiss.bg.ac.rs (M.P.)

2 CBIOS (Research Center for Biosciences and Health Technologies), Universidade Lusófona de Humanidades e Tecnologias, Campo Grande 376, 1749-024 Lisbon, Portugal; p5999@ulusofona.pt

3 Department of Chemistry, University of Coimbra, P-3004-535 Coimbra, Portugal

* Correspondence: patricia.rijo@ulusofona.pt (P.R.); camala@ibiss.bg.ac.rs (M.P.)

† These authors contributed equally to this work.

**Abstract:** Multidrug resistance in cancer is often mediated by P-glycoprotein. Natural compounds have been suggested as a fourth generation of P-glycoprotein inhibitors. Coleon U, isolated from *Plectranthus mutabilis* Codd., was reported to modulate P-glycoprotein activity but the underlying mechanism has not yet been revealed. Therefore, the effects of Coleon U on cell viability, proliferation, and cell death induction were studied in a non-small-cell lung carcinoma model comprising sensitive and multidrug-resistant cells with P-glycoprotein overexpression. P-glycoprotein activity and mitochondrial membrane potential were assessed by flow cytometry upon Coleon U, sodium-orthovanadate (an ATPase inhibitor), and verapamil (an ATPase stimulator) treatments. SwissADME was used to identify the pharmacokinetic properties of Coleon U, while P-glycoprotein expression was studied by immunofluorescence. Our results showed that Coleon U is not a P-glycoprotein substrate and is equally efficient in sensitive and multidrug-resistant cancer cells. A decrease in P-glycoprotein activity observed with Coleon U and verapamil after 72 h is antagonized in combination with sodium-orthovanadate. Coleon U induced a pronounced effect on mitochondrial membrane depolarization and showed a tendency to decrease P-glycoprotein expression. In conclusion, Coleon U-delayed effect on the decrease in P-glycoprotein activity is due to P-glycoprotein's functioning dependence on ATP production in mitochondria.

**Keywords:** Coleon U; P-glycoprotein; cancer multidrug resistance

## 1. Introduction

Non-small-cell lung cancer (NSCLC) is a common and highly lethal tumor worldwide [1]. Most NSCLC patients are diagnosed at an advanced stage of the disease; however, despite extensive research, the intrinsic and acquired chemotherapy resistance, particularly multidrug resistance (MDR), remains a significant challenge in improving treatment outcomes, often leading to treatment failure and disease relapse [2]. Statistical data indicate that drug resistance is a major contributor to over 90% of cancer patient mortality [3].

During chemotherapy, MDR can occur through a variety of mechanisms, including enhanced drug efflux, genetic factors (such as gene mutations, amplifications, and epigenetic changes), growth factors, increased DNA repair capacity, and increased metabolism of xenobiotics [3–7]. Each of these mechanisms decrease the effectiveness of administered drugs, making tumor treatment more challenging. P-glycoprotein (P-gp), an ATP-binding cassette (ABC) transporter, plays a crucial role in drug resistance [8]. P-gp is located in the cell membrane and regulates the distribution, absorption, and excretion of various chemical compounds [9]. It is highly expressed in cancer cells and functions as an efflux pump, actively removing drugs from the cells, leading to decreased effectiveness and contributing to MDR [3]. Developing new inhibitors that can inhibit P-gp function or discovering antitumor drugs that are not P-gp substrates is essential to prevent drug efflux and increase the sensitivity of cancer cells to drugs.

Natural products have gained attention as agents with the potential to overcome MDR [10]. Various compounds of natural origin, including flavonoids, alkaloids, coumarins, and terpenoids, have been investigated as potential modulators of P-gp activity [11–14]. Terpenoids, specifically abietane-type diterpenoids found in the *Plectranthus* genus, have been identified as inhibitors of P-gp activity [15]. *Plectranthus mutabilis* Codd. is a succulent plant from South Africa that is threatened with extinction, and its extract contains compounds with cytotoxic activity against various cancer cell lines [15–17]. Examining this rare plant with significant medicinal potential can raise awareness of its protection and conservation. Our work in the last few years is partly dedicated to this goal. Coleon U, an abietane diterpenoid, is the prominent diterpenoid in *Plectranthus mutabilis* Codd. extract (Coleon U is the most abundant compound with 96 μg/mg in the acetone extract from the leaves of *Plectranthus mutabilis* Codd.) [17]. The available literature on the Coleon U mode of action in cancer is limited and this motivated us to study its anticancer mechanism of action. It was discovered that Coleon U can impact the immune system's functional status by inhibiting the proliferation of human lymphocytes, making Coleon U a promising candidate for the development of immunomodulatory drugs [18]. Coleon U has also been shown to inhibit the growth of several human cancer cell lines: breast, central nervous system, kidney, and melanoma [19]. Although its precise mechanisms of action are not fully understood, Coleon U selectively activates a specific isoform of protein kinase C (PKC-δ) involved in apoptotic pathways, making it a promising candidate for further evaluation as an anticancer drug [20]. Additionally, Coleon U has shown the potential to reverse resistance to doxorubicin (DOX) in MDR NSCLC cells and increase its efficacy [17].

This study aims to elucidate the mechanism underlying the Coleon U effect against P-gp activity in MDR NSCLC cells. As a model system, we used sensitive NCI-H460 cells without P-gp expression and their MDR counterpart–NCI-H460/R cells with P-gp expression. Our investigation was directed towards revealing cellular processes that explain the delayed decrease in P-gp activity after Coleon U treatment because we showed that Coleon U does not directly interact with P-gp [17].

## 2. Materials and Methods

### 2.1. Reagents

RPMI 1640 medium and DMEM (Dulbecco's Modified Eagle Medium) were acquired from Capricorn Scientific GmbH (Ebsdorfergrund, Germany). L-glutamine, trypsin, and penicillin–streptomycin solution were from Biowest (Nuaillé, France). Thiazolyl blue tetrazolium bromide (MTT), fetal bovine serum (FBS), DMSO, Hoechst 33342, and carboxyfluorescein succinimidyl ester (CFSE) were purchased from Thermo Fisher Scientific (Waltham, MA, USA). An Apoptosis Detection Kit based on Annexin-V-FITC (AV) and propidium iodide (PI) staining was acquired from Abcam (Cambridge, UK). Bovine serum albumin (BSA) was obtained from Serva (Heidelberg, Germany). Rhodamine 123 (Rho123) and tetramethylrhodamine ethyl ester (TMRE) were purchased from Sigma (St Louis, MO, USA). Carbonyl cyanide m-chlorophenyl hydrazine (CCCP, C2759) was purchased from Sigma-Aldrich (Darmstadt, Germany). FITC-conjugated anti-P-gp antibody was obtained from BD Biosciences (Winnersh, Berkshire, UK). Anti-P-gp mouse monoclonal antibody (Abcam, Cambridge, UK; ab10333) and secondary antibodies Alexa Fluor 555 goat anti-mouse secondary antibody (#A-21422, Thermo Fisher Scientific, Waltham, MA, USA) were obtained.

### 2.2. Drugs

Coleon U compound, isolated from *Plectranthus mutabilis* Codd. [17], was dissolved in dimethyl sulfoxide (DMSO) as a 20 mM aliquot and kept at room temperature until use. Verapamil (Sigma-Aldrich Chemie Gmbh, Hamburg, Germany) and Sodium Vanadate ($Na_3VO_4$) (Promega, Madison, WI, USA) were stored as 10 mM aliquots at −20 °C. Paclitaxel—PTX (Sigma-Aldrich Chemie Gmbh, Hamburg, Germany) was diluted in absolute ethanol and 1 mM aliquots were stored at −20 °C. Before treatment, all drugs were freshly diluted in sterile water.

### 2.3. Cell Culture

NSCLC NCI-H460 cell line was purchased from the American Type Culture Collection (Rockville, MD, USA). P-gp overexpressing MDR NCI-H460/R cells were selected by continuous exposure to stepwise increasing concentrations of DOX from NCI-H460 cells [21]. NCI-H460/R and their sensitive counterparts were maintained in RPMI 1640 medium supplemented with 10% FBS, 2 mM L-glutamine, and 10,000 U/mL penicillin, 10 mg/mL streptomycin, 25 mg/mL amphotericin B solutions. Normal human embryonic pulmonary fibroblasts-MRC-5 cells were cultured in DMEM supplemented with 10% FBS, 4 g/L glucose, 2 mM L-glutamine, 5000 U/mL penicillin, and 5 mg/mL streptomycin solution. All cell lines were subcultured at 72 h intervals using 0.25% trypsin/EDTA and seeded into a fresh medium at 16,000 cells/$cm^2$. All cell lines were maintained at 37 °C in a humidified 5% $CO_2$ atmosphere.

### 2.4. MTT Assay

Cell viability was assessed by the MTT assay (AppliChem GmbH, Darmstadt, Germany). NCI-H460, NCI-H460/R, and MRC-5 cells grown in 25 $cm^2$ tissue flasks were trypsinized, and 2000 cells/well were seeded into flat-bottomed 96-well tissue culture plates following the overnight incubation. Subsequently, the cells were treated with increasing concentrations of Coleon U (2, 5, 10, 20, and 50 μM) for 72 h. At the end of the incubation period, MTT was added to each well in a final concentration of 0.2 mg/mL for 4 h. MTT-containing medium was then removed and formazan product was dissolved in 200 μL of dimethyl sulfoxide, and the absorbance was measured at 570 nm using an automatic Multiskan Sky reader (Thermo Scientific, Waltham, MA, USA).

### 2.5. CFSE Proliferation Assay

CFSE proliferation assay is based on the ability of the carboxyfluorescein succinimidyl ester (CFSE) dye to covalently bind free amines of the intracellular molecules via its succinimidyl group. CFSE is cell permeable and has no significant effect on their proliferative capacity. It enables the monitoring of cells over a period of 15 divisions [22]. The fluorescence intensity of CFSE, which gradually declines during cell divisions, enables the assessment of cell proliferation rates in treated versus untreated cells. For CFSE labeling, NCI-H460 and NCI-H460/R were incubated for 15 min in 0.1% FBS/PBS solution containing 1 μM CFSE at 37 °C, after which the cells were washed three times with PBS. Next, the CFSE-treated cells were seeded in 6-well plates, accommodated overnight, and then treated with Coleon U (7.5 μM, 15 μM, and 30 μM). PTX (100 nM in NCI-H460 and 2 μM in NCI-H460/R) was used as a positive control. After 48 h and 72 h, the cells were trypsinized and washed in ice-cold PBS and finally resuspended in 500 μL PBS. The fluorescence intensities of at least 20,000 cells per sample were detected on CytoFLEX flow cytometer (Beckman Coulter, Indianapolis, IN, USA) using fluorescence channel 1 (FL1) at 525 nm. The results were analyzed in Summit v4.3 software (Dako Colorado Inc., Fort Collins, CO, USA) using subtraction analysis.

### 2.6. Cell Death Analysis by Flow Cytometry

Induction of cell death after Coleon U treatment was assessed by Abcam Apoptosis Detection Kit through dual staining with Annexin V—FITC/Propidium Iodide (AV/PI) according to the manufacturer's instructions. NCI-H460 and NCI-H460/R cells were seeded in adherent 6-well plates at a density of 50,000 cells/well and incubated overnight. Both cell lines were treated with Coleon U (15, and 30 μM). Treatment with 500 nM PTX was used as a positive control in both cell lines. After 72 h, both attached and floating cells were collected and washed in PBS. After centrifugation, cells were resuspended in 100 μL of binding buffer containing AV and PI (1:1 ratio), and incubated for 15 min at room temperature in the dark. Cells were then analyzed on a CytoFLEX flow cytometer (Beckman Coulter, Indianapolis, IN, USA). The fluorescence intensities of AV and PI were measured in green fluorescence channel 1 (FL1) at 525 nm and red fluorescence channel 2 (FL2) at 585 nm, respectively. In each sample, the acquisition of 20,000 cells was performed, and the percentages of viable (AV− PI−), early apoptotic (AV+ PI−), late apoptotic (AV+ PI+), and necrotic (AV− PI+) cells were analyzed by CytExpert 2.4 software (Beckman Coulter, Indianapolis, IN, USA).

### *2.7. Rho123 Accumulation Assay*

The accumulation of Rho123 was analyzed by flow cytometry utilizing the ability of Rho123 as a P-gp substrate to emit fluorescence. The intensity of the fluorescence is proportional to Rho123 accumulation [23]. NCI-H460/R cells were seeded in 6-well plates and grown overnight. Cells were treated with Coleon U (5 μM and 10 μM), verapamil (5 μM), or $Na_3VO_4$ (1 μM), in single treatments or in combined treatments of Coleon U (5 μM and 10 μM) or Verapamil (5 μM) with $Na_3VO_4$ (1 μM). The cells were incubated for 30 min and 72 h. At the end of incubation periods, Rho123 (2.5 μM) was added and the cells were further incubated for 30 min at 5% $CO_2$ atmosphere at 37 °C. After the accumulation period, the cells were pelleted by centrifugation, washed with cold PBS, and kept on ice in the dark until the analysis. The samples were analyzed on a CytoFLEX flow cytometer (Beckman Coulter, Indianapolis, IN, USA). The orange fluorescence of Rho123 was assessed on fluorescence channel 1 (FL1) at 525 nm. A minimum of 10,000 events were assayed for each sample. The obtained mean fluorescence intensities were analyzed in Summit v4.3 software (Dako Colorado Inc., Fort Collins, CO, USA).

### *2.8. Mitochondrial Membrane Potential Analysis*

To analyze the effect of Coleon U on mitochondrial membrane potential, an early marker for apoptosis, TMRE dye, was used [24]. TMRE is a cell membrane permeable, fluorescent dye that accumulates in intact mitochondria. Depolarized or inactive mitochondria exhibit decreased membrane potential, resulting in reduced TMRE accumulation. NCI-H460/R cells were seeded in 6-well plates at a density of 50,000 cells/well. After 24 h, cells were treated with Coleon U (5 μM and 10 μM), verapamil (5 μM), and $Na_3VO_4$ (1 μM). Cells were incubated for an additional period of 72 h. CCCP was used as a positive control due to its notable depolarizing effect caused by the uncoupling of the proton gradient in the inner mitochondrial membrane [25]. The cells were treated with 10 μM CCCP 30 min prior to TMRE staining. For TMRE staining the cells were trypsinized, resuspended in RPMI media containing 500 nM TMRE, and incubated for 30 min at 37 °C in the dark. After washing twice in PBS, the red fluorescence emission of TMRE was immediately detected in the FL2 channel on a CytoFLEX flow cytometer (Beckman Coulter, Indianapolis, IN, USA). A minimum of 20,000 events were assayed per sample. The obtained data were analyzed in Summit v4.3 software (Dako Colorado Inc., Fort Collins, CO, USA), using subtraction analysis.

### *2.9. Flow Cytometric Analysis of P-gp Expression*

P-gp expression level in NCI-H460/R cells was assessed by flow cytometry. Cells were seeded in adherent 6-well plates and treated with Coleon U (5 and 10 μM) for 72 h. At the end of the incubation period, cells were collected by trypsinization, washed with PBS, and directly immunostained with FITC-conjugated anti-P-gp antibody according to the manufacturer's protocol. The samples were kept in the dark until the analysis on a

CytoFLEX flow cytometer (Beckman Coulter, Indianapolis, IN, USA). The fluorescence of FITC-conjugated anti-P-gp was detected on fluorescence channel 1 (FL1) at 525 nm. A minimum of 10,000 events were collected for each sample (the gate excluded cell debris and dead cells). The obtained results were analyzed by Summit v4.3 software (Dako Colorado Inc., Fort Collins, CO, USA) using subtraction analysis.

### 2.10. Immunocytochemistry

Immunocytochemistry was used for the quantification of the P-gp expression. For immunostaining, NCI-H460/R cells were seeded into 8-well chamber slides (Nunc, Nalgene, Roskilde, Denmark) at a density of 25,000 cells/chamber. Cells were allowed to attach to the surface overnight before the treatment with 30 μM Coleon U. In addition, NCI-H460/R cells were treated with PTX (500 nM), which served as a positive control for the induction of P-gp expression. After 72 h, cells were fixed with 4% paraformaldehyde for 20 min at RT and washed three times with PBS. Cells were then blocked with 2% bovine serum albumin (BSA) in PBS for 1 h at room temperature (RT). The anti-P-gp mouse monoclonal antibody was diluted 1:1000 in 2% BSA in PBS and incubated with the cells at 4 °C overnight. The cells were washed three times with PBS before adding Alexa Fluor 555 goat anti-mouse secondary antibody diluted 1:1000 in 2% BSA in PBS. The secondary antibody was incubated with the cells at RT for 2 h in the dark. To label the nuclei, the cells were incubated for 2 h in the dark with Hoechst 33342 at a final concentration of 1 μg/mL at RT. Finally, cells were washed three times with PBS, mounted in Mowiol, and stored at 4 °C in the dark before imaging. Fluorescently labeled cells were imaged using the ImageXpress® Pico Automated Cell Imaging System (Molecular Devices, San Jose, CA, USA). Analysis of the obtained images was performed using CellReporterXpress 2.9 software (Molecular Devices) with the Cell Scoring Analysis Protocol.

### 2.11. SwissADME Online Tool

WLOGP (water partition coefficient, a measure of the lipophilicity of a molecule) and TPSA (topological polar surface area, a measure of the surface area of a molecule that is polar) values for Coleon U and verapamil were generated using the SwissADME website [26,27]. The generated "boiled-egg" illustrates the position of the molecules in the WLOGP-versus-TPSA plot and enables the evaluation of passive gastrointestinal absorption and brain penetration of small molecules such as Coleon U and verapamil. The white region indicates a high probability of passive absorption by the human intestine (HIA), and the yellow region (yolk) indicates a high probability of blood–brain barrier (BBB) penetration. Yolk and white areas are not mutually exclusive. Points colored in blue show molecules predicted to be actively extruded by P-gp (PGP+) and are in red if predicted as nonsubstrate of P-gp (PGP−).

### 2.12. Statistical Analyses

For the results obtained by MTT assay, a nonparametric Kruskal–Wallis multiple comparisons test was applied, while $IC_{50}$ values were calculated by nonlinear regression analysis using GraphPad Prism 8.0.2 for Windows (San Diego, CA, USA). The results obtained by CFSE, Rho123 accumulation, mitochondrial membrane potential, and immunocytochemistry were analyzed by GraphPad Prism 8.0.2 (San Diego, CA, USA) using two-way-ANOVA Sidak's multiple comparisons test. For the cell death analysis, the results were analyzed by GraphPad Prism 8.0.2 (San Diego, CA, USA) using two-way-ANOVA Dunnett's multiple comparisons test.

## 3. Results and Discussion

### 3.1. Sensitivity of NSCLC Cells to Coleon U (Effects on Cell Viability, Proliferation, and Cell Death Induction)

The effects of Coleon U on cell viability were evaluated by MTT assay, which detects viable mitochondria (Figure 1). Therefore, the cell viability sensing of the MTT assay is due to preserved mitochondrial membrane electron transport chain [28]. The results obtained after 0 h and 72 h treatments in NSCLC NCI-H460, NCI-H460/R cell lines, and normal human embryonic pulmonary fibroblasts (MRC-5) show that the greatest reduction in cancer cell viability in the NCI-H460 and NCI-H460/R cell lines was observed with three concentrations of Coleon U (10, 20, and 50 μM), while the greatest reduction in cell viability in normal MRC-5 cells was achieved with the two highest concentrations of Coleon U (20 and 50 μM), as shown in Figure 1A. These concentrations that were at or below the threshold (half maximum absorbance, Figure 1A) were considered cytotoxic for each tested cell line.

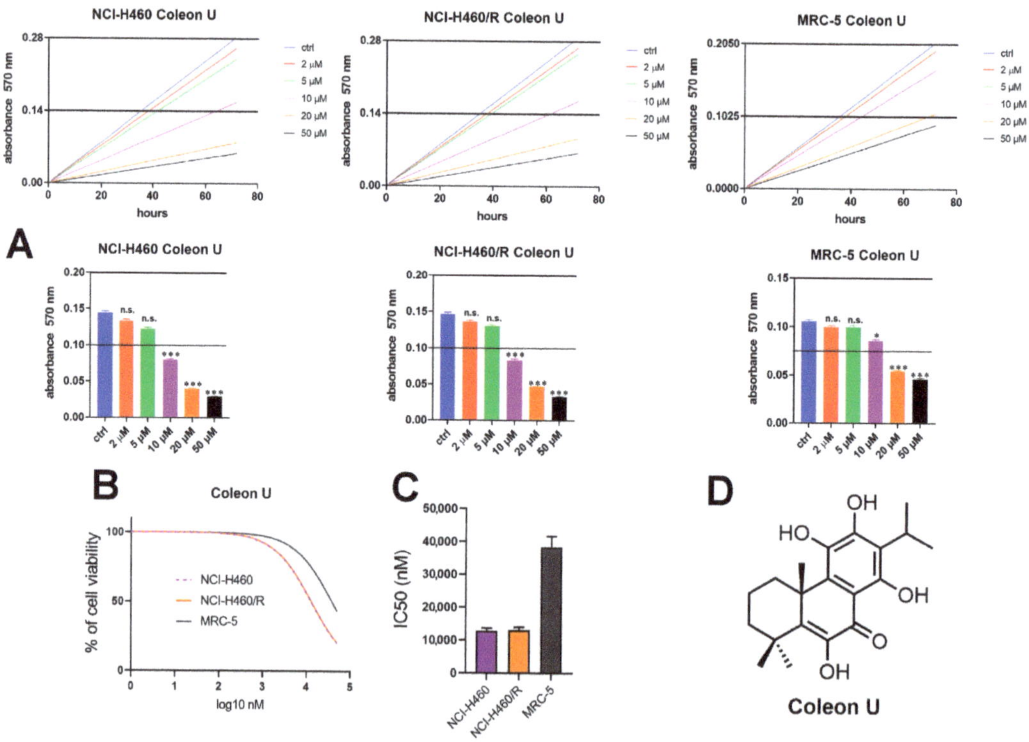


**Figure 1.** Concentration-dependent effect on cell viability upon Coleon U treatment. (**A**) The effects of different Coleon U concentrations (2, 5, 10, 20, and 50 μM) on the absorbance increase detected in NCI-H460, NCI-H460/R, and MRC-5 between time points 0 h and 72 h assessed by MTT presented as linear graphs and corresponding bar graphs with statistical analysis. The experiments were performed three times (n = 3). The statistically significant difference compared to the corresponding untreated controls: n.s. (nonsignificant), * ($p < 0.05$), *** ($p < 0.001$). (**B**) The cell viability was illustrated by nonlinear regression analysis; (**C**) the $IC_{50}$ values obtained in NCI-H460, NCI-H460/R, and MRC-5. Mean values ± SEM were calculated from three independent experiments (n = 3). (**D**) Coleon U chemical structure.

Importantly, Coleon U exerted a similar inhibitory effect on cell viability in sensitive and MDR NSCLC cell lines with similar $IC_{50}$ values (Figure 1B,C). This indicates that the MDR phenotype of NCI-H460/R cells did not reduce the effectiveness of Coleon U, which should be considered as a favorable feature of the potential anticancer compound. Additionally, the $IC_{50}$ value determined for the normal MRC-5 cells was threefold higher compared with the $IC_{50}$ values obtained in the cancer cell lines. Therefore, Coleon U has a greater inhibitory effect on the cancer cells compared to the normal cells, implying its selectivity towards cancer cells.

Other authors identified Coleon U as a selective activator of protein kinase C-δ [20], whose activity is related to antiproliferative effects [29]. Therefore, we assessed the antiproliferative effects of Coleon U after 48 h and 72 h by CSFE in NCI-H460 and NCI-H460/R cells (the live cell population was gated during acquisition on the flow cytometer). Cells were treated with three concentrations of Coleon U identified according to MTT assay: a noncytotoxic concentration of 7.5 μM, and two cytotoxic concentrations of 15 μM (approximately $IC_{50}$) and 30 μM (2 × $IC_{50}$). The results showed that treatment with 7.5 μM Coleon U led to a significant increase in CFSE fluorescence in both cell lines after 72 h (Figure 2A,B). Other treatments (15 μM and 30 μM) significantly sustained the proliferation of the non-small-cell lung carcinoma cells after 48 h and 72 h (Figure 2A,B). Obtained results also showed that there was a significant decrease in cell proliferation in both cell lines after treatment with PTX (100 nM in NCI-H460 and 2 μM in NCI-H460/R). Thus, Coleon U at 7.5 μM and 15 μM showed a time-dependent antiproliferative effect, while the effect of 30 μM was similar after 48 h and 72 h. Likewise, the PTX effect was not time-dependent (Figure 2A,B).

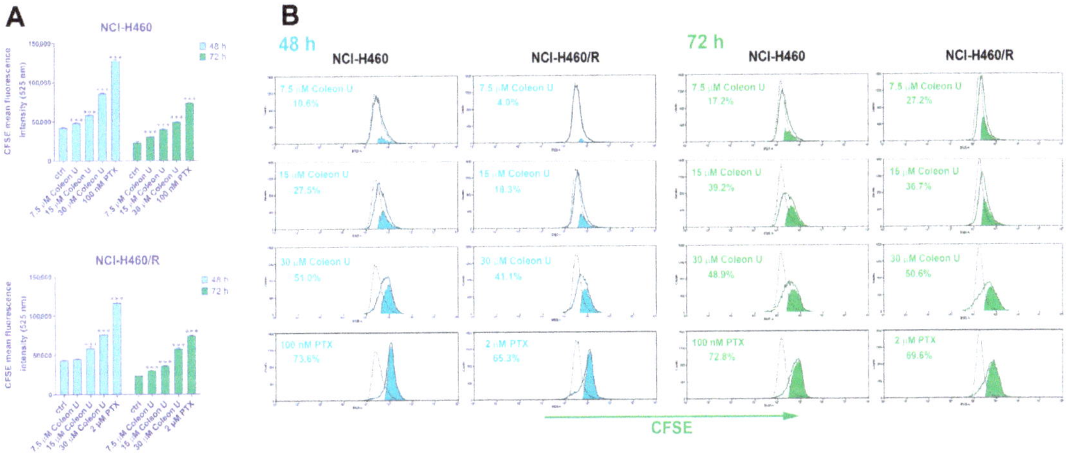


**Figure 2.** The antiproliferative effect of Coleon U. (**A**) The histograms show the increases in CFSE fluorescence intensity as a measure of sustained cell proliferation. The effect of different Coleon U concentrations (7.5, 15, and 30 μM) was evaluated after 48 h and 72 h in NCI-H460 and NCI-H460/R cells. PTX (100 nM in NCI-H460 and 2 μM in NCI-H460/R) was used as a positive control. The dotted line represents untreated control. Values are presented as mean ± SEM (n = 3). The statistically significant difference between the treated and control groups is shown as *** ($p < 0.001$). (**B**) Representative flow cytometric profiles of CFSE-labeled NCI-H460 and NCI-H460/R cells treated with Coleon U and PTX, evaluated after 48 h and 72 h, are shown.

To examine whether the induction of cell death contributes to the NSCLC cells' sensitivity to Coleon U, 15 μM and 30 μM treatments in NCI-H460 and NCI-H460/R were assessed by Annexin-V-FITC/Propidium Iodide staining after 72 h (Figure 3). The results showed that treatment with 30 μM Coleon U significantly increased the proportion of necrotic, late apoptotic, and early apoptotic NCI-H460 and NCI-H460/R cells compared to controls (Figure 3A,B). We reconfirmed that the potential of Coleon U to induce cell death was not compromised by the MDR phenotype of NCI-H460/R. However, the effect of 15 μM Coleon U was less pronounced, indicating that this concentration is not cytotoxic, as was previously suggested by the MTT assay. Therefore, the MTT can be considered the most sensitive method for the evaluation of Coleon U effects. It should be noted that the percentage of dying cells in the MDR NCI-H460/R cell line was significantly higher after 30 μM Coleon U treatment, compared with 500 nM PTX, and that the resistance to PTX was clearly mirrored in different efficacy in NCI-H460 and NCI-H460/R cells (Figure 3A,B).

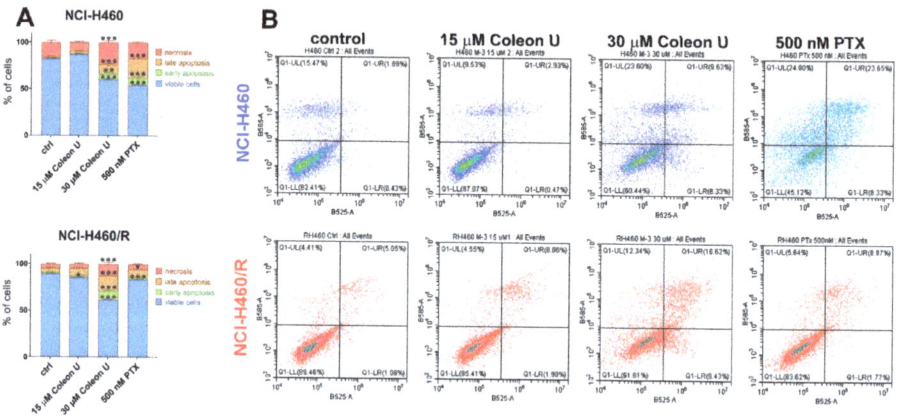


**Figure 3.** Cell death induction by Coleon U. AV/PI staining was used to assess cell death in NCI-H460 and NCI-H460/R cells following 72 h treatments with Coleon U (15 and 30 μM). PTX (500 nM) was used as a positive control. (**A**) Histograms show the percentage of viable (AV− PI−), early apoptotic (AV+ PI−), late apoptotic (AV+ PI+), and necrotic (AV− PI+) cells after treatments. The experiments were performed three times (n = 3). The statistically significant difference compared to the corresponding untreated controls: * ($p < 0.05$), ** ($p < 0.01$), *** ($p < 0.001$). (**B**) Representative flow cytometric profiles of AV/PI-labeled cells are shown for each treatment. Coleon U is nominated as M-3.

### *3.2. Coleon U Mechanisms Involved in P-gp Activity Modulation*

P-gp protein, found in MDR cancer cells, reduces the concentration of certain drugs inside the cells due to its transporter activity mediated by ATP hydrolysis. Therefore, the P-gp overexpression and activity significantly enhance the cellular ATP requirements [30]. In our previous study, Coleon U exerted a delayed effect on P-gp activity by decreasing it after 72 h and reversing doxorubicin resistance in the subsequent treatment [17]. To investigate how Coleon U affects P-gp activity, we conducted experiments using well-known inhibitors of this pump, sodium orthovanadate ($Na_3VO_4$), and verapamil. Sodium orthovanadate directly inhibits plasma membrane ATPases, including P-gp, and has shown anticancer activity against several types of cancer [31,32]. On the other hand, verapamil, a first-generation P-gp inhibitor, is a P-gp substrate that stimulates the ATPase activity of P-gp [33]. Verapamil promotes intracellular accumulation of the drug when applied with chemotherapeutic agents, as shown in various cancer cell lines: NSCLC, colorectal cancer, leukemia, and neuroblastoma [34–36]. To evaluate the ability of Coleon U to inhibit the P-gp transporter, a Rho123 accumulation assay was performed. Resistant NSCLC cells (NCI-H460/R) expressing P-gp, were treated with Coleon U (5 μM and 10 μM) and with verapamil (5 μM), in single treatments or in combination with sodium orthovanadate (1 μM). The treatments were conducted at two different time points: 30 min for the evaluation of the direct interaction with P-gp and 72 h for the evaluation of the indirect effect on P-gp activity such as effects on the P-gp expression [37], related proteins involved in the P-gp expression regulation [38], changes in the intracellular pH [39], or availability of ATP as a fuel molecule for the P-gp functioning [36]. Verapamil at 5 μM significantly inhibited the efflux of Rho123 in NCI-H460/R cells after 30 min, while Coleon U at both concentrations of 5 μM and 10 μM and sodium orthovanadate at 1 μM were not effective, as shown in Figure 4A. Sodium orthovanadate significantly decreased the level of Rho123 in combination with verapamil and Coleon U (10 μM) (Figure 4A). After 72 h, sodium orthovanadate (1 μM), verapamil (5 μM), and Coleon U (5 μM and 10 μM) increased the accumulation of Rho123 in NCI-H460/R cells. This suggests that verapamil directly interacts with P-gp as a P-gp substrate and sustains this effect during 72 h as an ATPase stimulator, while sodium orthovanadate and Coleon U effects on Rho123 intracellular accumulation are evident later, after 72 h. When NCI-H460/R cells were treated with combinations of verapamil or Coleon U with sodium orthovanadate, a decrease in the fluorescent Rho123 signal was observed after 72 h (Figure 4A). This suggests an antagonistic effect between verapamil (an ATPase stimulator) and sodium orthovanadate (an ATPase inhibitor) as well as Coleon U and sodium orthovanadate. Therefore, we assumed that Coleon U interferes with the ATP metabolism, probably its production in mitochondria. This is also supported by the fact that the MTT assay, which detects viable mitochondria, was more powerful in sensing Coleon U effects than AV/PI assay, which discriminates viable from dead cells.

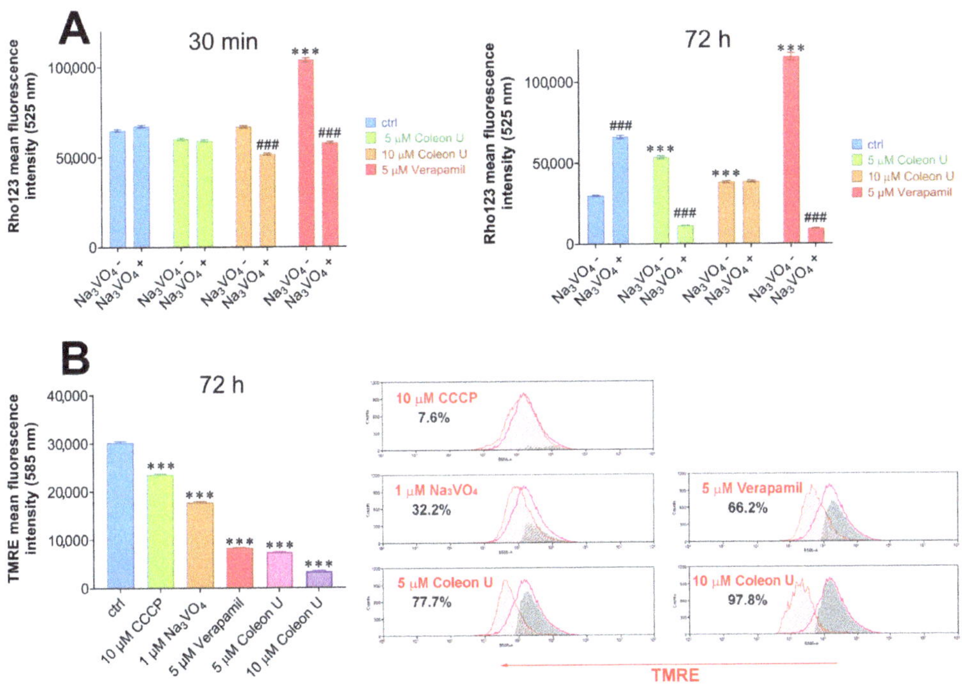


**Figure 4.** The effects of Coleon U on P-gp function and mitochondrial membrane potential in NCI-H460/R cells. (**A**) P-gp function in MDR NCI-H460/R cells was evaluated after 30 min and 72 h in individual treatments with Coleon U (5 and 10 μM), verapamil (5 μM), and sodium orthovanadate-$Na_3VO_4$ (1 μM) or in combined treatments of Coleon U or verapamil with sodium orthovanadate-$Na_3VO_4$. The histograms show Rho123 accumulation in treated NCI-H460/R cells. Values are presented as mean ± SEM (n = 3). The statistically significant difference between the treated and control groups is shown as *** ($p < 0.001$). The statistically significant difference between single treatment and corresponding combined treatment is shown as ### ($p < 0.001$). (**B**) Coleon U triggers loss of mitochondrial membrane potential in MDR NCI-H460/R cells. Mitochondrial membrane potential in NCI-H460/R cells was evaluated after 72 h treatments with Coleon U (5 and 10 μM), verapamil (5 μM), and sodium orthovanadate-$Na_3VO_4$ (1 μM) using TMRE staining. CCCP was used as a positive control. Histograms in the left panel represent the average fluorescence intensity of TMRE from three independent experiments. The statistically significant difference between the treated and control groups is shown as *** ($p < 0.001$). Representative flow cytometric profiles for each condition are shown in the right panel.

The inner mitochondrial membrane is essential for the generation of ATP via the mitochondrial respiratory chain. When the mitochondrial membrane potential is depolarized, the respiratory chain is blocked, leading to a reduced efficiency of ATP production [30]. Decreased intracellular ATP levels also can affect the functioning of the P-gp, as an ATP-dependent membrane transporter [30]. Thus, the compound Coleon U was examined for its effect on the disruption of mitochondrial potential in MDR NCI-H460/R cells. The TMRE labeling was used to assess changes in mitochondrial membrane potential. The results showed that 72 h after treatment with Coleon U (5 μM and 10 μM), TMRE fluorescence intensities were significantly decreased compared with the control (Figure 4B). This suggests that Coleon U induced notable depolarization of the mitochondrial membrane in NCI-H460/R. Treatments with sodium orthovanadate (1 μM) and verapamil (5 μM) also significantly lowered the TMRE signal in NCI-H460/R cells. The obtained results are consistent with previous findings in the literature that sodium orthovanadate decreases mitochondrial membrane potential and ATPase activity in sorafenib-resistant hepatocellular carcinoma cells, ultimately promoting apoptosis [32]. Our results suggest that Coleon U induces depolarization of the mitochondrial membrane potential in MDR NCI-H460/R cancer cells that can cause a decrease in ATP production, indirectly affecting P-gp function. Furthermore, the disruption of mitochondrial function caused by Coleon U can potentially contribute to the activation of cell death pathways.

Although the results obtained by MTT, CFSE, and AV/PI showed no significant difference in Coleon U efficacy between NCI-H460 and NCI-H460/R cells, which means that Coleon U efficacy is not affected by MDR and P-gp activity, we performed SwissADME analysis to ensure that Coleon U is not a P-gp substrate. SwissADME results illustrated as "boiled egg" [27] showed that Coleon U can be easily absorbed by the human intestine and confirmed that Coleon U is not a P-gp substrate, while verapamil is a P-gp substrate (Figure 5A). To examine the above mechanisms that may lead to a decrease in P-gp expression and contribute to the observed delayed effect of Coleon U on P-gp activity, we studied P-gp expression in NCI-H460/R cells after 72 h treatment with Coleon U (5 and 10 μM), as shown in Figure 5B,C. Analysis of P-gp expression by flow cytometry revealed that Coleon U tended to decrease P-gp expression in NCI-H460/R cells. Specifically, treatment with 5 μM resulted in an approximate 9% decrease in P-gp expression, while treatment with 10 μM led to an approximately 12% decrease in P-gp expression (Figure 5B). We also examined whether the cytotoxic concentration of Coleon U (30 μM) affects P-gp expression in NCI-H460/R cells (Figure 5C). To that end, we used fluorescent labeling of P-gp and analysis performed on the ImageXpress® Pico Automated Cell Imaging System. The results obtained by CellReporterXpress software revealed that 30 μM Coleon U significantly reduced the portion of NCI-H460/R cells expressing P-gp without changing the level of expression per cell (Figure 5C). This suggests that Coleon U exerts its cytotoxic effect on P-gp-positive cells and does not induce P-gp expression. Therefore, Coleon U has valuable characteristics for the treatment of MDR cancers. In contrast, treatment of NCI-H460/R cells with PTX resulted in a significant increase in P-gp fluorescence intensity per cell, while the portion of P-gp cells was not changed (Figure 5C). This suggests a continuation of the P-gp upregulation upon PTX treatment, while Coleon U may hold promise in overcoming MDR by modulating both P-gp activity and expression.

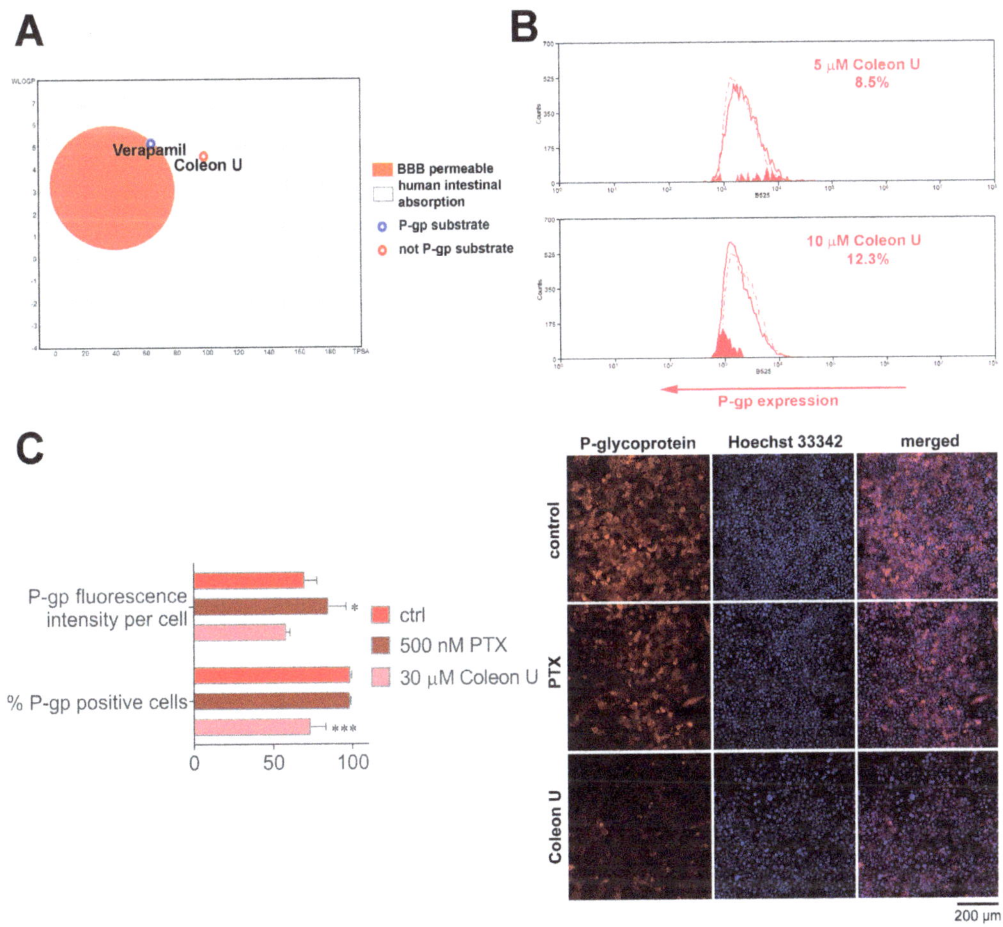


**Figure 5.** The effects of Coleon U on P-gp expression. (**A**) The "boiled-egg" plot – Brain or IntestinaL EstimateD permeation predictive model (plot of WLOGP against TPSA) of Coleon U and verapamil from SwissADME online tool. The points in yellow (boiled-egg's yolk) represent molecules predicted to passively permeate through the blood–brain barrier (BBB), while the points in white (boiled-egg's white) represent molecules predicted to be passively absorbed by the gastrointestinal tract (HIA). Blue dots represent molecules predicted to be P-gp substrates, while red dots are molecules predicted not to be substrates for P-gp. (**B**) Flow cytometric profiles of P-gp expression in NCI-H460/R cells treated with Coleon U for 72 h treatment. (**C**) The histograms in the left panel show P-gp fluorescence intensity per cell and percent of P-gp positive NCI-H460/R cells treated with Coleon U and PTX for 72 h. Values are presented as mean ± SEM (n = 3). The statistically significant difference between the treated and control groups is shown as * ($p < 0.05$) and *** ($p < 0.001$). Representative immunofluorescence micrographs of anti-P-gp labeled NCI-H460/R cells are shown in the right panel. Nuclei were counterstained with Hoechst 33342. Scale bar = 200 μm.

## 4. Conclusions

Coleon U is a natural compound with a significant anticancer potential whose activity against MDR cancers has not been truly investigated. Although the activation of protein kinase C-δ (that promotes antiproliferative and proapoptotic effects) was identified as the Coleon U mechanism of anticancer action [20], the literature data on Coleon U as an anticancer agent is rare. It seems that the anticancer potential of Coleon U has not attracted the attention of the scientific community enough. Herein, we showed that Coleon U is not a P-gp substrate and does not directly interact with P-gp (an ATP-dependent transporter). However, it decreases P-gp activity after 72 h and this delayed effect is accompanied by a significant mitochondrial membrane potential decrease, suggesting the exhaustion of the ATP necessary for the P-gp activity. Moreover, we showed that sodium orthovanadate, which is an ATPase inhibitor, antagonizes the effect of verapamil, an ATPase stimulator, while the same antagonizing effect of sodium orthovanadate is achieved in combination with Coleon U. An important finding is the tendency of Coleon U to decrease the P-gp expression, which also qualifies Coleon U as a valuable modulator of MDR mediated by P-gp expression and activity. In our recent study, we showed the sensitizing effect of Coleon U in the subsequent combination with doxorubicin in MDR cancer cells [17]. Our work may stimulate other studies on different natural compounds that interfere with ATP production for the identification of novel P-gp modulators.

**Author Contributions:** Conceptualization, M.P. (Milica Pešić) and P.R.; methodology, S.J.S., E.N.N., J.D., A.P.-R. and M.P. (Milica Pajović); formal analysis, S.J.S. and M.P. (Milica Pešić); investigation, S.J.S., E.N.N., J.D., A.P.-R. and M.P. (Milica Pajović); resources, S.J.S. and M.P. (Milica Pešić); writing—original draft preparation, S.J.S., E.N.N. and M.P. (Milica Pešić); writing—review and editing, M.P. (Milica Pešić) and P.R.; supervision, M.P. (Milica Pešić); funding acquisition, M.P. (Milica Pešić). All authors have read and agreed to the published version of the manuscript.

**Funding:** This research was funded by the Ministry of science, technological development and innovation of the Republic of Serbia, grant number 451-03-47/2023-01/200007.

**Institutional Review Board Statement:** Not applicable.

**Informed Consent Statement:** Not applicable.

**Data Availability Statement:** The data presented in this study are available on request from the corresponding author. The data are not publicly available because our institutional repository is not yet ready to collect experimental data.

**Acknowledgments:** This work was performed within the framework of COST Action CA17104 "New diagnostic and therapeutic tools against multidrug resistant tumours" and COST Innovator Grant IG17104 "Pan-European Educational Platform on Multidrug Resistant Tumours and Personalised Cancer Treatment". This research was also funded by financial support from national funds (FCT/MCTES, Fundação para a Ciência e Tecnologia and Ministério da Ciência, Tecnologia e Ensino Superior) through CBIOS (UIDB/04567/2020, UIDP/04567/2020).

**Conflicts of Interest:** The authors declare no conflict of interest.

## References

1. Herbst, R.S.; Morgensztern, D.; Boshoff, C. The Biology and Management of Non-Small Cell Lung Cancer. *Nature* **2018**, *553*, 446–454. [CrossRef] [PubMed]
2. Wang, X.; Zhang, H.; Chen, X. Drug Resistance and Combating Drug Resistance in Cancer. *Cancer Drug Resist.* **2019**, *2*, 141. [CrossRef] [PubMed]
3. Bukowski, K.; Kciuk, M.; Kontek, R. Mechanisms of Multidrug Resistance in Cancer Chemotherapy. *Int. J. Mol. Sci.* **2020**, *21*, 3233. [CrossRef]
4. Muriithi, W.; Macharia, L.W.; Heming, C.P.; Echevarria, J.L.; Nyachieo, A.; Filho, P.N.; Neto, V.M. ABC Transporters and the Hallmarks of Cancer: Roles in Cancer Aggressiveness beyond Multidrug Resistance. *Cancer Biol. Med.* **2020**, *17*, 253–269. [CrossRef] [PubMed]
5. Zhao, Z.; Shilatifard, A. Epigenetic Modifications of Histones in Cancer. *Genome Biol.* **2019**, *20*, 245. [CrossRef]
6. Rocha, C.R.R.; Silva, M.M.; Quinet, A.; Cabral-Neto, J.B.; Menck, C.F.M. DNA Repair Pathways and Cisplatin Resistance: An Intimate Relationship. *Clinics* **2018**, *73*, e478s. [CrossRef]
7. Pathania, S.; Bhatia, R.; Baldi, A.; Singh, R.; Rawal, R.K. Drug Metabolizing Enzymes and Their Inhibitors' Role in Cancer Resistance. *Biomed. Pharmacother.* **2018**, *105*, 53–65. [CrossRef]
8. Wang, J.Q.; Wu, Z.X.; Yang, Y.; Teng, Q.X.; Li, Y.D.; Lei, Z.N.; Jani, K.A.; Kaushal, N.; Chen, Z.S. ATP-Binding Cassette (ABC) Transporters in Cancer: A Review of Recent Updates. *J. Evid. Based Med.* **2021**, *14*, 232–256. [CrossRef]

9. Kim, Y.; Chen, J. Molecular Structure of Human P-Glycoprotein in the ATP-Bound, Outward-Facing Conformation. *Science* **2018**, *359*, 915–919. [CrossRef]
10. Dewanjee, S.; Dua, T.K.; Bhattacharjee, N.; Das, A.; Gangopadhyay, M.; Khanra, R.; Joardar, S.; Riaz, M.; De Feo, V.; Zia-Ul-Haq, M. Natural Products as Alternative Choices for P-Glycoprotein (P-Gp) Inhibition. *Molecules* **2017**, *22*, 871. [CrossRef]
11. Cui, J.; Liu, X.; Chow, L.M.C. Flavonoids as P-Gp Inhibitors: A Systematic Review of SARs. *Curr. Med. Chem.* **2019**, *26*, 4799–4831. [CrossRef] [PubMed]
12. Joshi, P.; Vishwakarma, R.A.; Bharate, S.B. Natural Alkaloids as P-Gp Inhibitors for Multidrug Resistance Reversal in Cancer. *Eur. J. Med. Chem.* **2017**, *138*, 273–292. [CrossRef] [PubMed]
13. Yoshida, N.; Koizumi, M.; Adachi, I.; Kawakami, J. Inhibition of P-Glycoprotein-Mediated Transport by Terpenoids Contained in Herbal Medicines and Natural Products. *Food Chem. Toxicol.* **2006**, *44*, 2033–2039. [CrossRef]
14. Kasaian, J.; Mosaffa, F.; Behravan, J.; Masullo, M.; Piacente, S.; Ghandadi, M.; Iranshahi, M. Reversal of P-Glycoprotein-Mediated Multidrug Resistance in MCF-7/Adr Cancer Cells by Sesquiterpene Coumarins. *Fitoterapia* **2015**, *103*, 149–154. [CrossRef] [PubMed]
15. Garcia, C.; Isca, V.M.S.; Pereira, F.; Monteiro, C.M.; Ntungwe, E.; Sousa, F.; Dinic, J.; Holmstedt, S.; Roberto, A.; Díaz-Lanza, A.; et al. Royleanone Derivatives From Plectranthus Spp. as a Novel Class of P-Glycoprotein Inhibitors. *Front. Pharmacol.* **2020**, *11*, 557789. [CrossRef] [PubMed]
16. Śliwiński, T.; Sitarek, P.; Skała, E.; Isca, V.M.S.; Synowiec, E.; Kowalczyk, T.; Bijak, M.; Rijo, P. Diterpenoids from Plectranthus Spp. as Potential Chemotherapeutic Agents via Apoptosis. *Pharmaceuticals* **2020**, *13*, 123. [CrossRef] [PubMed]
17. Ntungwe, E.N.; Stojanov, S.J.; Duarte, N.M.; Candeias, N.R.; Díaz-Lanza, A.M.; Vágvölgyi, M.; Hunyadi, A.; Pešić, M.; Rijo, P. C20- nor-Abietane and Three Abietane Diterpenoids from Plectranthus Mutabilis Leaves as P-Glycoprotein Modulators. *ACS Med. Chem. Lett.* **2022**, *13*, 674–680. [CrossRef]
18. Cerqueira, F.; Cordeiro-Da-Silva, A.; Gaspar-Marques, C.; Simões, F.; Pinto, M.M.M.; Nascimento, M.S.J. Effect of Abietane Diterpenes from Plectranthus Grandidentatus on T- and B-Lymphocyte Proliferation. *Bioorg. Med. Chem.* **2004**, *12*, 217–223. [CrossRef]
19. Marques, C.G.; Pedro, M.; Simões, M.F.A.; Nascimento, M.S.J.; Pinto, M.M.M.; Rodriguez, B. Effect of Abietane Diterpenes from Plectranthus Grandidentatus on the Growth of Human Cancer Cell Lines. *Planta Med.* **2002**, *68*, 839–840. [CrossRef]
20. Coutinho, I.; Pereira, G.; Simões, M.F.; Côrte-Real, M.; Gonçalves, J.; Saraiva, L. Selective Activation of Protein Kinase C-Delta and -Epsilon by 6,11,12,14-Tetrahydroxy-Abieta-5,8,11,13-Tetraene-7-One (Coleon U). *Biochem. Pharmacol.* **2009**, *78*, 449–459. [CrossRef]
21. Pesic, M.; Markovic, J.Z.; Jankovic, D.; Kanazir, S.; Markovic, I.D.; Rakic, L.; Ruzdijic, S. Induced Resistance in the Human Non Small Cell Lung Carcinoma (NCI-H460) Cell Line in Vitro by Anticancer Drugs. *J. Chemother.* **2006**, *18*, 66–73. [CrossRef]
22. Kang, W.; Nielsen, O.; Fenger, C.; Leslie, G.; Holmskov, U.; Reid, K.B.M. Induction of DMBT1 Expression by Reduced ERK Activity during a Gastric Mucosa Differentiation-like Process and Its Association with Human Gastric Cancer. *Carcinogenesis* **2005**, *26*, 1129–1137. [CrossRef] [PubMed]
23. Jouan, E.; Le Vée, M.; Mayati, A.; Denizot, C.; Parmentier, Y.; Fardel, O. Evaluation of P-Glycoprotein Inhibitory Potential Using a Rhodamine 123 Accumulation Assay. *Pharmaceutics* **2016**, *8*, 12. [CrossRef]
24. Crowley, L.C.; Christensen, M.E.; Waterhouse, N.J. Measuring Mitochondrial Transmembrane Potential by TMRE Staining. *Cold Spring Harb. Protoc.* **2016**, *2016*, 1092–1096. [CrossRef]
25. Zhang, L.; Zheng, X.C.; Huang, Y.Y.; Ge, Y.P.; Sun, M.; Chen, W.L.; Liu, W.B.; Li, X.F. Carbonyl Cyanide 3-Chlorophenylhydrazone Induced the Imbalance of Mitochondrial Homeostasis in the Liver of Megalobrama Amblycephala: A Dynamic Study. *Comp. Biochem. Physiol. C Toxicol. Pharmacol.* **2021**, *244*, 109003. [CrossRef]
26. Daina, A.; Michielin, O.; Zoete, V. SwissADME: A Free Web Tool to Evaluate Pharmacokinetics, Drug-Likeness and Medicinal Chemistry Friendliness of Small Molecules. *Sci. Rep.* **2017**, *7*, 42717. [CrossRef]
27. Daina, A.; Zoete, V. A BOILED-Egg To Predict Gastrointestinal Absorption and Brain Penetration of Small Molecules. *ChemMedChem* **2016**, *11*, 1117–1121. [CrossRef]
28. Urra, F.; Weiss-López, B.; Araya-Maturana, R. Determinants of Anti-Cancer Effect of Mitochondrial Electron Transport Chain Inhibitors: Bioenergetic Profile and Metabolic Flexibility of Cancer Cells. *Curr. Pharm. Des.* **2016**, *22*, 5998–6008. [CrossRef] [PubMed]
29. Hernández-Maqueda, J.G.; Luna-Ulloa, L.B.; Santoyo-Ramos, P.; Castañeda-Patlán, M.C.; Robles-Flores, M. Protein Kinase C Delta Negatively Modulates Canonical Wnt Pathway and Cell Proliferation in Colon Tumor Cell Lines. *PLoS ONE* **2013**, *8*, e58540. [CrossRef] [PubMed]
30. Seelig, A. P-Glycoprotein: One Mechanism, Many Tasks and the Consequences for Pharmacotherapy of Cancers. *Front. Oncol.* **2020**, *10*, 576559. [CrossRef]
31. Duran, G.E.; Derdau, V.; Weitz, D.; Philippe, N.; Blankenstein, J.; Atzrodt, J.; Sémiond, D.; Gianolio, D.A.; Macé, S.; Sikic, B.I. Cabazitaxel Is More Active than First-Generation Taxanes in ABCB1(+) Cell Lines Due to Its Reduced Affinity for P-Glycoprotein. *Cancer Chemother. Pharmacol.* **2018**, *81*, 1095–1103. [CrossRef] [PubMed]
32. Jiang, W.; Li, G.; Li, W.; Wang, P.; Xiu, P.; Jiang, X.; Liu, B.; Sun, X.; Jiang, H. Sodium Orthovanadate Overcomes Sorafenib Resistance of Hepatocellular Carcinoma Cells by Inhibiting Na+/K+-ATPase Activity and Hypoxia-Inducible Pathways. *Sci. Rep.* **2018**, *8*, 9706. [CrossRef]

33. Ledwitch, K.V.; Gibbs, M.E.; Barnes, R.W.; Roberts, A.G. Cooperativity between Verapamil and ATP Bound to the Efflux Transporter P-Glycoprotein. *Biochem. Pharmacol.* **2016**, *118*, 96. [CrossRef] [PubMed]
34. Jaferian, S.; Soleymaninejad, M.; Daraee, H. Verapamil (VER) Enhances the Cytotoxic Effects of Docetaxel and Vinblastine Combined Therapy Against Non-Small Cell Lung Cancer Cell Lines. *Drug Res.* **2018**, *68*, 146–152. [CrossRef] [PubMed]
35. Wang, X.; Wang, Z.; Wang, K.; Gao, M.; Zhang, H.; Xu, X. Metabolomics Analysis of Multidrug Resistance in Colorectal Cancer Cell and Multidrug Resistance Reversal Effect of Verapamil. *Biomed. Chromatogr.* **2021**, *35*, e4976. [CrossRef] [PubMed]
36. Zhao, L.; Zhao, Y.; Schwarz, B.; Mysliwietz, J.; Hartig, R.; Camaj, P.; Bao, Q.; Jauch, K.W.; Guba, M.; Ellwart, J.W.; et al. Verapamil Inhibits Tumor Progression of Chemotherapy-Resistant Pancreatic Cancer Side Population Cells. *Int. J. Oncol.* **2016**, *49*, 99. [CrossRef]
37. Wang, P.P.; Xu, D.J.; Huang, C.; Wang, W.P.; Xu, W.K. Astragaloside IV Reduces the Expression Level of P-Glycoprotein in Multidrug-Resistant Human Hepatic Cancer Cell Lines. *Mol. Med. Rep.* **2014**, *9*, 2131–2137. [CrossRef]
38. Dinić, J.; Podolski-Renić, A.; Jovanović, M.; Musso, L.; Tsakovska, I.; Pajeva, I.; Dallavalle, S.; Pešić, M. Novel Heat Shock Protein 90 Inhibitors Suppress P-Glycoprotein Activity and Overcome Multidrug Resistance in Cancer Cells. *Int. J. Mol. Sci.* **2019**, *20*, 4575. [CrossRef]
39. Podolski-Renić, A.; Dinić, J.; Stanković, T.; Jovanović, M.; Ramović, A.; Pustenko, A.; Žalubovskis, R.; Pešić, M. Sulfocoumarins, Specific Carbonic Anhydrase IX and XII Inhibitors, Interact with Cancer Multidrug Resistant Phenotype through PH Regulation and Reverse P-Glycoprotein Mediated Resistance. *Eur. J. Pharm. Sci.* **2019**, *138*, 105012. [CrossRef]

pharmaceutics

Article

# Development and Evaluation of a Self-Nanoemulsifying Drug Delivery System for Sinapic Acid with Improved Antiviral Efficacy against SARS-CoV-2

**Hani A. Alhadrami [1,2,3,†], Ahmed S.G. Srag El-Din [4,†], Hossam M. Hassan [5,6], Ahmed M. Sayed [5], Albaraa H. Alhadrami [7], Mostafa E. Rateb [7,*] and Demiana M. Naguib [8,*]**

1 Faculty of Applied Medical Sciences, Department of Medical Laboratory Technology, King Abdulaziz University, P.O. Box 80402, Jeddah 21589, Saudi Arabia; hanialhadrami@kau.edu.sa
2 King Fahd Medical Research Centre, King Abdulaziz University, P.O. Box 80402, Jeddah 21589, Saudi Arabia
3 Molecular Diagnostics Laboratory, King Abdulaziz University Hospital, P.O. Box 80402, Jeddah 21589, Saudi Arabia
4 Department of Pharmaceutics, Faculty of Pharmacy, Delta University for Science & Technology, Gamasa City 35712, Egypt; ahmed.serageldin@deltauniv.edu.eg
5 Department of Pharmacognosy, Faculty of Pharmacy, Nahda University, Beni-Suef 62513, Egypt; hossam.mokhtar@nub.edu.eg (H.M.H.); ahmed.mohamed.sayed@nub.edu.eg (A.M.S.)
6 Department of Pharmacognosy, Faculty of Pharmacy, Beni-Suef University, Beni-Suef 62514, Egypt
7 School of Computing, Engineering & Physical Sciences, University of the West of Scotland, Paisley PA1 2BE, UK
8 Department of Pharmaceutics, Faculty of Pharmacy, Nahda University (NUB), Beni-Suef 62513, Egypt
* Correspondence: mostafa.rateb@uws.ac.uk (M.E.R.); demiana.monier@nub.edu.eg (D.M.N.)
† These authors contributed equally to this work.

**Abstract:** This study aimed to develop a self-nanoemulsifying drug delivery system (SNE) for sinapic acid (SA) to improve its solubility and antiviral activity. Optimal components for the SA-SNE formulation were selected, including Labrafil as the oil, Cremophor EL as the surfactant, and Transcutol as the co-surfactant. The formulation was optimized using surface response design, and the optimized SA-SNE formulation exhibited a small globule size of 83.6 nm, high solubility up to 127.1 $\pm$ 3.3, and a 100% transmittance. In vitro release studies demonstrated rapid and high SA release from the formulation. Pharmacokinetic analysis showed improved bioavailability by 2.43 times, and the optimized SA-SNE formulation exhibited potent antiviral activity against SARS-CoV-2. The developed SA-SNE formulation can enhance SA's therapeutic efficacy by improving its solubility, bioavailability, and antiviral activity. Further in silico, modeling, and Gaussian accelerated molecular dynamics (GaMD)-based studies revealed that SA could interact with and inhibit the viral main protease ($M^{pro}$). This research contributes to developing effective drug delivery systems for poorly soluble drugs like SA, opening new possibilities for their application via nebulization in SARS-CoV-2 therapy.

**Keywords:** self-nanoemulsifying drug delivery system; sinapic acid; SARS-CoV-2; GaMD; $M^{pro}$

## 1. Introduction

Sinapic acid (SA) is a natural organic compound belonging to the phenolic acid family [1]. It is abundant in various plant sources, including fruits, vegetables, grains, and spices [2]. SA is synthesized by plants from the amino acid phenylalanine through the phenylpropanoid pathway [3]. It is involved in several biological processes in plants, such as UV protection, cell wall lignification, and defense against pathogens and oxidative stress [1,4]. In addition to its role in plants, SA has been studied extensively for its potential health benefits in humans. It exhibits potent antioxidant [5], anti-inflammatory [6], and anticancer properties [7]. It also shows promising effects on various health conditions, including diabetes [8], cardiovascular diseases [9], and neurodegenerative disorders [10]. It was also confirmed earlier during the coronavirus pandemic that SA selectively inhibited SARS-CoV-2 replication by targeting its envelope protein [11].

The body readily absorbs SA, and it can be found in the bloodstream after consuming foods containing it [12]. It can also be obtained through dietary supplements [12]. However, more research is needed to determine the optimal dose and long-term safety of SA supplementation. Also, SA has poor solubility in water, which limits its therapeutic efficacy and bioavailability [13]. A self-nanoemulsifying drug delivery system (SNEDDS) has been used to solve these problems.

An SNEDDS is a drug delivery system that uses a mixture of oils, surfactants, and co-surfactants to form a fine oil-in-water nano-emulsion when exposed to aqueous media, such as gastrointestinal fluids. This system is thermodynamically stable and can enhance the solubility and bioavailability of poorly soluble drugs, thereby improving their therapeutic efficacy [14]. An SA self-nano-emulsion (SA-SNE) is a promising drug delivery system that can enhance SA's solubility, absorption, and bioavailability. The development of an SA-SNE requires the optimal choice of oils, surfactants, and co-surfactants. The choice of these components is critical in achieving a stable and effective self-nanoemulsifying formulation. The oils, surfactants, and co-surfactants selected for this study were chosen based on data collected from the literature [15–17]. The selection of oils is pivotal in an SNEDDS as they play a central role in stabilizing the emulsion and determining its properties. The oils chosen for this study were Labrafil, Capryol-90, Labrafac, olive oil, anise oil, almond oil, soybean oil, and corn oil. These oils were selected because they are known to be effective emulsifiers capable of forming oil-in-water (O/W) emulsions upon contact with an aqueous phase [18]. Each of these oils has unique characteristics that can influence the emulsion's stability, viscosity, and compatibility with various active ingredients [19,20].

Surfactants are essential for emulsion stability, reducing interfacial tension between oil and water phases. This study employed hydrophilic non-ionic surfactants, including Cremophor El, Tween 20, and Tween 80. The choice of these surfactants is supported by their high hydrophilic–lipophilic balance (HLB) values, which indicate their ability to stabilize O/W emulsions effectively [21–23]. Furthermore, their low oral toxicity is crucial for pharmaceutical and cosmetic applications, where safety is a paramount concern [24,25]. Co-surfactants are often used with surfactants to enhance an emulsion's stability and modify its properties. The co-surfactants selected for this study were Labrasol, Transcutol, PEG 400, and propylene glycol. These co-surfactants were chosen based on their ability to improve an emulsion's droplet size distribution, viscosity, and overall stability [26,27]. Their compatibility with the chosen oils and surfactants makes them suitable for creating emulsions with specific desired characteristics.

The current study's aim was to formulate an SA-SNE with a careful selection of its components, oils, surfactants, and co-surfactants. The optimization process was performed using Design Expert software (Version 11.1.2.0). The safety and efficacy of the SA-SNE in clinical situations were evaluated via pharmacokinetic study in rabbits through nebulization of the prepared SA-SNE mist and in vitro antiviral activity study against SARS-CoV-2. SA's mode of action was investigated via in vitro-based and in silico-based experiments, including MD simulations. In this context, we present a novel SNE system that produces nano-emulsion mists for pulmonary SA delivery via nebulization.

## 2. Materials and Methods

### 2.1. Screening of Components for SA-SNE Development

The equilibrium solubility of SA was studied in various oils, surfactants, and cosurfactants. Oils of different saturation degrees (medium- or long-chain triglycerides) were utilized to create an SNE to test their solubilization and nano-emulsification ability. The oil with the highest ability to solubilize SA was chosen due to its significant impact on drug solubilization ability and absorption. The tested oils included Labrafil, Capryol-90, Labrafac, olive oil, anise oil, almond oil, soybean oil, and corn oil. The screened surfactants were hydrophilic non-ionic surfactants (Cremophor El, Tween 20, and Tween 80); they have high HLB values and low oral toxicity. Labrasol, Transcutol, PEG 400, and propylene glycol were the screened cosurfactants. An isothermal method previously reported in the literature [15] was utilized for the solubility determination. In brief, 300 mg of SA was added to 1 mL of each component in screw-capped glass vials and mixed using a vortex mixer. The mixtures were then transferred to a thermodynamic water bath shaker (JULABO™, Julabo Labortechnik GMBH, Seelbach, Germany) for continuous shaking at a speed of 100 rpm at room temperature for three days, followed by centrifugation at 4000 rpm for 25 min. SA concentration was measured spectrophotometrically at 322 nm after the supernatant was diluted with methanol using a magnetic stirrer at 500 rpm for 2 min at room temperature [28]. The results of this study provide important insights into the solubility of SA in various oils, surfactants, and cosurfactants, which is crucial for the formulation of SA in a self-nanoemulsifying dosage form.

### 2.2. Construction of Pseudo-Ternary Phase Diagram

Based on the highest solubility of SA in different components, Labrafil, Cremophor El, and Transcutol were selected as the oil phase, surfactant, and co-surfactant, respectively. Various mixtures with varying concentrations of oil, surfactant, and co-surfactant were prepared, and a pseudo-ternary diagram was plotted using ProSim software (Labège, France, https://www.prosim.net/produit/prosimplus-simulation-optimisation-procede-industriel/ (accessed on 10 June 2023)). Labrafil, Cremophor El, and Transcutol levels varied from 10 to 30% *w*/*w*, 20 to 80 % *w*/*w*, and 10 to 70% *w*/*w*, respectively. All compositions were visually examined for nano-emulsion formation after diluting each mixture 100 times with deionized water [29].

### 2.3. Computer-Aided Optimization of SA-SNE Formulation Using Mixture I-Optimal Design

Recently, experimental design has been used in formulation development. This method yields ideal data with the fewest possible experiments and additional information regarding the effect of components on responses [30]. Pre-optimization research determined the proportion of each component (Labrafil, Cremophor El, and Transcutol). These concentrations were optimized using Design Expert (Version 11.1.2.0) software from Stat-Ease, Inc., Minneapolis, MN, USA. Labrafil, Cremophor El, and Transcutol concentrations were established as independent variables or factors. According to the pre-optimization research, the limits of every independent variable were identified and prescribed in the previous section. The sum of all components in a formulation always equals 100 percent. The parameters chosen included transmittance, solubility, and globule size. The optimization batches were chosen following the desirability function. Formulations were selected with a desired factor close to 1.0.

### 2.4. Globule Size, Size Distributions and Zeta Potential

SA-SNEs were diluted with deionized water and subjected to particle size examination. The average particle size, polydispersity index (PDI), and zeta potential were determined using dynamic light scattering on a Malvern Zetasizer (Malvern Instruments, Malvern, UK) [31].

### 2.5. Dispersibility Studies

Self-emulsification time was assessed through dispersibility studies. First, 1 mL of SA-SNE was added dropwise to 100 mL of isotonic phosphate buffer (pH 7.4) with gentle agitation using USP Type II (paddle) dissolution apparatus rotating at 50 rpm at $37 \pm 0.5$ °C. The process of self-emulsification was visually checked. Precipitation was assessed through visual monitoring of the resultant emulsion after 24 h storage at room temperature [14].

### 2.6. Solubility Studies

Solubility studies were conducted by adding excess SA to 1 mL of each SA-SNE mixture. The mixtures were moved to a thermodynamic water bath shaker (JULABO™,

Julabo Labortechnik GMBH, Seelbach, Germany) for continuous shaking at a shaking speed of 100 rpm at room temperature for three days, then centrifuged at 5000 rpm for 25 min. The supernatant was diluted with methanol using a magnetic stirrer at 500 rpm for 2 min at room temperature and then subjected to the quantification of SA content spectrophotometrically at 322 nm [32].

### 2.7. *Transmittance*

Using distilled water to reconstitute the SA-SNE, the resulting nano-emulsion was examined visually for any turbidity. A UV-vis spectrophotometer was then used to assess its percent transmittance at 638.2 nm, with distilled water serving as the blank. The experiments were performed following a 100-fold dilution [33].

### 2.8. *Differential Scanning Calorimetry Studies*

Thermal analysis of SA, Labrafil, Cremophor El, Transcutol, and the optimized SA-SNE mixture was carried out using a differential scanning calorimeter (DSC) (Shimadzu, DSC 60 TSW 60, Kyoto, Japan). The study was conducted within the 50–200 °C range at a scanning rate of 10 °C/min. An empty pan was used as a reference [34].

### 2.9. *Transmission Electron Microscopy*

A transmission electron microscope (JEOL JEM-1400, Tokyo, Japan) examination of the optimized SA-SNE was performed for visual observation. A copper grid was stained for five minutes at room temperature with a 1 percent $w/v$ solution of phosphotungstic acid before being treated with a drop of diluted SA-SNE. A transmission electron microscope was used to capture images at a 100 kV accelerated voltage [35].

### 2.10. *In Vitro Release Study and Kinetic Analysis*

An in vitro drug release was carried out to compare the release profiles of SA from optimized SA-SNE and its suspension formulation, containing 10 mg of SA each. The in vitro dissolution tests on SA SNE were carried out using the dialysis method. A glass cylinder open at both ends was attached to the shaft of United States Pharmacopoeia (USP) dissolution apparatus (Erweka DT-720, Erweka GmbH, Heusenstamm, Germany), and a dialysis membrane (MWt of 12,000 kDa) was tied at one end of the glass cylinder [31,35]. In vitro drug release tests were conducted in a beaker containing 50 mL of isotonic phosphate buffer (pH 7.4) (dissolution medium). These experiments were conducted at 100 rpm with the dissolving media kept at a constant 37 ± 0.5 °C. At regular time intervals, three milliliters of sample from each formulation were removed, and the same volume of SA-free new isotonic phosphate buffer (pH 7.4) was added instead. At 322 nm, the concentration of SA in each sample was measured spectrophotometrically. The relationship between spectrophotometric absorbance and SA concentration was plotted using the calibration curve. The amount of SA in the media was calculated at various time intervals from the calibration curve. The release kinetics data for sinapic acid suspension were gathered using three different models: Zero Order, First Order, and Diffusion. These models provide insights into the drug release behavior and offer valuable information for the design and formulation of sinapic-acid-based pharmaceutical products.

### 2.11. *Pharmacokinetic Study in Rabbits*

The pharmacokinetic study was carried out to evaluate the bioavailability of SA from the optimized SA-SNE nebulized dosage form, and it was compared with an oral SA suspension. The ultimate success of any formulation depends on its in vivo performance, so an in vivo study was conducted in a rabbit model.

Albino New Zealand rabbits obtained from the Nahda University animal center (Beni-Suef, Egypt), of either sex, 3 months old, and weighing between 1.35 and 1.72 kg, were used in this study. They were maintained on a standard pellet diet and water. A total of 12 rabbits were divided into two groups, each containing 6 ($n$ = 6) animals. The animals of group A were fed with 1 mL of drug suspension containing 10 mg/Kg of SA by mouth using an oral feeding needle. The animals of group B were given optimized SA-SNE containing the same amount of SA via nebulization using an ultrasonic nebulizer (PARI GmbH, Starnberg, Germany) [36,37].

The ultrasonic nebulizer was configured and connected to a power source for operation. Within the nebulizer, the medication chamber was filled with optimized SA-SNE. Upon activation, the ultrasonic nebulizer emitted high-frequency vibrations (ultrasonic waves) that were transmitted to the liquid medication within the chamber, resulting in the transformation of the medication into a fine mist or aerosol comprising minute medication droplets suspended in the surrounding air. This aerosol was then directed to the rabbits for inhalation, typically facilitated within a suitable chamber, thereby enabling the aerosolized medication to be absorbed directly into the rabbits' respiratory system, including the lungs and airways [38,39].

The rabbits were anesthetized using diethyl ether, and 0.5 mL of blood was withdrawn from the tip of the ear using a 19–23-gauge butterfly needle away from the base of the ear. Blood flow was stopped by applying pressure with sterile gauze placed at the blood sampling site for approximately 2 min to achieve hemostasis. Blood samples were withdrawn from the marginal ear vein of each rabbit at 0, 1, 2, 3, 4, 6, and 24 h, then collected in 1.5 mL capacity EDTA Eppendorf tubes. The tubes were centrifuged at 5000 rpm for 5 min, and the plasma was separated and stored at −80 °C until analysis. The animal experiments conducted were approved by the Institutional Animal Ethics Committee of Nahda University (Regd. No. NUB-032-022).

### *2.12. Pharmacokinetic Analysis of SA in Plasma through HPLC-UV Method*

#### 2.12.1. HPLC Method

HPLC separation was carried out on a stationary-phase ZORBAX Eclipse Plus® C18 column (250 × 4.6 mm, 5 μm), using a mobile phase consisting of acetonitrile and water in the proportion of (98:2, %*v*/*v*) at a flow rate of 1.0 mL/min using ciprofloxacin as an internal standard. The overall run time was 7 min, and 20.0 μL of each sample was injected in triplicates. The UV detection was conducted at 320.0 nm, and the temperature was adjusted to 25 °C.

#### 2.12.2. Sample Preparation

The samples of frozen plasma previously collected from rabbits were thawed to room temperature, 100.0 μL of the collected plasma samples was accurately transferred into a 2 mL centrifuge tube, then 40.0 μL (IS) was added from its stock solution (1.00 mg/mL). The volume was then adjusted to 1.0 mL with acetonitrile. Samples were then vortex-mixed for one minute and centrifuged for 20 min at 4000 rpm; the clear supernatant of each sample was completely taken and evaporated to dryness. The dried samples were then reconstituted with 1.0 mL acetonitrile, and 20.0 μL was injected for analysis.

### *2.13. $M^{pro}$ Enzyme Assay*

Top-scoring compounds were assessed for their in vitro enzyme inhibition activities using a 3CL Protease, tagged (SARS-CoV-2) Assay Kit (Catalogue #: 79955-1, BPS Bioscience, Inc., Allentown, PA, USA) according to the manufacturer's protocol. The in vitro FRET assay was monitored at an emission wavelength of 460 nm with an excitation at 360 nm using a Flx800 fluorescence spectrophotometer (BioTek Instruments, Winooski, VT, USA).

### *2.14. Antiviral Assay*

The antiviral screening was performed using SARS-CoV-2 (wild strain) virus at the Egyptian Company for Production of Vaccines (VACSERA) using strain hCoV-19/Egypt/NRC-3/2020 isolate, which was isolated in the Ministry of Health Laboratories for diagnostic reasons in May 2020. The cytotoxicity and the antiviral activities were investigated using Vero E6 cells (Sigma Aldrich, St. Louis, MO, USA). Then, serial concentrations of the test compounds were incubated with Vero E6 and then infected with SARS-CoV-2 (hCoV-19/Egypt/NRC-3/2020 isolate) at an MOI (multiplicity of infection) of approximately 0.5. DMSO was used as a solvent control. The antiviral screening was performed in a Biological Safety Cabinet Class III BCBS-502 (Biolab Scientific Ltd., Suite 300, Toronto, ON M1V 0B8, Canada) at VACSERA Co., Giza, Egypt. The detailed quantitative real-time PCR antiviral assay results can be found in the Supplementary Materials.

### 2.15. In Silico Study

Docking, molecular dynamics (MD) simulations, and Gaussian accelerated molecular dynamics (GaMD) simulations were carried out according to the previously described methods [40]. The methods are described in detail in the Supplementary Materials.

## 3. Results and Discussion

### 3.1. Screening of Components for SA-SNE Development

SNEDDSs are used to improve the solubility, bioavailability, and therapeutic efficacy of poorly water-soluble drugs [41]. The selection of surfactant, oil, and co-surfactant components can impact the physical properties and performance of the resulting nano-emulsion, such as particle size, stability, and drug release profile [42]. The solubility of pharmaceutical molecules in various components is the most essential requirement for screening components [43]. In this study, a systematic approach was employed to choose these components. Different oils with varying saturation degrees, specifically medium- or long-chain triglycerides, were evaluated to assess their solubilization and nano-emulsification capabilities. The oil demonstrating the highest proficiency in solubilizing SA was selected due to its significant influence on drug solubility and absorption. Additionally, the surfactants chosen for this study, namely Cremophor El, Tween 20, and Tween 80, were carefully screened based on their hydrophilic, non-ionic properties, which are advantageous for promoting emulsification and maintaining stability [25]. These surfactants are known for their high hydrophilic–lipophilic balance values, indicating their suitability for forming stable nano-emulsions [44]. Moreover, their low oral toxicity profile is a crucial consideration to ensure the safety of the delivery system [45]. Furthermore, the preference for oils with medium-length carbon chains over longer-chain oils, such as Labrafil, Capryol-90, and Labrafac, was based on empirical evidence indicating that medium-chain triglycerides exhibit superior solubilization and emulsification capabilities [24]. This choice aligns with the goal of optimizing drug delivery efficiency, underscoring the importance of selecting components that enhance drug solubility and bioavailability. These choices were made to design an efficient self-nanoemulsifying drug delivery system that could improve the solubility and absorption of SA, ultimately enhancing its therapeutic efficacy.

The solubility of SA was determined in various oils, surfactants, and cosurfactants using an isothermal method (Figure 1), and Tukey's post hoc test and one-way ANOVA were applied to check the significance of the obtained results. Among the oils, Labrafil demonstrated the highest solubilization concentration (29.98 mg/mL) by a significant margin at a P value less than 0.05. It was rated adequate, indicating its potential as an excellent carrier oil for drug delivery applications. Capryol 90 and Labrafac had moderately high concentrations, making them suitable alternatives to Labrafil in various formulations. Olive oil, anise oil, almond oil, soybean oil, and corn oil had lower concentrations, indicating their suitability for use in formulations where low oil concentration is preferred.

Surfactants are used in formulations to improve the solubility and bioavailability of poorly soluble drugs [46]. The results showed that Cremophore El had the highest concentration (138 mg/mL), followed by Tween 20 (122 mg/mL) and Tween 80 (78 mg/mL), while Labrasol had the lowest concentration (60 mg/mL). The choice of surfactant depends on the specific drug and formulation requirements. Cremophore El is often used in intravenous formulations, while Tween 20 and Tween 80 are commonly used in oral formulations due to their low toxicity [47]. Co-surfactants are added to formulations to enhance the solubilization and absorption of drugs. The results indicated that Transcutol had the highest concentration (192 mg/mL) by a significant margin at $p < 0.05$, followed by PEG 400 (102 mg/mL) and propylene glycol (58 mg/mL). The choice of co-surfactant depends on the specific formulation requirements and the compatibility with the surfactant [48]. The differences in the solubility of SA in different oils could be attributed to their physicochemical properties, such as their polarity, molecular weight, and viscosity [49]. The higher the solubility of SA in an oil, the better it can be used as a solvent or carrier for SA in various applications. These findings were consistent with previous studies reporting the high solubility of SA in Labrafil due to its high lipophilicity and low hydrophilicity [15]. Moreover, the high solubility of SA in Cremophor El and Transcutol might be attributed to their ability to form micelles and solubilize hydrophobic drugs [28]. Based on these results, Labrafil, Cremophor EL, and Transcutol, as the oil, surfactant, and cosurfactant, respectively, were further utilized for constructing a ternary phase diagram.

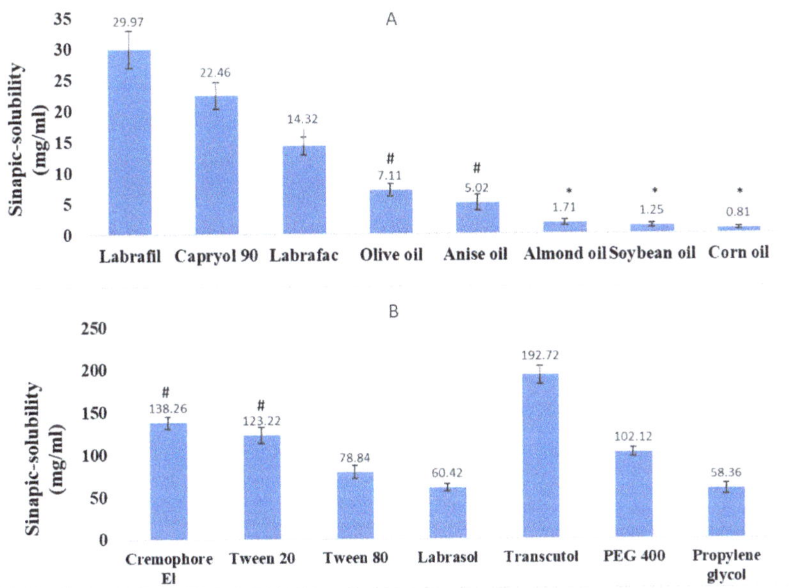


**Figure 1.** The saturated solubility of sinapic acid in: (**A**) different oils, (**B**) different surfactants and cosurfactants (* and # represent insignificance at $p < 0.05$ using ANOVA and SPSS 16.0 software).

### *3.2. Construction of Pseudo-Ternary Phase Diagram*

A pseudo-ternary phase diagram provided a useful tool for selecting the optimal composition of oil, surfactant, and cosurfactant for the formulation of a stable and efficient SA-SNE. Based on the results obtained from the solubility study, the pseudo-phase diagram was constructed using Labrafil as oil, Cremophor EL as surfactant, and Transcutol as cosurfactant. The ternary phase diagrams (Figure 2) were constructed to determine the concentration range of components required to form a stable nano-emulsion. The darker region in the phase diagram represented the self-emulsification area [50]. It was observed that the addition of surfactant–cosurfactant improved the efficiency of emulsification due to their higher hydrophilicity properties [51]. A stable system was seen with an oil concentration of up to 30% *w*/*w*. The addition of a cosurfactant, Transcutol, was found to improve the self-emulsification property of the system. Furthermore, it was observed that spontaneous emulsion formation was not efficient, with less than 40% *w*/*w* of the surfactant in the system. The results also showed that the drug solubility was influenced by the oil concentration, surfactant, and co-surfactant. As the concentration of the oil increases, the drug solubility decreases. Conversely, as the concentration of the surfactant and co-surfactant increases, the drug solubility increases. This is because the surfactant and co-surfactant help in solubilizing the drug in the microemulsion [52]. The results of this study provided a basis for optimizing the formulation of the SA-SNE.

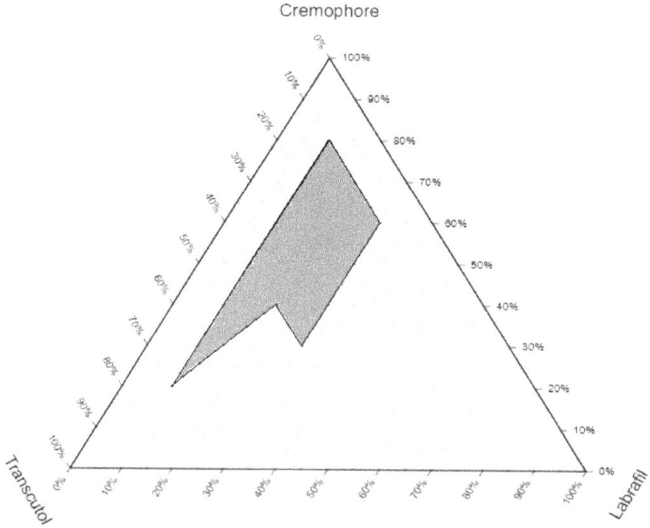


**Figure 2.** Pseudo-ternary phase diagram created using ProsSim software 1.0 for the selected system (Labrafil as the oil, Cremophor EL as the surfactant, and Transcutol as the co-surfactant.), showing the nano-emulsification area in gray.

### *3.3. Computer-Aided Optimization of SA-SNE Formulation Using Mixture I-Optimal Design*

Computer-aided optimization of an SA-SNE is an efficient approach to develop an optimal formulation with desirable characteristics. One approach to optimize an SA-SNE formulations is the use of mixture I-optimal design. Following this approach, the formulation variables, such as oil, surfactant, and cosurfactant concentrations, were varied according to a designed experimental plan. The experimental plan was based on a statistical approach that minimized the number of experiments required while providing a comprehensive understanding of the effect of each variable on the response of interest. Computer-aided optimization using mixture I-optimal design had several advantages over traditional experimental approaches [34,53]. It provided a comprehensive understanding of the formulation variables' effects. This approach also allowed for the optimization of multiple response variables simultaneously, leading to a more efficient formulation process.

The pre-optimization studies helped to identify the ranges of concentrations for the critical components of the SA-SNE formulation, i.e., oil, surfactant, and co-surfactant. To optimize the SA-SNE formulation, the concentrations of oil, SA, and co-surfactant were chosen as independent variables, while globule size, solubility, and percent transmittance were selected as responses. Globule size is an important parameter that determines the physical stability of the SA-SNE [54], while solubility is a critical factor that affects the bioavailability of the drug [31]. Percent transmittance is a measure of the clarity and transparency of the formulation.

A total of 15 formulations were constructed using Design Expert software (Version 11.1.2.0 of Stat-Ease, Inc., Minneapolis, MN, USA) based on the pre-optimization studies. The design matrix of the 15 formulations is presented in Table 1. This design aimed to find the optimal formulation to achieve the desired responses. The design matrix was analyzed using the Design Expert software, and the responses were fitted to a second-order polynomial equation. The model was evaluated using ANOVA, and significant factors were identified.

**Table 1.** The dependent and independent variables for the prepared SNEDDS using a mixture of experimental designs.

| Sample No. | Surfactant (%) (X1) | Co-Surfactant (%) (X3) | Oil (%) (X2) | Transmittance % (Y1) | Solubility (mg/mL) (Y2) | Size nm (Y3) | PDI | Emulsification Time (s) | Zeta Potential (mV) |
|---|---|---|---|---|---|---|---|---|---|
| 1 | 20 | 70 | 10 | 94 | 230.5 ± 5.2 | 191.2 | 0.214 | 41 | −17.1 |
| 2 | 30 | 60 | 10 | 93 | 210.2 ± 7.3 | 184.4 | 0.291 | 48 | −13.3 |
| 3 | 40 | 50 | 10 | 92 | 199.4 ± 4.6 | 203.1 | 0.232 | 55 | −14.6 |
| 4 | 40 | 40 | 20 | 100 | 174.3 ± 3.1 | 94.93 | 0.176 | 67 | −14.2 |
| 5 | 40 | 30 | 30 | 100 | 128.6 ± 1.4 | 102.4 | 0.137 | 75 | −17.8 |
| 6 | 50 | 40 | 10 | 95 | 153.5 ± 3.6 | 189.5 | 0.162 | 46 | −16.5 |
| 7 | 50 | 30 | 20 | 100 | 134.2 ± 4.2 | 95.35 | 0.219 | 45 | −12.1 |
| 8 | 50 | 20 | 30 | 94 | 117.3 ± 2.3 | 169.5 | 0.241 | 64 | −12.4 |
| 9 | 30 | 40 | 30 | 93 | 129.8 ± 1.9 | 195.3 | 0.263 | 72 | −14.7 |
| 10 | 60 | 30 | 10 | 100 | 184.2 ± 3.7 | 93.7 | 0.231 | 43 | −16.9 |
| 11 | 60 | 20 | 20 | 100 | 156.4 ± 6.2 | 97.4 | 0.156 | 46 | −15.7 |
| 12 | 60 | 10 | 30 | 99 | 101.2 ± 2.7 | 125.6 | 0.175 | 61 | −16.1 |
| 13 | 70 | 20 | 10 | 100 | 118.7 ± 3.1 | 88.8 | 0.310 | 35 | −13.4 |
| 14 | 70 | 10 | 20 | 100 | 111.9 ± 4.2 | 95.3 | 0.134 | 37 | −14.6 |
| 16 | 80 | 10 | 10 | 100 | 127.1 ± 3.3 | 83.6 | 0.164 | 42 | −15.1 |

After developing the models, optimization was conducted using the desirability function. The desirability function combines the responses into a single value between 0 and 1, where 1 is the ideal response. The formulations with the highest desirability factor were selected for further analysis. The R-squared and adjusted $R^2$ values provide an indication of how well the model fits the data. In this study, the R-squared and adjusted $R^2$ values were relatively close, suggesting that the model fit was adequate. The adequate precision (AP) ratio measures the signal-to-noise ratio and indicates whether the experimental region is well represented by the model. The AP ratios for all responses were greater than 4, which is sufficient for routing the design space. Therefore, the results suggested that the models developed were reliable and suitable for the optimization of the SA-SNE formulation [55].

The results of the design matrix (Table 1) showed that as the percentage of surfactant (X1) increased, the transmittance percentage (Y1) also increased, indicating better transparency of the SA-SNE. However, there was a trade-off between transmittance and particle size (Y3), as the size decreased with increasing surfactant percentage. The co-surfactant percentage (X3) and oil percentage (X2) also affected the solubility (Y2) and size (Y3) of the SA-SNE. The results showed that the transmittance values ranged from 92 to 100%, showing that the samples had good clarity. The solubility values varied from 101.2 to 230.5 mg/mL, which suggested that the SNEDDS effectively enhanced the drug's solubility. The size of the SA-SNE ranged from 83.6 to 203.1 nm, indicating that the samples were in the nano-emulsion range. The PDI values ranged from 0.134 to 0.310, indicating a low to moderate level of polydispersity. The emulsification time ranged from 35 to 72 s, indicating that the preparation of SA-SNE was relatively easy. The zeta potential values ranged from −17.8 to −12.1 mV, indicating that the samples had a negative charge on their surface.

Table 2 summarizes the regression analysis for the three response variables: globule size, solubility, and % transmittance. The model for all three responses was linear, as indicated by the "Linear" in the "Model" column. In general, the linear model is the most appropriate for the mixture design since the relations between the mixture ingredients are shown by directly proportional responses [56]. In addition, linear models are less complex than other models [57]. The *p*-values for solubility and % transmittance were highly significant at 0.0001 and 0.0185, respectively, indicating that the model terms were significant for these responses. The R-squared values for solubility and % transmittance were high at 0.8931 and 0.4857, respectively, indicating that the model could explain a large proportion of the total variation in these responses. The R-squared value for globule size was lower at 0.5231, indicating that the model explained a lower proportion of the variation in this response. The adjusted R-squared values considered that the number of terms in the model were lower than the R-squared values. However, the adjusted R-squared values for solubility and % transmittance were still relatively high at 0.8753 and 0.4000, respectively, indicating a good model fit.

**Table 2.** Data of regression analysis for responses.

| | Globule Size (Y1) | Solubility (Y2) | % Transmittance (Y3) |
|---|---|---|---|
| Model | Linear | Linear | Linear |
| *p*-value | 0.0118 | 0.0001 | 0.0185 |
| R-squared | 0.5231 | 0.8931 | 0.4857 |
| Adjusted $R^2$ | 0.4436 | 0.8753 | 0.4000 |
| Predicted $R^2$ | 0.3114 | 0.8470 | 0.2459 |
| Adequate precision | 7.523 | 20.140 | 6.990 |
| %CV | 26.53 | 9.18 | 2.52 |
| PRESS | 21,902.93 | 3339.41 | 105.38 |
| Y1 = +136.49 X1 + 239.65 X2 + 6.38 X3 | | | (Equation (1)) |
| Y2 = +81.77 X1 + 221.32 X2 + 160.34 X3 | | | (Equation (2)) |
| Y3 = +100.49 X1 + 91.56 X2 + 95.35 X3 | | | (Equation (3)) |

The adjusted R-squared value for globule size was lower at 0.4436, indicating that the model might be overfitting the data. The predicted R-squared values were lower than the R-squared values, indicating that the model could have limited predictive ability for new observations. The adequate precision ratios for all three responses were above the recommended threshold of 4, indicating that the model could navigate the design space [58]. The coefficients of variation (%CV) for solubility and % transmittance were low, indicating good precision and reproducibility of the measurements. The %CV for globule size was higher, indicating more variability in the measurements. The PRESS values for all three responses were relatively high, indicating that the model might not fit the data and could lead to residual error. This suggested that the model could be further improved by including additional terms or refining the experimental design.

The regression equations (Table 2) represent mathematical models that predicted the response of the dependent variable, Y, based on the values of the independent variables, X1, X2, and X3. The dependent variable could be a drug delivery system's physical properties, such as viscosity, surface tension, or solubility. In contrast, the independent variables were the compositions of the system, represented by the amounts of Cremophor EL (X1), Labrafil (X2), and Transcutol (X3). The general form of the equations is Y = aX1 + bX2 + cX3, where a, b, and c are coefficients that represent the contribution of each independent variable to the dependent variable. The specific values of these coefficients were obtained through statistical analysis of experimental data, such as regression analysis, and were unique to the specific system and dependent variable studied.

From the regression data of the linear models, the equations mentioned in Table 2 were obtained. These equations showed positive values for the three independent variables, showing their positive effect on the three responses. In Equation (1), the Y value was predicted based on the specific coefficients of 136.49 for X1 (Cremophor EL), 239.65 for X2 (Labrafil), and 6.38 for X3 (Transcutol). In Equation (2), the coefficients were 81.77 for X1, 221.32 for X2, and 160.34 for X3, and in Equation (3), the coefficients were 100.49 for X1, 91.56 for X2, and 95.35 for X3. These equations could be used to predict the response of the dependent variable, such as the solubility of a drug, based on the independent variables, such as the amounts of surfactant, oil, and co-surfactant in a self-nanoemulsifying drug delivery system. The equations could also be used to optimize the system's composition to achieve a desired response or to identify the independent variables with the most influence on the dependent variable. Equations (1) and (2) had a similar range of coefficients for X1 and X2, but Equation (2) had a much higher coefficient for X3. This suggested that X3 (Transcutol) might be more critical in predicting the response in Equation (2) than in Equation (1). Equation (3) had lower coefficients overall compared to Equations (1) and (2), which suggested that the independent variables might have a minor impact on the dependent variable in this case.

Figures 3–6 illustrate model equations and show the effects of independent variables on responses. They were created to interpret the mixture region. In the response surface, each factor (pure mixture component) is represented in a corner of an equilateral triangle; each point within this triangle refers to a different proportion of components in the mixture. The maximum percentage of each ingredient considered in the regression is placed in the corresponding corner, while the minimum is positioned in the middle of the opposite side of the triangle. The 2D contour plots show the effect of the independent variables, surfactant percentage (X1), oil percentage (X2), and co-surfactant percentage (X3), on the % of transmittance (Y1) (Figure 3A). The contour plot indicates that the % of transmittance was primarily influenced by the surfactant percentage (X1) and oil percentage (X2). The co-surfactant percentage (X3) had a relatively weaker effect on the transmittance. The contour lines on the plot are elliptical, which suggests an interactive effect of the two variables on the transmittance. The maximum % of transmittance is observed at the center of the plot, which corresponds to the formulation with a higher surfactant percentage and a lower oil percentage. Based on the contour plot, it could be inferred that a surfactant percentage of around 40–50% and an oil percentage of around 10–20% could provide optimum transmittance for the formulation. The contour plot provided a graphical representation of the relationship between the independent and response variables and helped identify the optimum formulation conditions.

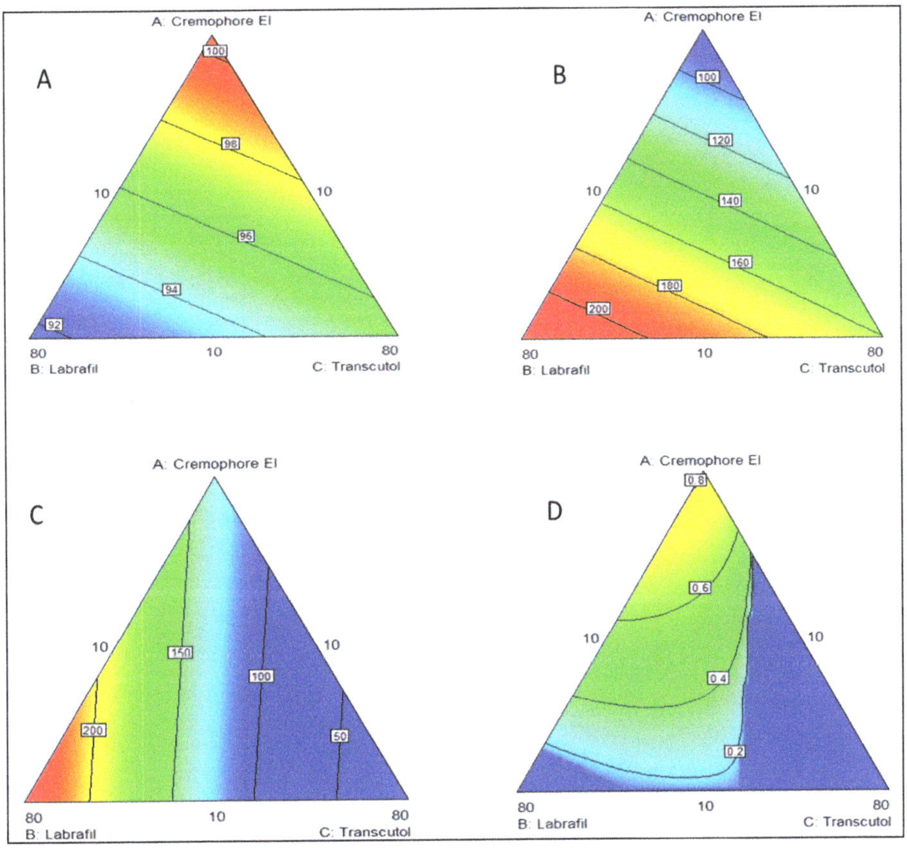


**Figure 3.** 2D contour plots for the effect of the independent variables (concentration of oil, surfactant, and cosurfactant) on (**A**) enhancing the % of transmittance up to 100%, (**B**) lowering the globule size from 200 to 100 nm, (**C**) enhancing the solubility, and (**D**) increasing the desirability.

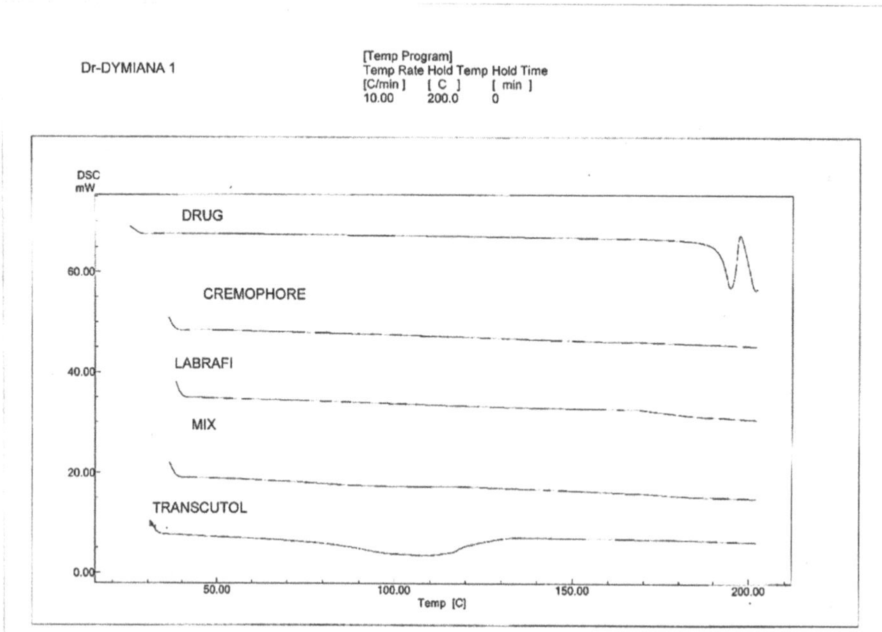


**Figure 4.** DSC thermograms of sinapic acid (drug), Cremophore El, Transcutol, Labrafil, and the optimum SNEDDS mix.

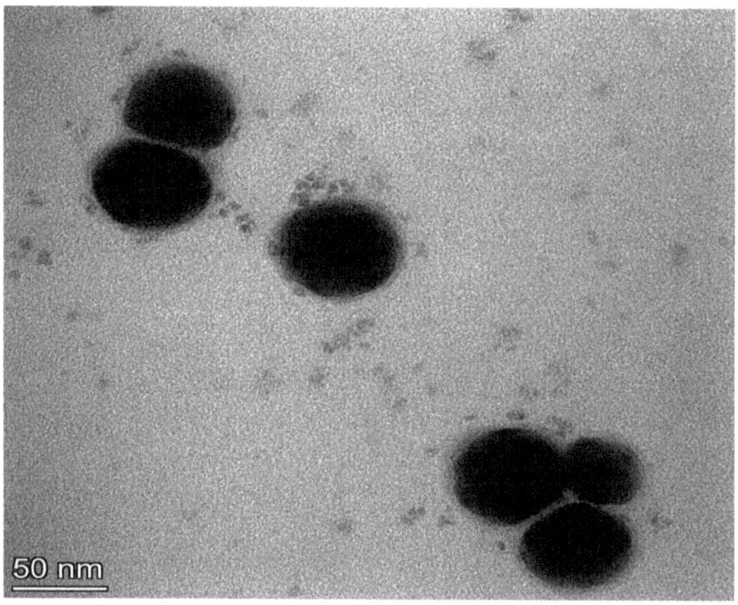


**Figure 5.** Photomicrograph of the optimum SNE, taken using TEM (50,000×).

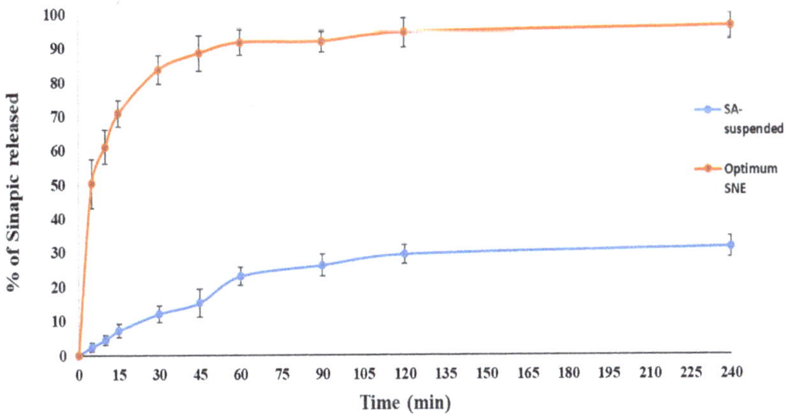


**Figure 6.** In vitro release of sinapic acid from the optimized SNE compared with the sinapic suspension.

The effect of the independent variables on the globule size Figure 3B reveals that the surfactant percentage and oil percentage had opposite effects on the globule size. As the surfactant percentage increased, the globule size decreased, while increasing the oil percentage resulted in an increase in globule size. This trend is consistent with emulsions' general behavior, where surfactants stabilize oil droplets, preventing them from coalescing into larger droplets [59]. In addition to the surfactant and oil percentages, the contour plot also reveals that the lowest values of globule size occurred when the surfactant percentage was high (above 40%) and the oil percentage was low (below 20%). This result indicated that the choice of surfactant and oil types, as well as their relative concentrations, could have a significant impact on the globule size of an emulsion. The contour plot could also be used to estimate the optimal values of the independent variables that resulted in the desired globule size. For example, if the target globule size was around 100 nm, the contour plot suggests that this could be achieved by using a high surfactant percentage (above 40%) and a low oil percentage (below 20%).

Figure 3C illustrates the effect of the independent variables on SA solubility. The contour plot shows that the co-surfactant percentage had a strong positive correlation with SA solubility. As the co-surfactant percentage increased, SA solubility also increased. This trend is depicted by the contour lines that move upwards and to the right, indicating higher solubility values. Therefore, it could be inferred that a higher co-surfactant percentage might help improve the solubility of SA. However, the effect of emulsification time on SA solubility was less clear. There was a slight trend of increasing solubility with increasing emulsification time, but the effect was not as strong as the effect of co-surfactant percentage. The contour lines for emulsification time are more horizontal, indicating that the effect of this variable on SA solubility was less significant compared to the co-surfactant percentage.

Based on the regression models, optimal factor settings were selected using Design-Expert® in order to identify experimental settings in which all desirability criteria were met as much as possible. The optimization process of the design was based on minimizing globule size and maximizing both the transmittance and solubility. According to the optimization parameter input, the software generated an optimum formulation (F 16), which was produced and analyzed. The composition of this optimized formula is described in Table 3. The effect of formulation factors on desirability is shown in Figure 3D. The optimized formula had a droplet size of 83.6 nm with a small PDI OF 0.164, SA solubility of 127.1 $\pm$ 3.3, and transmittance of 100%. Experimental and predicted values of the optimum SNE (F 16) were compared and showed close resemblance with a small relative prediction error (less than 9%), suggesting the validity of the generated regression equations.

**Table 3.** Composition of the optimized SNE system.

| Surfactant | Oil | Co-Surfactant |
|---|---|---|
| Cremophore El<br>80% | Labrafil<br>10% | Transcutol<br>10% |

### 3.4. Differential Scanning Calorimetry Studies

Thermograms of SA, Labrafil, Cremophor El, Transcutol, and optimized SA-SNE physical mixture were obtained using a differential scanning calorimeter, as shown in Figure 4. In this figure, each curve represents a different material. The figure allowed us to observe each material's melting point and other thermal properties, which are important parameters that affect the stability and bioavailability of drugs. The sharp endothermic peaks observed for SA, Cremophore El, and Transcutol at around 150 °C, 60 °C, and 50 °C, respectively, indicate that these materials had high melting points. This means it might be more difficult for these to be dispersed in water and absorbed by the body.

In contrast, Labrafil and the optimized SA-SNE mix showed broad endothermic peaks at around 40–50 °C, indicating their lower melting points. These materials are more suitable for drug delivery systems as they can be easily dispersed in water and absorbed by the body [31,60,61]. The change in the melting behavior of SA could be attributed to the inhibition of its crystallization and solubilization of SA in the SA-SNE. Therefore, it could be concluded that SA in the SA-SNE was amorphous [62]. It is known that transforming the physical state of a drug to the amorphous or partially amorphous state leads to a high-energy state and high disorder, resulting in enhanced solubility [63,64].

### 3.5. Transmission Electron Microscopy

TEM photographs of the SA-SNE formulae after dilution with distilled water are shown in Figure 5 and were interpreted for surface morphology and globule size. From the presented figures, it is apparent that globules of all formulae were well dispersed, and no globule aggregation occurred. TEM analysis revealed that most formulae showed spherical and homogeneous droplets with a small size less than 100 nm, which satisfies the criteria of the nanometric size range required for the nanoemulsifying formula. These results support the results obtained through dynamic light scattering for the optimized formula, which had a droplet size of 83.6 nm with a small PDI OF 0.164.

### 3.6. In Vitro Release

In vitro release profiles of the SA-SNE optimized formulation were compared with the SA suspension. A rapid initial release of SA from the SA-SNE was observed. As shown in Figure 6, the SA suspension released less than 25% within 24 min, with much lower drug release than the optimized-formula SA-SNE, which released more than 92% of SA simultaneously. The highest release of SA from the formulation SA-SNE was possible due to its lowest droplet size (83.6 nm), lowest PDI value (0.164), and high Cremophore El concentration (80%). The higher drug release rate from the formula is because the quantitative drug release from developed nano-emulsions is droplet-size-dependent. This led to a greater interfacial area in the nano-emulsion with small drops, promoting rapid drug release [53]. These results could be explained by the fact that the SA-SNE was composed of a mixture of oil, surfactant, and co-surfactant, which spontaneously form nano-emulsions when exposed to an aqueous environment. The nano-emulsion droplets' small size increases the drug's surface area in contact with the dissolution medium, which enhances the SA dissolution and release rate [65]. Additionally, the surfactant and co-surfactant components of the SA-SNE can help solubilize the SA and keep it in solution, further enhancing the release rate [66,67]. In contrast, an SA suspension was composed of solid SA particles suspended in a liquid medium. The kinetics analysis of the release data began by assessing the goodness of fit for each model, as indicated by the R-squared ($R^2$) values (Table 4). The $R^2$ value quantifies how well each model aligns with the experimental data. It was observed that the Diffusion model exhibited the highest $R^2$ value of 0.9226, signifying an excellent fit to the release kinetics of sinapic acid. This result suggests that the Diffusion model is particularly well suited for describing the release behavior of the drug from the suspension. In the Diffusion model for the suspension, a steep negative slope (−2.4448) was observed, indicating a rapid release rate of sinapic acid. This suggests that sinapic acid is released quickly from the suspension. Conversely, in the First Order model for the nano-emulsion, the negative slope (−0.0051) suggests a gradual decrease in sinapic acid concentration over time.

**Table 4.** Kinetics analysis of the release data.

| | Model | | |
|---|---|---|---|
| | **Zero** | **First** | **Diffusion** |
| Slope | −0.13688 | −0.00073 | −2.44477 |
| Intercept | 93.24012 | 1.969396 | 100.7459 |
| $R^2$ | 0.745386 | 0.773053 | 0.922613 |
| dF | 8 | 8 | 8 |
| F | 23.42007 | 27.25046 | 95.37716 |
| SE slope | 0.028285 | 0.000139 | 0.250331 |
| SE int | 2.637903 | 0.012988 | 1.963146 |
| SE y | 6.270964 | 0.030876 | 3.457209 |
| SS reg | 920.9942 | 0.025979 | 1139.976 |
| SS resid | 314.5999 | 0.007627 | 95.61835 |
| C.V.% | 7.393113 | 1.604223 | 53.07682 |

The release rate of the SA from the suspension was dependent on the rate of dissolution of the SA particles in the liquid medium. Since SA is poorly soluble [68], the dissolution rate from the suspension might be slow, resulting in a lower release rate than the SA-SNE. In summary, when comparing the release kinetics of sinapic acid in the suspension and the self-nanoemulsifying drug delivery system, it is evident that the nano-emulsion formulation offers several advantages. The nano-emulsion exhibited a good fit, statistical significance, and lower variability in parameter estimation. The gradual release profile observed in the nano-emulsion suggests enhanced solubility and controlled release behavior compared to the rapid release from the suspension. These findings underscore the efficacy of the nanoemulsifying system in improving the solubility and release kinetics of sinapic acid, making it a promising approach for pharmaceutical product development.

### *3.7. Pharmacokinetic Analysis of SA in Plasma through HPLC-UV Method*

The pharmacokinetic parameters for SA in the oral pure SA suspension and the nebulized optimum SA-SNE are presented in Table 5. These parameters provide information on the body's drug absorption, distribution, metabolism, and excretion. The results show that the nebulized optimized SA-SNE formulation resulted in higher $C_{max}$ (maximum plasma concentration) by 2.65 times and $AUC_{0–24}$ (area under the plasma concentration–time curve from 0 to 24 h) by 2.43 times compared to the oral SA suspension, Figure 7. This suggests that the optimized SA SNE formulation improved the bioavailability of SA by enhancing its absorption and reducing its first-pass metabolism. The higher $C_{max}$ also indicates that the drug reached its peak concentration faster in the blood with the optimized SA-SNE formulation, as shown by the similar $T_{max}$ (time to reach $C_{max}$) values for both formulations. The shorter half-life ($t_{1/2}$) and mean residence time (MRT) of SA with the optimized SA-SNE formulation suggests that the drug was eliminated from the body more quickly compared to the oral suspension [69,70]. This could be due to the increased absorption of SA in the optimized SA-SNE formulation, which could result in faster clearance and metabolism of the drug. Overall, the results of this study suggest that the nebulized optimized SA-SNE formulation could improve the pharmacokinetic profile of SA, which could have implications for its clinical use as a therapeutic agent.

**Table 5.** Pharmacokinetic parameters for sinapic acid following administration of oral pure sinapic acid suspension and the nebulized optimized SNEDDS.

| Parameters | Oral Sinapic Acid Suspension | Optimized SNEDDS |
|---|---|---|
| $C_{max}$ (μg/mL) | 119.02 ± 13.2 | 315.31 ± 20.3 |
| $T_{max}$ (h) | 2 | 2 |
| $AUC_{0-24}$ (μg·h/mL) | 766.48 ± 130 | 1865.17 ± 185 |
| $t_{1/2}$ (h) | 8.38 ± 0.62 | 5.76 ± 0.35 |
| $AUC_{0-\infty}$ (μg·h/mL) | 904.84 ± 27.2 | 1993.57 ± 43.2 |
| $MRT_{0-\infty}$ (h) | 11.36 ± 1.8 | 8.34 ± 1.4 |

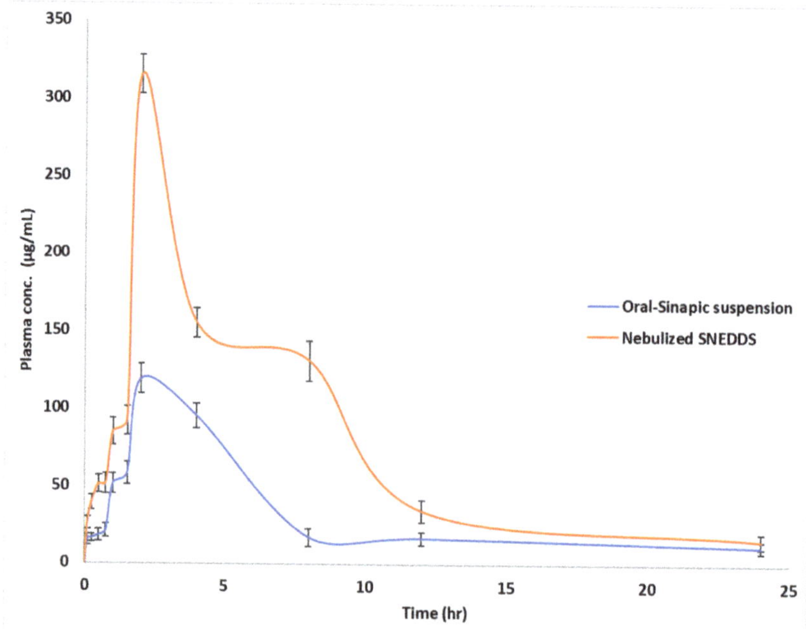


**Figure 7.** Mean sinapic acid plasma concentration–time curves following administration of oral sinapic acid suspension and nebulized SNE optimized formula to Wistar albino rabbits (n = 6) at ($p < 0.05$).

### 3.8. In Vitro Antiviral Activity Study

The $IC_{50}$ for SA's in vitro antiviral activity against SARS-CoV-2 was determined to be 2.53 ± 0.065 μg/mL. This value is considered 64-fold higher than the reference antiviral drug remdesivir ($IC_{50}$ = 0.036 ± 0.002 μg/mL). The $CC_{50}$ values for SA and the reference antiviral medication remdesivir against the host cells were very low (196.23 ± 15.1 and 170.8 ± 9.8 μg/mL, respectively). It is worth noting that the SA-free SNEDDS did not show any cytotoxicity up to 300 μg/mL ($CC_{50}$ > 300 μg/mL).

Compared to the unformulated SA, the newly prepared nano-formulation significantly decreased viral multiplication ($IC_{50}$ = 0.095 ± 0.0005 μg/mL). Accordingly, the new formulation's antiviral efficacy became comparable to remdesivir's. This significant bioactivity improvement might be a direct result of the enhanced cellular delivery of SA.

### 3.9. In Vitro $M^{pro}$ Inhibitory Activity

In the present study, we tested SA for its inhibitory activity against the viral $M^{pro}$. The reason for selecting this specific target for testing was the recently published study about baicalein that proved the ability of this natural product to inhibit $M^{pro}$ activity [71]. As shown in Figure 8, the baicalin scaffold is significantly close to SA, with a Tanimoto similarity index of 0.46. The key difference between the two structures is that SA has a terminal hydrophilic carboxylate moiety, while the reference inhibitor baicalin has a hydrophobic benzene ring. Accordingly, both compounds showed comparable in vitro $M^{pro}$ inhibitory activity ($IC_{50}$ = 6.74 ± 0.95 and 1.12 ± 0.35 μM, respectively) [71].

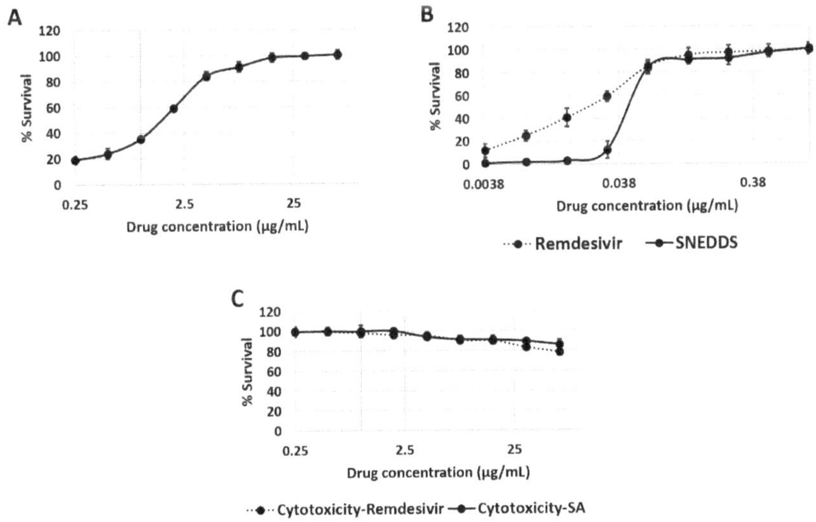


**Figure 8.** Antiviral activity of SA and its nano-formulation (i.e., SNEDDS) ((**A**,**B**), respectively) against SARS-CoV-2. The antiviral activity of SNEDDS was compared to that of remdesivir in the same Figure (**B**). Cellular cytotoxicities of both SA and remdesivir against Vero E6 cells (**C**). The assays were carried out in triplicate (n = 3).

### *3.10. Investigation of SA's Mode of Interaction with SARS CoV-2 $M^{pro}$*

Being an $M^{pro}$ inhibitor, SA's modeled structure was docked into the enzyme active site crystal structure (PDB ID: 6M2N). The resulting poses (10 poses) were of convergent orientations, where the RMSD between the pose of the best score and that of the worst one was 1.79 Å. Accordingly, we chose the best-scoring pose for the subsequent investigations.

As shown in Figure 9, both SA and baicalein structures were aligned inside the $M^{pro}$ active site, where they shared some interactions, particularly the hydrophilic ones (e.g., H-bonds with HIS-41 and ASN-142). The main difference between the interactions of each structure was that the carboxylate moiety of SA was involved in H-binding with GLN-192. In contrast, baicalein's extended aromatic ring (i.e., ring B) was involved in hydrophobic interactions with LEU-167 and PRO-168.

Accordingly, we tried to validate the SA interaction inside the $M^{pro}$ active site by (i) calculating the absolute binding free energy ($\Delta G_{Binding}$) of SA; (ii) running 100 ns long molecular dynamics (MD) simulations to study the degree of SA stability inside the $M^{pro}$ active site; and (iii) running Gaussian accelerated molecular dynamics (GaMD) simulations to reveal the binding pathway of SA to the $M^{pro}$ active site.

First, the docking poses of SA and baicalein were subjected to 100 ns long MD simulations. As shown in Figure 10, both structures achieved stable binding throughout MD simulation, with an average RMSD of 2.23 Å and 2.16 Å for baicalein and SA, respectively. Hence, their calculated interaction energies in terms of electrostatic and van der Waals energies were convergent (−66.56 and −61.45 kcal/mol, respectively). In addition, their calculated absolute binding free energies ($\Delta G_{Binding}$) were also convergent (−9.36 and −8.78 kcal/mol, respectively).

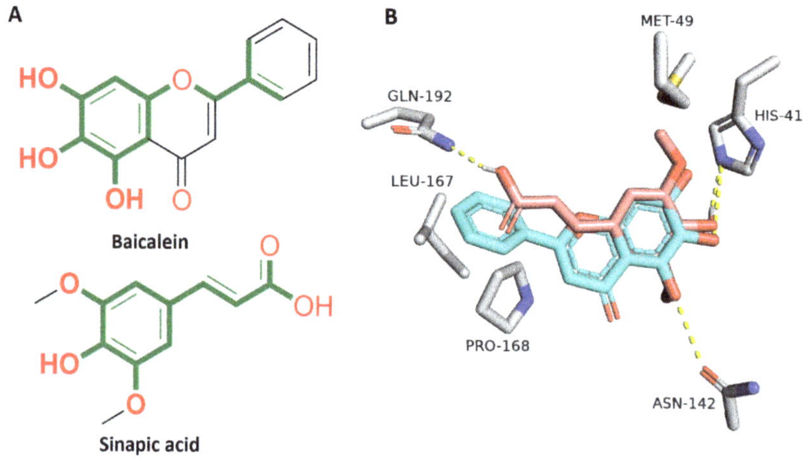


**Figure 9.** (**A**) Chemical structures of both baicalein and SA showing the structural similarity in green. (**B**) Binding modes of both baicalein and SA in alignment with each other inside the $M^{pro}$ active site. The green color indicates the structural similarity between both molecules.

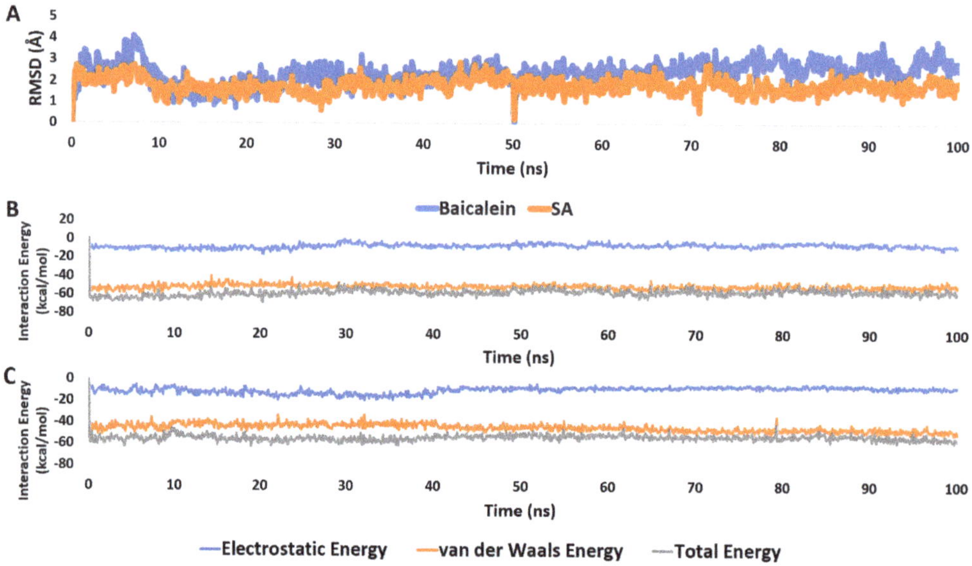


**Figure 10.** (**A**): RMSDs of SA and baicalein structures inside the $M^{pro}$ active site throughout 100 ns long MD simulation. (**B**,**C**) Interaction energies (i.e., electrostatic and van der Waals energies) of SA and baicalein structures inside the $M^{pro}$ active site over the course of 100 ns long MD simulation.

Second, we tried to re-produce the docking pose of SA inside the $M^{pro}$ active site by running five independent GaMD simulations. Five copies of the SA structure were simulated for 200 ns each, with each copy located at least 20Å away from the modeled $M^{pro}$ structure in the solvent box. As shown in Figure 11, ligand binding was observed in one of the five simulations after ~40 ns and remained stable in the $M^{pro}$ active site until the completion of the simulation. The binding state (Figure 10B) was highly similar to the docking pose chosen earlier (RMSD = 2.61 Å), suggesting this is the most likely binding pose.

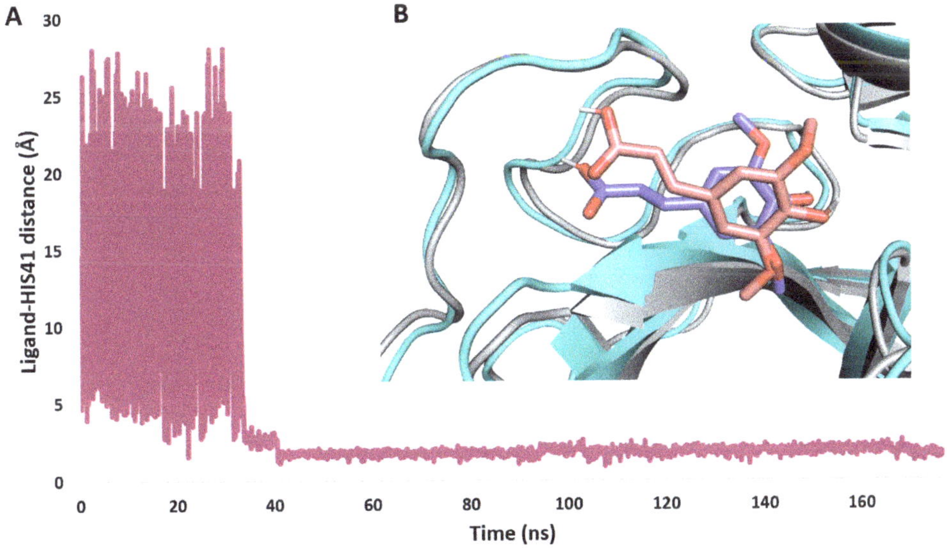


**Figure 11.** (**A**): Simulation of SA binding event inside the $M^{pro}$ active site observed in GaMD simulation after ~40 ns. (**B**): Structure alignment of the SA docking pose (blue structure) on its most populated structure after binding inside the active site of $M^{pro}$ during the GaMD simulation (brick-red structure). RMSD value between the two states = 2.61 Å. A video showing the binding event of SA inside the active site of $M^{pro}$ can be found on the Zenodo website: https://zenodo.org/record/8175715 (accessed on 10 June 2023).

These results indicate that SA exhibited a binding profile comparable to that of the co-crystalized inhibitor baicalein.

## 4. Conclusions

In 2021, we proved that sinapic acid (SA) selectively inhibited SARS-CoV-2 viral replication in vitro by targeting its envelope protein. However, it lacked druggability properties, with low water solubility and poor pharmacokinetic properties. This study successfully developed and optimized a self-nanoemulsifying drug delivery system (SNE) for SA to overcome its poor solubility and enhance its therapeutic efficacy. Our in-depth screening identified Labrafil, Cremophor EL, and Transcutol as suitable choices for the formulation. The construction of a pseudo-ternary phase diagram provided valuable insights into the optimal composition of the SA-SNE formulation, ensuring stability and efficient emulsification. Mixture I-optimal design facilitated computer-aided optimization, considering key factors such as globule size, solubility, and transmittance. The optimized SA-SNE formulation exhibited promising characteristics, including small globule size, high solubility, and rapid drug release. These properties are crucial for improving the bioavailability of SA. Pharmacokinetic analysis revealed that the optimized SA-SNE formulation via nebulization resulted in higher maximum plasma concentration ($C_{max}$) and area under the plasma concentration–time curve ($AUC_{0–24}$) compared to the oral SA suspension. This indicated enhanced absorption and reduced first-pass metabolism of SA. Furthermore, the SA-SNE formulation demonstrated potent antiviral activity against SARS-CoV-2, with an $IC_{50}$ comparable to the reference antiviral drug remdesivir. In silico, modeling, and Gaussian accelerated molecular dynamics (GaMD) experiments demonstrated that SA could interact with and inhibit the viral main protease ($M^{pro}$). This suggests the potential of the SA-SNE formulation as a promising candidate for antiviral therapy with multiple modes of action.

**Supplementary Materials:** The following supporting information can be downloaded at: https://www.mdpi.com/article/10.3390/pharmaceutics15112531/s1, Figure S1: Initial model used in the GaMD simulations. Sinapic acid is shown as green spheres and $M^{pro}$ as cyan ribbons. Golden-yellow balls represent the $Na^+$ and $Cl^-$ ions.

**Author Contributions:** Conceptualization, D.M.N., M.E.R., H.M.H. and H.A.A.; methodology, D.M.N., A.S.G.S.E.-D., A.H.A., A.M.S. and H.A.A.; validation, D.M.N., A.S.G.S.E.-D., M.E.R., H.A.A. and H.M.H.; data curation, D.M.N., A.S.G.S.E.-D., H.A.A., A.H.A., H.M.H., M.E.R. and A.M.S.; writing—original draft preparation, D.M.N. and A.S.G.S.E.-D.; writing—review and editing, All authors; funding acquisition, H.A.A. and M.E.R. All authors have read and agreed to the published version of the manuscript.

**Funding:** This study was funded by the Jameel Fund for Infectious Disease Research and Innovation in Saudi Arabia (grant number (8-8)2120191).

**Institutional Review Board Statement:** The animal study protocol was approved by the animal ethical committee of Nahda University (Regd. No. NUB-032-022) on 15 May 2022.

**Informed Consent Statement:** Not applicable.

**Data Availability Statement:** The data presented in this study are available in this article and Supplementary Materials. All references related to the experimental data in the SI are listed below [72–82].

**Acknowledgments:** The authors therefore gratefully acknowledge King Abdulaziz University and Community Jameel (grant number (8-8)2120191) for their financial and technical support. The authors thank Essam Rashwan, Head of the Confirmatory Diagnostic Unit, VACSERA, Egypt, for carrying out the in vitro study on human coronavirus.

**Conflicts of Interest:** The authors declare no conflict of interest.

## References

1. Nguyen, V.T.; Stewart, J.D.; Ioannou, I.; Allais, F. Sinapic acid and sinapate esters in Brassica: Innate accumulation, biosynthesis, accessibility via chemical synthesis or recovery from biomass, and biological activities. *Front. Chem.* **2021**, *9*, 664602. [CrossRef] [PubMed]
2. Reungoat, V.; Allais, F.; Ducatel, H.; Ioannou, I. Extraction and Purification Processes of Sinapic Acid Derivatives from Rapeseed and Mustard Seed By-Products. *Sep. Purif. Rev.* **2022**, *51*, 521–544. [CrossRef]
3. Oliva, M.; Guy, A.; Galili, G.; Dor, E.; Schweitzer, R.; Amir, R.; Hacham, Y. Enhanced production of aromatic amino acids in tobacco plants leads to increased phenylpropanoid metabolites and tolerance to stresses. *Front. Plant Sci.* **2021**, *11*, 604349. [CrossRef] [PubMed]
4. Šamec, D.; Karalija, E.; Šola, I.; Vujčić Bok, V.; Salopek-Sondi, B. The role of polyphenols in abiotic stress response: The influence of molecular structure. *Plants* **2021**, *10*, 118. [CrossRef] [PubMed]
5. Yang, J.; Chen, J.; Hao, Y.; Liu, Y. Identification of the DPPH radical scavenging reaction adducts of ferulic acid and sinapic acid and their structure-antioxidant activity relationship. *LWT* **2021**, *146*, 111411. [CrossRef]
6. Olszewska, M.A.; Granica, S.; Kolodziejczyk-Czepas, J.; Magiera, A.; Czerwińska, M.E.; Nowak, P.; Rutkowska, M.; Wasiński, P.; Owczarek, A. Variability of sinapic acid derivatives during germination and their contribution to antioxidant and anti-inflammatory effects of broccoli sprouts on human plasma and human peripheral blood mononuclear cells. *Food Funct.* **2020**, *11*, 7231–7244. [CrossRef]
7. Pandi, A.; Kalappan, V.M. Mechanisms involved in the anticancer effects of sinapic acid. *Bull. Natl. Res. Cent.* **2022**, *46*, 1–9. [CrossRef]
8. Altındağ, F.; Rağbetli, M.Ç.; Özdek, U.; Koyun, N.; Alhalboosi, J.K.I.; Elasan, S. Combined treatment of sinapic acid and ellagic acid attenuates hyperglycemia in streptozotocin-induced diabetic rats. *Food Chem. Toxicol.* **2021**, *156*, 112443. [CrossRef]
9. Aldubayan, M.A.; Ahmed, A.S.; Emara, A.M.; Ahmed, A.A.; Elgharabawy, R.M. Sinapic acid attenuates cardiovascular disorders in rats by modulating reactive oxygen species and angiotensin receptor expression. *Oxidative Med. Cell. Longev.* **2020**, *2020*, 1436858. [CrossRef]
10. Lee, S.-B.; Yang, H.O. Sinapic acid ameliorates REV-ERB $\alpha$ modulated mitochondrial fission against MPTP-induced Parkinson's disease model. *Biomol. Ther.* **2022**, *30*, 409–417. [CrossRef]
11. Koriem, K.M.; Gad, I.B. Sinapic acid restores blood parameters, serum antioxidants, and liver and kidney functions in obesity. *J. Diabetes Metab. Disord.* **2022**, *21*, 293–303. [CrossRef] [PubMed]
12. Bento-Silva, A.; Koistinen, V.M.; Mena, P.; Bronze, M.R.; Hanhineva, K.; Sahlstrøm, S.; Kitrytė, V.; Moco, S.; Aura, A.-M. Factors affecting intake, metabolism and health benefits of phenolic acids: Do we understand individual variability? *Eur. J. Nutr.* **2020**, *59*, 1275–1293. [CrossRef]

13. Ahad, A.; Jardan, Y.A.B.; Raish, M.; Al-Mohizea, A.M.; Al-Jenoobi, F.I. Ternary Inclusion Complex of Sinapic Acid with Hydroxypropyl-β-cyclodextrin and Hydrophilic Polymer Prepared by Microwave Technology. *Processes* **2022**, *10*, 2637. [CrossRef]
14. Nair, A.B.; Singh, B.; Shah, J.; Jacob, S.; Aldhubiab, B.; Sreeharsha, N.; Morsy, M.A.; Venugopala, K.N.; Attimarad, M.; Shinu, P. Formulation and evaluation of self-nanoemulsifying drug delivery system derived tablet containing sertraline. *Pharmaceutics* **2022**, *14*, 336. [CrossRef] [PubMed]
15. Salem, H.F.; Kharshoum, R.M.; Halawa, A.K.A.; Naguib, D.M. Preparation and optimization of tablets containing a self-nano-emulsifying drug delivery system loaded with rosuvastatin. *J. Liposome Res.* **2018**, *28*, 149–160. [CrossRef]
16. Bansal, T.; Mustafa, G.; Khan, Z.I.; Ahmad, F.J.; Khar, R.K.; Talegaonkar, S. Solid self-nanoemulsifying delivery systems as a platform technology for formulation of poorly soluble drugs. *Crit. Rev. Ther. Drug Carr. Syst.* **2008**, *25*, 63–116. [CrossRef] [PubMed]
17. Negi, J.S. Nanolipid materials for drug delivery systems: A comprehensive Review. In *Characterization and Biology of Nanomaterials for Drug Delivery*; Elsevier: Amsterdam, The Netherlands, 2019; pp. 137–163.
18. Kim, H.-J.; Yoon, K.A.; Hahn, M.; Park, E.-S.; Chi, S.-C. Preparation and in vitro evaluation of self-microemulsifying drug delivery systems containing idebenone. *Drug Dev. Ind. Pharm.* **2000**, *26*, 523–529. [CrossRef]
19. Balakrishnan, P.; Lee, B.-J.; Oh, D.H.; Kim, J.O.; Lee, Y.-I.; Kim, D.-D.; Jee, J.-P.; Lee, Y.-B.; Woo, J.S.; Yong, C.S. Enhanced oral bioavailability of Coenzyme Q10 by self-emulsifying drug delivery systems. *Int. J. Pharm.* **2009**, *374*, 66–72. [CrossRef]
20. Palamakula, A.; Khan, M.A. Evaluation of cytotoxicity of oils used in coenzyme Q10 Self-emulsifying Drug Delivery Systems (SEDDS). *Int. J. Pharm.* **2004**, *273*, 63–73. [CrossRef]
21. Weerapol, Y.; Limmatvapirat, S.; Nunthanid, J.; Sriamornsak, P. Self-nanoemulsifying drug delivery system of nifedipine: Impact of hydrophilic–lipophilic balance and molecular structure of mixed surfactants. *AAPS Pharmscitech* **2014**, *15*, 456–464. [CrossRef]
22. Smail, S.S.; Ghareeb, M.M.; Omer, H.K.; Al-Kinani, A.A.; Alany, R.G. Studies on surfactants, cosurfactants, and oils for prospective use in formulation of ketorolac tromethamine ophthalmic nanoemulsions. *Pharmaceutics* **2021**, *13*, 467. [CrossRef] [PubMed]
23. Pavoni, L.; Perinelli, D.R.; Bonacucina, G.; Cespi, M.; Palmieri, G.F. An overview of micro-and nanoemulsions as vehicles for essential oils: Formulation, preparation and stability. *Nanomaterials* **2020**, *10*, 135. [CrossRef] [PubMed]
24. Buya, A.B.; Beloqui, A.; Memvanga, P.B.; Préat, V. Self-nano-emulsifying drug-delivery systems: From the development to the current applications and challenges in oral drug delivery. *Pharmaceutics* **2020**, *12*, 1194. [CrossRef] [PubMed]
25. Bandivadekar, M.; Pancholi, S.; Kaul-Ghanekar, R.; Choudhari, A.; Koppikar, S. Single non-ionic surfactant based self-nanoemulsifying drug delivery systems: Formulation, characterization, cytotoxicity and permeability enhancement study. *Drug Dev. Ind. Pharm.* **2013**, *39*, 696–703. [CrossRef]
26. Pires, P.C.; Paiva-Santos, A.C.; Veiga, F. Antipsychotics-Loaded Nanometric Emulsions for Brain Delivery. *Pharmaceutics* **2022**, *14*, 2174. [CrossRef]
27. Yousef, S.A.; Mohammed, Y.H.; Namjoshi, S.; Grice, J.E.; Benson, H.A.; Sakran, W.; Roberts, M.S. Mechanistic evaluation of enhanced curcumin delivery through human skin in vitro from optimised nanoemulsion formulations fabricated with different penetration enhancers. *Pharmaceutics* **2019**, *11*, 639. [CrossRef]
28. Shakeel, F.; Haq, N.; Alam, P.; Jouyban, A.; Ghoneim, M.M.; Alshehri, S.; Martinez, F. Solubility of sinapic acid in some (ethylene glycol+ water) mixtures: Measurement, computational modeling, thermodynamics, and preferential solvation. *J. Mol. Liq.* **2022**, *348*, 118057. [CrossRef]
29. Patel, P.; Solanki, S.; Mahajan, A.; Mehta, F.; Shah, K. Preparation and characterization of novel self nano emulsifying drug delivery system of allopurinol. *Res. J. Pharm. Technol.* **2021**, *14*, 2108–2114. [CrossRef]
30. Narayanan, H.; Luna, M.F.; von Stosch, M.; Cruz Bournazou, M.N.; Polotti, G.; Morbidelli, M.; Butté, A.; Sokolov, M. Bioprocessing in the digital age: The role of process models. *Biotechnol. J.* **2020**, *15*, 1900172. [CrossRef]
31. Gardouh, A.R.; El-Din, A.S.S.; Salem, M.S.; Moustafa, Y.; Gad, S. Starch Nanoparticles for Enhancement of Oral Bioavailability of a Newly Synthesized Thienopyrimidine Derivative with Anti-Proliferative Activity against Pancreatic Cancer. *Drug Des. Devel. Ther.* **2021**, *15*, 3071. [CrossRef]
32. Agarry, I.E.; Wang, Z.; Cai, T.; Kan, J.; Chen, K. Chlorophyll encapsulation by complex coacervation and vibration nozzle technology: Characterization and stability study. *Innov. Food Sci. Emerg. Technol.* **2022**, *78*, 103017. [CrossRef]
33. Feng, Z.; Li, L.; Zhang, Y.; Li, X.; Liu, C.; Jiang, B.; Xu, J.; Sun, Z. Formation of whey protein isolate nanofibrils by endoproteinase GluC and their emulsifying properties. *Food Hydrocoll.* **2019**, *94*, 71–79. [CrossRef]
34. Gardouh, A.R.; El-Din, A.S.; Mostafa, Y.; Gad, S. Starch Nanoparticles Preparation and Characterization by in situ combination of Sono-precipitation and Alkali hydrolysis under Ambient Temperature. *Res. J. Pharm. Technol.* **2021**, *14*, 3543–3552. [CrossRef]
35. El-Menshawe, S.F.; Ali, A.A.; Halawa, A.A.; Srag El-Din, A.S. A novel transdermal nanoethosomal gel of betahistine dihydrochloride for weight gain control: In-vitro and in-vivo characterization. *Drug Des. Devel. Ther.* **2017**, *11*, 3377–3388. [CrossRef]
36. Khan, I.; Needham, R.; Yousaf, S.; Houacine, C.; Islam, Y.; Bnyan, R.; Sadozai, S.K.; Elrayess, M.A.; Elhissi, A. Impact of phospholipids, surfactants and cholesterol selection on the performance of transfersomes vesicles using medical nebulizers for pulmonary drug delivery. *J. Drug Deliv. Sci. Technol.* **2021**, *66*, 102822. [CrossRef]
37. Xu, J.; Zhang, S.; Li, H.; Bao, Y.; Du, Y.; Zhou, Y.; Zhao, D.; Liu, F. LncRNA-AK007111 affects airway inflammation in asthma via the regulation of mast cell function. *Int. Immunopharmacol.* **2023**, *121*, 110341. [CrossRef]
38. Khairnar, S.V.; Jain, D.D.; Tambe, S.M.; Chavan, Y.R.; Amin, P.D. Nebulizer systems: A new frontier for therapeutics and targeted delivery. *Ther. Deliv.* **2022**, *13*, 31–49. [CrossRef]

39. Kole, E.; Jadhav, K.; Sirsath, N.; Dudhe, P.; Verma, R.K.; Chatterjee, A.; Naik, J. Nanotherapeutics for pulmonary drug delivery: An emerging approach to overcome respiratory diseases. *J. Drug Deliv. Sci. Technol.* **2023**, *81*, 104261. [CrossRef]
40. Abd El-Aleam, R.H.; Sayed, A.M.; Taha, M.N.; George, R.F.; Georgey, H.H.; Abdel-Rahman, H.M. Design and synthesis of novel benzimidazole derivatives as potential Pseudomonas aeruginosa anti-biofilm agents inhibiting LasR: Evidence from comprehensive molecular dynamics simulation and in vitro investigation. *Eur. J. Med. Chem.* **2022**, *241*, 114629. [CrossRef]
41. Parveen, N.; Sheikh, A.; Abourehab, M.A.; Karwasra, R.; Singh, S.; Kesharwani, P. Self-nanoemulsifying drug delivery system for pancreatic cancer. *Eur. Polym. J.* **2023**, *190*, 111993. [CrossRef]
42. Ameta, R.K.; Soni, K.; Bhattarai, A. Recent Advances in Improving the Bioavailability of Hydrophobic/Lipophilic Drugs and Their Delivery via Self-Emulsifying Formulations. *Colloids Interfaces* **2023**, *7*, 16. [CrossRef]
43. Faraz, O.; Poustchi, M.; Denyani, E.N.; Movahedi, P.; Kouchi, F.R.; Shahriari, R. Thermodynamic modeling of pharmaceuticals solubility in pure, mixed and supercritical solvents. *J. Mol. Liq.* **2022**, *353*, 118809. [CrossRef]
44. Lo, Y.-L. Relationships between the hydrophilic–lipophilic balance values of pharmaceutical excipients and their multidrug resistance modulating effect in Caco-2 cells and rat intestines. *J. Control. Release* **2003**, *90*, 37–48. [CrossRef] [PubMed]
45. Patel, K.; Patil, A.; Mehta, M.; Gota, V.; Vavia, P. Medium chain triglyceride (MCT) rich, paclitaxel loaded self nanoemulsifying preconcentrate (PSNP): A safe and efficacious alternative to Taxol®. *J. Biomed. Nanotechnol.* **2013**, *9*, 1996–2006. [CrossRef] [PubMed]
46. Chaudhari, S.P.; Dugar, R.P. Application of surfactants in solid dispersion technology for improving solubility of poorly water soluble drugs. *J. Drug Deliv. Sci. Technol.* **2017**, *41*, 68–77. [CrossRef]
47. Strickley, R.G. Solubilizing excipients in oral and injectable formulations. *Pharm. Res.* **2004**, *21*, 201–230. [CrossRef]
48. Sripriya, R.; Raja, K.M.; Santhosh, G.; Chandrasekaran, M.; Noel, M. The effect of structure of oil phase, surfactant and co-surfactant on the physicochemical and electrochemical properties of bicontinuous microemulsion. *J. Colloid Interface Sci.* **2007**, *314*, 712–717. [CrossRef]
49. Wang, Q.; Cui, S.W. Understanding the physical properties of food polysaccharides. In *Food Carbohydrates: Chemistry, Physical Properties, and Applications*; CRC Press: Boca Raton, FL, USA, 2005; pp. 162–214.
50. Galatage, S.T.; Trivedi, R.; Bhagwat, D.A. Oral self-emulsifying nanoemulsion systems for enhancing dissolution, bioavailability and anticancer effects of camptothecin. *J. Drug Deliv. Sci. Technol.* **2022**, *78*, 103929. [CrossRef]
51. Jain, S.; Dongare, K.; Nallamothu, B.; Dora, C.P.; Kushwah, V.; Katiyar, S.S.; Sharma, R. Enhanced stability and oral bioavailability of erlotinib by solid self nano emulsifying drug delivery systems. *Int. J. Pharm.* **2022**, *622*, 121852. [CrossRef]
52. He, C.-X.; He, Z.-G.; Gao, J.-Q. Microemulsions as drug delivery systems to improve the solubility and the bioavailability of poorly water-soluble drugs. *Expert Opin. Drug Deliv.* **2010**, *7*, 445–460. [CrossRef]
53. Gardouh, A.R.; Nasef, A.M.; Mostafa, Y.; Gad, S. Design and evaluation of combined atorvastatin and ezetimibe optimized self-nano emulsifying drug delivery system. *J. Drug Deliv. Sci. Technol.* **2020**, *60*, 102093. [CrossRef]
54. Sulthana, S.; Chary, P.S.; Bhavana, V.; Pardhi, E.; Singh, S.B.; Mehra, N.K. Development and evaluation emulgel for effective management of the imiquimod-induced psoriasis. *Inflammopharmacology* **2023**, *31*, 301–320. [CrossRef] [PubMed]
55. Yeom, D.W.; Song, Y.S.; Kim, S.R.; Lee, S.G.; Kang, M.H.; Lee, S.; Choi, Y.W. Development and optimization of a self-microemulsifying drug delivery system for ator vastatin calcium by using D-optimal mixture design. *Int. J. Nanomed.* **2015**, *10*, 3865.
56. Cornell, J.A. *Experiments with Mixtures: Designs, Models, and the Analysis of Mixture Data*; John Wiley & Sons: Hoboken, NJ, USA, 2011.
57. Faraway, J.J. *Linear Models with R*; CRC Press: Boca Raton, FL, USA, 2014.
58. Parthiban, A.; Sathish, S.; Suthan, R.; Sathish, T.; Rajasimman, M.; Vijayan, V.; Jayaprabakar, J. Modelling and optimization of thermophilic anaerobic digestion using biowaste. *Environ. Res.* **2023**, *220*, 115075. [CrossRef] [PubMed]
59. Hsieh, C.-M.; Yang, T.-L.; Putri, A.D.; Chen, C.-T. Application of Design of Experiments in the Development of Self-Microemulsifying Drug Delivery Systems. *Pharmaceuticals* **2023**, *16*, 283. [CrossRef] [PubMed]
60. Le Dévédec, F.; Strandman, S.; Hildgen, P.; Leclair, G.g.; Zhu, X. PEGylated bile acids for use in drug delivery systems: Enhanced solubility and bioavailability of itraconazole. *Mol. Pharm.* **2013**, *10*, 3057–3066. [CrossRef] [PubMed]
61. Chiou, W.L.; Riegelman, S. Pharmaceutical applications of solid dispersion systems. *J. Pharm. Sci.* **1971**, *60*, 1281–1302. [CrossRef]
62. Urbanetz, N.A. Stabilization of solid dispersions of nimodipine and polyethylene glycol 2000. *Eur. J. Pharm. Sci.* **2006**, *28*, 67–76.
63. Yan, Y.-D.; Kim, J.A.; Kwak, M.K.; Yoo, B.K.; Yong, C.S.; Choi, H.-G. Enhanced oral bioavailability of curcumin via a solid lipid-based self-emulsifying drug delivery system using a spray-drying technique. *Biol. Pharm. Bull.* **2011**, *34*, 1179–1186. [CrossRef]
64. Won, D.-H.; Kim, M.-S.; Lee, S.; Park, J.-S.; Hwang, S.-J. Improved physicochemical characteristics of felodipine solid dispersion particles by supercritical anti-solvent precipitation process. *Int. J. Pharm.* **2005**, *301*, 199–208. [CrossRef]
65. Choradiya, B.R.; Patil, S.B. A comprehensive review on nanoemulsion as an ophthalmic drug delivery system. *J. Mol. Liq.* **2021**, *339*, 116751. [CrossRef]
66. Omari-Siaw, E.; Zhu, Y.; Wang, H.; Peng, W.; Firempong, C.K.; Wang, Y.W.; Cao, X.; Deng, W.; Yu, J.; Xu, X. Hypolipidemic potential of perillaldehyde-loaded self-nanoemulsifying delivery system in high-fat diet induced hyperlipidemic mice: Formulation, in vitro and in vivo evaluation. *Eur. J. Pharm. Sci.* **2016**, *85*, 112–122. [CrossRef] [PubMed]

67. Mehanna, M.M.; Mneimneh, A.T.; Abed El Jalil, K. Levofloxacin-loaded naturally occurring monoterpene-based nanoemulgel: A feasible efficient system to circumvent MRSA ocular infections. *Drug Dev. Ind. Pharm.* **2020**, *46*, 1787–1799. [CrossRef] [PubMed]
68. Pandi, A.; Kalappan, V.M. Pharmacological and therapeutic applications of Sinapic acid—An updated review. *Mol. Biol. Rep.* **2021**, *48*, 3733–3745. [CrossRef]
69. Sharma, A.N.; Upadhyay, P.K.; Dewangan, H.K. Development, evaluation, pharmacokinetic and biodistribution estimation of resveratrol-loaded solid lipid nanoparticles for prostate cancer targeting. *J. Microencapsul.* **2022**, *39*, 563–574. [CrossRef] [PubMed]
70. Qi, S.; Guan, H.; Wang, Y.; Fang, Q.; Cheng, X.; Liu, P.; Wei, H.; Liu, W.; Wang, C. Simultaneous determination of cordycepin and its metabolite 3′-deoxyinosine in rat whole blood by ultra-high-performance liquid chromatography coupled with Q Exactive hybrid quadrupole orbitrap high-resolution accurate mass spectrometry and its application to accurate pharmacokinetic studies. *J. Sep. Sci.* **2023**, *46*, 2200602.
71. Su, H.-X.; Yao, S.; Zhao, W.-F.; Li, M.-J.; Liu, J.; Shang, W.-J.; Xie, H.; Ke, C.-Q.; Hu, H.-C.; Gao, M.-N. Anti-SARS-CoV-2 activities in vitro of Shuanghuanglian preparations and bioactive ingredients. *Acta Pharmacol. Sin.* **2020**, *41*, 1167–1177. [CrossRef]
72. Chow, E.; Rendleman, C.A.; Bowers, K.J.; Dror, R.O.; Hughes, D.H.; Gullingsrud, J.; Shaw, D.E. *Desmond Performance on a Cluster of Multicore Processors*; DE Shaw Research Technical Report DESRES/TR (2008) 2008-01; DE Shaw Research: New York, NY, USA, 2008.
73. Bergdorf, M.; Baxter, S.; Rendleman, C.A.; Shaw, D.E. *Desmond/GPU Performance as of October 2015*; DE Shaw Research: New York, NY, USA, 2015.
74. Wang, L.; Wu, Y.; Deng, Y.; Kim, B.; Pierce, L.; Krilov, G.; Lupyan, D.; Robinson, S.; Dahlgren, M.K.; Greenwood, J.; et al. Accurate and reliable prediction of relative ligand binding potency in prospective drug discovery by way of a modern free-energy calculation protocol and force field. *J. Am. Chem. Soc.* **2015**, *137*, 2695–2703. [CrossRef]
75. Phillips, J.C.; Braun, R.; Wang, W.; Gumbart, J.; Tajkhorshid, E.; Villa, E.; Chipot, C.; Skeel, R.D.; Kalé, L.; Schulten, K. Scalable molecular dynamics with NAMD. *J. Comput. Chem.* **2014**, *26*, 1781–1802. [CrossRef]
76. Kim, S.; Oshima, H.; Zhang, H.; Kern, N.R.; Re, S.; Lee, J.; Roux, B.; Sugita, Y.; Jiang, W.; Im, W. CHARMM-GUI free energy calculator for absolute and relative ligand solvation and binding free energy simulations. *J. Chem. Theory Comput.* **2020**, *16*, 7207–7218. [CrossRef]
77. Ngo, S.T.; Tam, N.M.; Pham, M.Q.; Nguyen, T.H. Benchmark of popular free energy approaches revealing the inhibitors binding to SARS-CoV-2 Mpro. *J. Chem. Inf. Model* **2021**, *61*, 2302–2312. [CrossRef] [PubMed]
78. Pang, Y.T.; Miao, Y.; Wang, Y.; McCammon, J.A. Gaussian accelerated molecular dynamics in NAMD. *J. Chem. Theory Comput.* **2017**, *13*, 9–19. [CrossRef] [PubMed]
79. Mosmann, T. Rapid colorimetric assay for cellular growth and survival: Application to proliferation and cytotoxicity assays. *J. Immunol. Methods* **1983**, *65*, 55–63. [CrossRef]
80. Boom, R.C.J.A.; Sol, C.J.; Salimans, M.M.; Jansen, C.L.; Wertheim-van Dillen, P.M.; Van der Noordaa, J.P.M.E. Rapid and simple method for purification of nucleic acids. *J. Clin. Microbiol.* **1990**, *28*, 495–503. [CrossRef] [PubMed]
81. Drosten, C.; Günther, S.; Preiser, W.; van der Werf, S.; Brodt, H.R.; Becker, S.; Rabenau, H.; Panning, M.; Kolesnikova, L.; Fouchier, R.A.; et al. Identification of a novel coronavirus in patients with severe acute respiratory syndrome. *N. Engl. J. Med.* **2003**, *348*, 1967–1976. [CrossRef] [PubMed]
82. Asper, M.; Sternsdorf, T.; Hass, M.; Drosten, C.; Rhode, A.; Schmitz, H.; Günther, S. Inhibition of different Lassa virus strains by alpha and gamma interferons and comparison with a less pathogenic arenavirus. *J. Virol.* **2004**, *78*, 3162–3169. [CrossRef] [PubMed]

*pharmaceutics*

MDPI

*Article*

# Bio-Fabricated Silver Nanoparticles from the Leaf Extract of the Poisonous Plant, *Holigarna arnottiana*: Assessment of Antimicrobial, Antimitotic, Anticancer, and Radical-Scavenging Properties

Anthyalam Parambil Ajaykumar [1,*], Ovungal Sabira [1], Valiyaparambil Sivadasan Binitha [2], Sudhir Rama Varma [3], Anjaly Mathew [4], Kodangattil Narayanan Jayaraj [5,*], Pandikkadan Ayyappan Janish [1], Koladath Vasu Zeena [1], Padannappurath Sheena [1], Veena Venugopal [1], Priyanka Palakkapparambil [1] and Aswathi [1]

1 Division of Biomaterial Sciences, Department of Zoology, Sree Neelakanta Government Sanskrit College, Pattambi, Palakkad 679303, Kerala, India; sabiraovungal@gmail.com (O.S.); janishpa@sngscollege.org (P.A.J.); zeenasalim@gmail.com (K.V.Z.); sheenapnarayan@gmail.com (P.S.); veenarun86@gmail.com (V.V.); priyapkd1995@gmail.com (P.P.); achuthy7@gmail.com (A.)

2 Department of Zoology, Sree Narayana College, Nattika, Thrissur 680566, Kerala, India; binisncn@gmail.com

3 Clinical Sciences Department, Centre for Medical and Bio-Allied Health Sciences Research, Ajman University, Ajman P.O. Box 346, United Arab Emirates; s.varma@ajman.ac.ae

4 Department of Chemistry, Sree Neelakanta Government Sanskrit College, Pattambi, Palakkad 679303, Kerala, India; anjalymathew4@gmail.com

5 Basic Sciences Department, Centre for Medical and Bio-Allied Health Sciences Research, Ajman University, Ajman P.O. Box 346, United Arab Emirates

* Correspondence: ajaykumar@sngscollege.org (A.P.A.); j.narayanan@ajman.ac.ae (K.N.J.)

**Abstract:** This study presents a novel approach to fabricate silver nanoparticles (AgNPs) using the poisonous plant, *Holigarna arnottiana* leaf extract. The formation of AgNPs was confirmed by a color change from green to dark brown and validated by UV analysis. FTIR analysis identified functional groups on the AgNPs, while Zeta potential analysis assessed their stability. TEM analysis established an average diameter of 18 nm and a spherical morphology for the nanoparticles. LC MS analysis coupled with database searches revealed the presence of diverse bioactive compounds, including flavonoids, nucleotides, dipeptides, enzymes, and glycosides. These compounds are postulated to act as reducing agents in the leaf extract-mediated synthesis process. Moreover, the bio-fabricated AgNPs exhibited noteworthy anticancer properties against DLA cells. In addition, AgNPs displayed substantial antimitotic effects in an assay involving *Allium cepa* root cells. These findings underscore the potential of the AgNPs as cytotoxic agents. The biosynthesized AgNPs showed antimicrobial activity against various bacterial pathogens, including *Escherichia coli*, *Klebsiella pneumoniae*, and *Staphylococcus aureus*. Furthermore, the AgNPs exhibited outstanding radical-scavenging properties in the DPPH assay, suggesting their potential application in antioxidant therapies. The study collectively highlights the successful synthesis of AgNPs through a green, biocompatible approach, and demonstrates their promising potential for anticancer, antimitotic, and radical-scavenging applications.

**Keywords:** *Holigarna arnottiana*; silver nanoparticles; radical scavenging; anticancer activity; LC-MS analysis

## 1. Introduction

The nanotechnology field has recently experienced significant advancements in the synthesis of metal nanoparticles because of the broad range of applications for nanoparticles, from medicine to environmental remediation. Among these nanoparticles, silver nanoparticles (AgNPs) have attracted the most interest due to their unique physicochemical properties that make them suitable for biological, catalytic, and antibacterial applications. However, the environment is often compromised by the use of hazardous chemicals and energy-intensive processes in current chemical and other AgNP synthesis methods. To solve these issues and advance environmentally friendly nanoparticle synthesis, researchers have increasingly concentrated on green synthesis methods that utilize natural resources like plant extracts to produce AgNPs [1–4].

The endemic tree *Holigarna arnottiana* (Family: Anacardiaceae) is a large tree, that can reach a height of up to 35 m. Its bark exhibits fine fissures and pustules, while its branches are robust, circular, and devoid of hair. Its leaves are simple, alternate, arranged spirally and clustered at twig ends. The tree is endemic to The Western Ghats [5]. It is used as a medicinal plant and as a folk medicine by village people. The plant is used to treat a wide range of discomforts like fever, cough, skin diseases, rheumatism, snakebite, jaundice, dysentery, etc. [6]. *Holigarna* has a scanty exudate of colorless sap which becomes black on drying. The sap of the tree causes various severe anaphylactic skin responses [7]. Mane and co-workers made an investigation, aiming to obtain chromosomal information in *Holigarna* without previous cytological information and find out the karyotype formula as 21 m + 8 sm [8]. sm [8]. GC-MS analysis of leaf and bark extracts uncovered the presence of around 20 distinct compounds known for their beneficial biological properties. Additionally, phytochemical investigations confirmed the existence of compounds like coumarins, saponins, tannins, reducing sugars, and others. Notably, the extract was also found to contain urushiol, a substance with the potential to induce allergic reactions [9]. *Holigarna* shows strong antibacterial activity towards several Gram positive and Gram negative pathogenic bacteria, and against human and shrimp pathogenic bacteria, and a GC-MS analysis was used to elucidate its antimicrobial principles. Based on the findings, it could be inferred that *H. arnottiana* could be used as a reliable source for developing bio-therapeutics for shrimp and humans in the future [10]. Contradictory to their strong antimicrobial activity, fungal isolates were resistant to the plant-derived extracts of *H. arnottiana* and showed specific activity against bacteria only. The organic extracts had more antibacterial activity as compared to the water extracts [11]. Recent studies unveiled their antioxidant activity using the superoxide radical scavenging method and the Hydroxy radical assay. Strong cytotoxic activity was detected using the MTT assay and DLA assay [12,13].

The plant kingdom provides a rich source of reducing and capping agents for the environmentally friendly synthesis of nanoparticles due to its vast variety of secondary metabolites. The flavonoids, terpenoids, and phenolic chemicals found in this plant are recognized for their wide range of bioactive substances, which have a variety of medicinal qualities. In the current study, the potential bioactivity of silver nanoparticles produced from *H. arnottiana* leaf extract is explored. The synthesis process makes use of the inherent reduction power of the leaf extract of *H. arnottiana* and is thus an appealing replacement for conventional methods because it is eco-friendly, free of dangerous chemicals, and energy-efficient. The natural capping agents included in the plant extract are expected to improve the stability and biocompatibility of the resulting silver nanoparticles, which may boost their potential use in biomedicine.

The primary objective of this research is to assess the synthesized silver nanoparticles by characterizing them and examining their potential in various aspects, including anticancer, antimicrobial, cytological aberration, and scavenging properties. In the realm of oncology, there is an increasing necessity for novel anticancer agents that offer enhanced effectiveness while minimizing side effects. Moreover, in light of the escalating challenge of antibiotic resistance, there is a growing demand for alternative antimicrobial agents, and silver nanoparticles have displayed promise in addressing this concern. Additionally, the antioxidant and free-radical-scavenging properties of nanoparticles hold the potential to combat disorders associated with oxidative stress, thus creating new avenues for their utilization in antioxidant therapy.

## 2. Materials and Methods

### 2.1. Preparation of Holigarna arnottiana Leaf Extract

Fresh leaves of *H. arnottiana* (Figure 1) were locally gathered from the Malappuram district (located at 75°58′25.7″ E) in the Indian state of Kerala. An expert from the Department of Botany at the University of Calicut, India, validated their identification to ensure accuracy. The leaves were subjected to thorough washing with deionized water through repeated cycles. Subsequently, 10 g of the sliced leaves was placed into a 250 mL beaker, and 200 mL of distilled water was added. Subsequently, the beaker underwent microwave irradiation for approximately 2 min, utilizing microwave radiation at a frequency of 2.45 GHz and a microwave power of 350 W. The pale-green filtrate was acquired by centrifuging the mixture at 6000× *g* for 2 min. Afterward, it was placed in a refrigerator at a controlled temperature of 5 °C to ensure its preservation for further investigation.

**Figure 1.** *H. arnottiana*.

#### LC MS Analysis of Leaf Extract of *H. arnottiana*

Metabolite identification from the methanol extract of *H. arnottiana* leaf was assessed using an LC-ESI-Q-TOF-MS system (Agilent Technologies 6550 i-Funnel, Santa Clara, CA, USA) with a G4220B pump, G4226A autosampler, and G1316C, along with a polydiode array detector (PAD). The elution solution was composed of a gradient mixture of 0.1% formic acid in water (A) and acetonitrile (B), operating at a flow rate of 0.30 mL/min. The gradient began at 95% A and 5% B, transitioning to 5% A and 95% B over 25 min, before returning to the initial composition (95% A: 5% B) within 10 min, maintained for 5 min. The mass spectrometry (MS) analysis was conducted using ESI in the positive ionization mode. The MS source settings were as follows: capillary voltage of 3500 V, gas temperature set at 250 °C, drying gas flow rate of 13 L/min, sheath gas temperature of 300 °C, sheath gas flow at 11 units, nebulizing gas pressure at 35 psig, fragment set at 175 V, skimmer at 65 V, octopole RF peak at 750 V, and a mass range for detection spanning m/z 50 to 1000. The Mass Hunter software, Version: Q-TOF B.05.01 (B5125.3), developed by Agilent Technologies, was employed to analyze the mass spectra and chromatograms.

### 2.2. Synthesis of Silver Nanoparticles Using H. arnottiana Leaf Extracts

Silver nitrate ($AgNO_3$) was purchased from Merck Chemicals (99.8%) and used for the biosynthesis of silver nanoparticles. Silver nanoparticles (AgNPs) were synthesized through a method involving the heating of a solution comprising $AgNO_3$ and *H. arnottiana* leaf extract in a common household microwave oven (LG-MS-2029 UW). The reaction mixture was prepared by mixing 10 mL of the leaf extract combined with 50 mL of a 0.001 M silver nitrate solution. Subsequently, the reaction mixture was put into a domestic microwave oven operating at 2.45 GHz and 350 W for approximately 10 min. An alteration in color, progressing from pale green to pale yellow and eventually to deep brown, was observed, indicating the successful formation of AgNPs. The AgNPs synthesized and mediated by *H. arnottiana* were isolated by being subjected to centrifugation at 12,000 rpm for 10 min. This procedure was repeated two times to eliminate any residual-free silver. Subsequently, the resultant green-synthesized AgNPs were dried and stored at a temperature of 4 °C until they were needed for future analysis.

### 2.3. Characterization of AgNPs

The initial confirmation of AgNP formation was assessed through UV–vis spectroscopy. The identification of functional groups linked to the *H. arnottiana*-mediated AgNPs was accomplished via FTIR (PerkinElmer, Waltham, MA, USA) analysis. The stability assessment of the AgNPs was conducted using zeta potential measurements. The dimensions and morphology of the AgNPs were determined through TEM (JEOLJEM-2100 microscope, Montgomery, AL, USA) imaging. Collectively, these analytical techniques contributed to an in-depth characterization of the synthesized AgNPs, offering insights into their optical, structural, surface chemical, and colloidal attributes.

### 2.4. Cyclic Voltammetry Analysis

The electrochemical behavior of *H. arnottiana* leaf extract was investigated through cyclic voltammetry (CV). The CV measurement was conducted using a digital Ivy Potentiostat (Model No. DY2000EN) equipped with DY 2000 software. A three-electrode electrochemical arrangement was employed, consisting of a glassy carbon electrode (GCE) as the working electrode, a platinum wire serving as the counter electrode, and an Ag/AgCl reference electrode for monitoring CV signals. The CV measurements were carried out in *H. arnottiana* leaf extract containing a solution of 0.1 M phosphate buffer (pH = 7.4) and 0.1 M KCl as a supporting electrolyte.

### 2.5. Antimicrobial Effects of AgNPs on Bacterial Pathogens

The antimicrobial properties of green synthesized AgNPs were tested against four different bacterial pathogens, such as *Escherichia coli*, *Klebsiella pneumoniae*, *Staphylococcus aureus*, and using the agar well diffusion method [14–16]. The four bacteria strains were applied on Mueller–Hinton agar using a sterilized cotton swab (MHA). A sterile blank antimicrobial susceptibility disc was used in the test. The discs were loaded with 10 µL of AgNPs as an experiment (represented as A); leaf extracts were utilized as a control (represented as C), and the antibiotic was tetracycline employed as a positive control (represented as B). The discs were then placed on the agar plate and incubated at 37 °C for 24 h. The inhibitory zone was examined after 24 h of incubation.

### 2.6. Antimitotic Assay

Various chromosomal abnormalities were studied in healthy young *Allium* bulbs. They were grown in synthesized AgNPs used for this purpose. The control was grown in distilled water. Using the squashing method, the squash preparation was performed, and using the camera-attached microscope, the chromosome images were taken at 40× magnification.

### 2.7. Cytotoxicity of AgNPs in Cancer Cell Lines

The evaluation of the cytotoxicity of biofabricated AgNPs involved an investigation into its impact on Dalton's lymphoma ascites cells (DLA cells). The analysis was carried out in Amala Cancer Research Institute, Thrissur, Kerala, India. The primary objective of this study was to establish cell viability, which was accomplished through the utilization of the trypan blue exclusion technique. Tubes containing varying concentrations of biosynthesized AgNPs (12.5 μg/mL, 25 μg/mL, 50 μg/mL, 100 μg/mL, and 150 μg/mL and 200 μg/mL) were meticulously prepared. Each tube was combined with a viable cell solution comprising $1 \times 10^6$ cells in 0.1 mL. In order to achieve a total volume of 1 mL, phosphate buffer solution was added to each tube, except for the control tube which solely contained the cell suspension. Following this, the resulting mixture was subjected to an incubation period exceeding three hours at a temperature of 37 °C.

Afterwards, the cell solution was merged with 0.1 mL of a 1% trypan blue solution. The mixture was allowed to settle for 2–3 min, after which it was transferred to a hemocytometer. Notably, the trypan blue imparted a blue hue to the deceased cells, while the viable cells remained devoid of staining. Subsequent to this, the quantities of stained and unstained cells were directly counted. This process facilitated the determination of both the total number of stained cells and the number of unstained, viable cells [17]. The percentage of cytotoxicity was calculated by using the following formula. Percentage of cytotoxicity = number of dead cells/number of live cells + number of dead cells × 100.

### 2.8. Antioxidant Activity

The DPPH free radical scavenging assay was used to assess the antioxidant properties of various substances. A brand-new DPPH solution was made in methanol, and the precise beginning concentration was determined spectrophotometrically from a calibration curve. At 517 nm, the absorbance reduction was seen. The following equation was used to compute the DPPH radical-scavenging capability. S% = [(A control − A sample)/A control] × 100.

The linear regression equation was used to determine the sample concentration needed to induce 50% scavenging activity (IC 50 value) from the graph.

## 3. Result and Discussion

### 3.1. Characterization of Metabolites in the Leaf Extract by LCMS Analysis

The LC MS analyses of the secondary metabolites present in the leaf extract of *H. arnottiana* are presented in the Figure 2. The results unveiled a diverse range of compounds in the analysis. These included fatty acids such as Propanoic acid and 2-Hydroxypelargonic acid, flavonoids like Quercetin, Quercetin 3-beta-L-arabinopyranoside, Lonchocarpene and Robinetin 7-glucoside, phenolic compounds such as Deshydroxypyrogallin-4-carboxylic acid, as well as peptides such as hydroxyphenylalanyl-aspartic acid. Furthermore, the leaf extract consist of an enzyme like 2-Hydroxymethylserine, specific amino acid derivatives such as 2-Hydroxymethylserine, nucleotide sugar like UDP-2-acetamido-4-amino-2,4,6-trideoxyglucose, and compounds like naphthoquinones (e.g., juglalin) and terpenes. Each of these identified molecules within the leaf extract of *H. arnottiana* possesses distinct chemical properties that could potentially play a role in reducing and stabilizing silver ions during the synthesis of nanoparticles.

### 3.2. UV Analysis

During the biological synthesis of AgNPs, a noticeable shift in solution color from pale green to brown serves as a prominent indicator of AgNP formation (Figure 3. The findings of the UV-Vis analysis are presented in Figure 4. The result showed a prominent peak at 448 nm in the spectrum, which signifies the Surface Plasmon Resonance feature resulting from the oscillation of surface electrons of silver metal within an electromagnetic environment, which indicates that the effective formation of AgNPs through the utilization of leaf extract is validated. Earlier studies with green synthetic silver nano from *Ocimum sanctum* and marigold extract found that the maximum absorbance ranged from 400 to 500 [18,19].

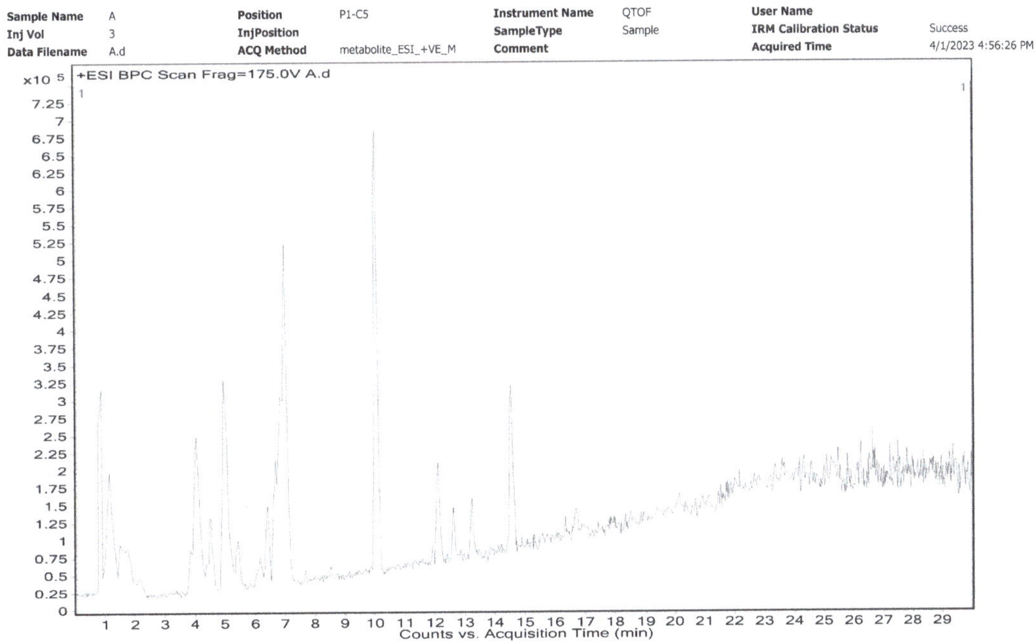


**Figure 2.** LC MS profile of leaf extract of *H. arnottiana*.

**Figure 3.** The successful synthesis of AgNPs was evidenced by a visual transition from a pale green to a pale yellow shade, followed by a deep brown coloration.

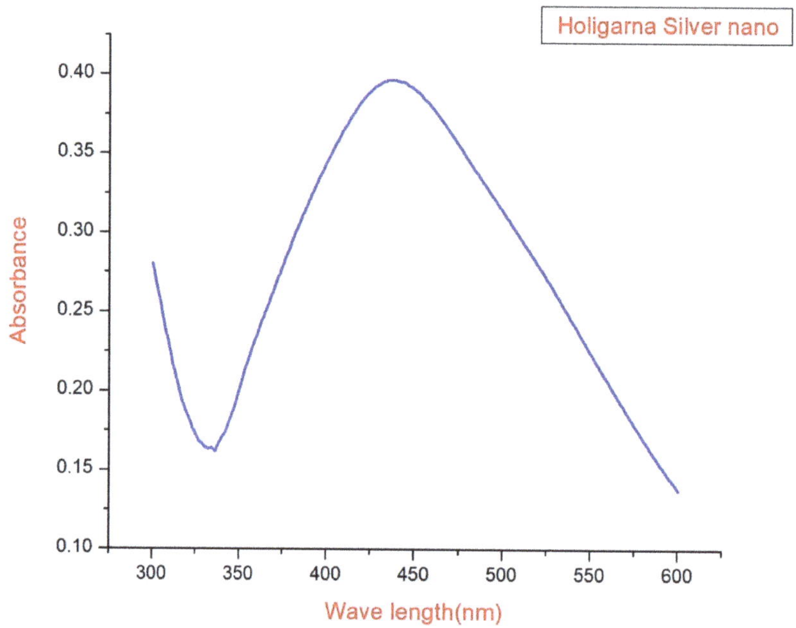


**Figure 4.** UV-Vis spectrum of AgNPs.

### 3.3. FTIR Analysis

Using FTIR spectroscopy, the dual function of leaf extract as a reducing and capping agent was established. The FTIR spectra of the powders of the aqueous extract of *H. arnottiana* and AgNPs were recorded using the KBr pellet method with a Perkin Elmer spectrophotometer in the range of 400–4000 $cm^{-1}$. Figure 5 shows the FTIR spectra of *H. arnottiana* leaf extract and AgNPs synthesized using *H. arnottiana*. The two spectra may be seen to have small differences in intensity, making them practically identical.

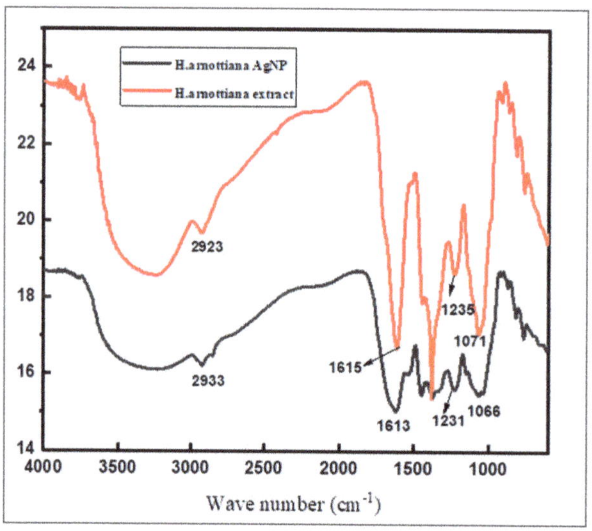


**Figure 5.** FTIR spectra of leaf extract and AgNPs.

A peak at 2933 $cm^{-1}$ of the leaf extract arises from the C–H stretching vibrations. An amide I band appears at 1615 $cm^{-1}$. The peak at 1383 $cm^{-1}$ is assigned to the –COO stretching from amino acid groups. The band at 1071 $cm^{-1}$ and 822 $cm^{-1}$ is assigned to C–C stretching vibration and the aromatic –C–H bending vibrations, respectively. This implies that terpenoids, flavonoids, phenols, and phytosterols found in the *H. arnottiana* leaf extract play a role in the synthesis and stabilization of nanoparticles [20]. These molecules have been identified in the *H. arnottiana* leaf extract through LC MS analysis (as shown in Figure 2 and Table 1). It can be inferred that biomolecules from the leaf extract containing active functional groups are capable of reducing silver ions into metallic AgNPs. Furthermore, there is a proposition that these functional groups may combine with silver, creating a protective capping layer around the AgNPs, preventing their aggregation. Previous research has demonstrated that carbonyl, ester, amino, hydroxyl, and phenolic groups effectively bind with the noble metals through coordination and polar interactions [21]. The small variation in the intensity and position of the peaks in the spectrum of nanoparticles is due to the coordination of phytochemicals with metals [22].

**Table 1.** Different compounds identified from the leaf extracts of *H. arnottiana* via LC MS analysis.

| SL. No | Name of Molecule | Retention Time (RT) | Mass | Chemical Formula | Class of Compound |
|---|---|---|---|---|---|
| 1 | 7-Deshydroxypyrogallin-4-carboxylicacid | 0.941 | 248.0343 | $C_{12}H_8O_6$ | Phenolic compound |
| 2 | 2-Hydroxymethylserine | 1.092 | 135.0525 | $C_4H_9NO_4$ | Derivative of the amino acid serine |
| 3 | ProLeu (prolylleucine) | 1.351 | 228.1447 | $C_{11}H_{20}N_2O_3$ | Dipeptide |
| 4 | 2-Hydroxypelargonic acid | 2.10 | 174.1256 | $C_9H_{18}O_3$ | Fatty acid |
| 5 | Lonchocarpenin | 5.024 | 448.1876 | $C_{27}H_{28}O_6$ | Flavonoid |
| 6 | UDP-2-acetamido-4-amino-2,4,6-trideoxyglucose | 5.396 | 590.1023 | $C_{17}H_{28}N_4O_{15}P_2$ | Nucleotide sugar |
| 7 | Robinetin7-glucoside | 6.408 | 464.0897 | $C_{21}H_{20}O_{12}$ | Flavonoid |
| 8 | Juglalin | 6.488 | 418.0895 | $C_{20}H_{18}O_{10}$ | Naphthoquinone, |
| 9 | Propanoic acid | 6.495 | 248.0676 | $CH_3CH_2COOH$ | Short-chain fatty acid |
| 10 | Quercetin 3-beta-L-arabinopyranoside | 6.706 | 434.079 | $C_{20}H_{18}O_{11}$ | Flavonoid glycoside |
| 11 | Quercetin | 6.779 | 302.0387 | $C_{15}H_{10}O_7$ | Flavonoid |
| 12 | Asn-Trp-OH | 7.014 | 426.1121 | $C_{20}H_{18}N_4O_7$ | Dipeptide |
| 13 | Dihydrophaseicacid4-O-beta-D-glucoside | 13.278 | 444.1992 | $C_{21}H_{32}O_{10}$ | Glycoside |

### 3.4. Zeta Potential Analysis

Zeta potential analysis serves as a pivotal factor that characterizes the surface electrostatic charge of nanoparticles. In this study, the AgNPs that were synthesized displayed a zeta potential value of −8.7 mV (Figure 6). The negative zeta potential value (−8.7 mV) implies a state of repulsion among the AgNPs, potentially bolstering their stability in suspension. This outcome contributes to the effective dispersion and compatibility of AgNPs within physiological surroundings. The findings enrich our comprehension of how AgNPs synthesized with the assistance of *H. arnottiana* behave in biological and environmental settings, and how they can be further applied in the biomedical field. According to a current study, AgNPs are more stable. The capping of the extract's polyphenolic contents may be the cause of the high negative potential value [23,24].

### 3.5. Cyclic Voltammetry Analysis

Cyclic voltammetry (CV) was utilized to explore the electrochemical behavior of the leaf extract derived from *H. arnottiana*. A single anodic peak current appeared in the cyclic voltagram of the plant extract at a potential of −0.318 V (Figure 7). CV analysis confirmed the effective reduction capacity of *H. arnottiana* leaf extract, facilitating the conversion of Ag ions into AgNPs. Silver nanoparticles synthesized via a green synthesis approach displayed no reducing properties, as shown in Supplementary Figure S1. This finding inspires additional investigation to fully realize the potential of natural resources for a range of applications and shows great promise for environmentally friendly and sustainable nanoparticle synthesis processes. Comparable outcomes were achieved via the CV analysis using *Uverianarum* and *Mimosa albida* leaf extract, as reported in a previous study [25,26].

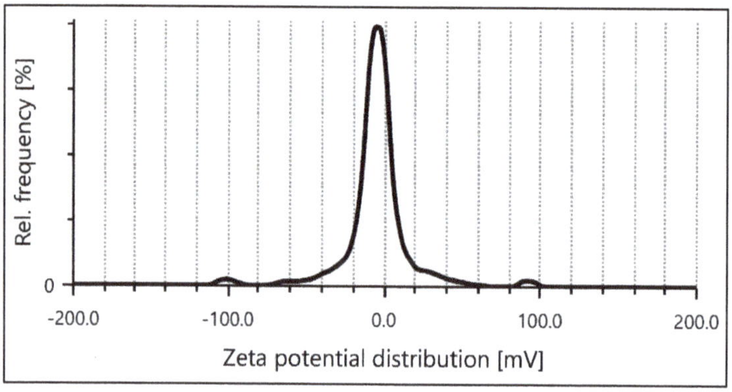


**Figure 6.** Zeta potential distribution of AgNPs.

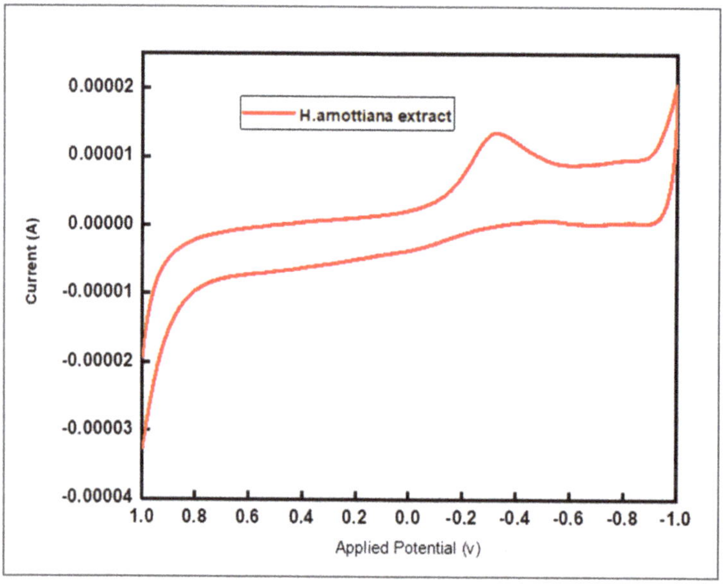


**Figure 7.** Cyclic voltammetry of leaf extract of *H. arnottiana*.

### *3.6. Transmission Electron Microscopy Analysis*

The leaf extract of *H. arnottiana* was harnessed for the synthesis of AgNPs, which were subsequently examined using Transmission Electron Microscopy to gain insights into their shape and size. These variables hold substantial influence over their attributes and potential applications. The investigation disclosed that the AgNPs created through the utilization of *H. arnottiana* leaf extract displayed a spherical shape, with an average diameter measuring 18 nm (Figure 8), consistent with earlier research findings [27]. This choice of size holds critical importance, as nanoparticle dimensions yield a substantial impact on a multitude of traits, encompassing optical characteristics, magnetic responses, and catalytic capabilities. Furthermore, due to their nano-scale dimensions, AgNPs possess the potential to permeate cellular structures, rendering them robust candidates for applications spanning targeted therapeutic interventions, imaging modalities, and drug delivery methods. Ecologically sound plant extracts contain biomolecules that serve as capping and reducing agents to create stable, shape-controlled nanoparticles. Biomolecules including phenolics, terpenoids, and polysaccharides are among the main substances that have been documented to alter the reduction and capping of nanoparticles [28,29].

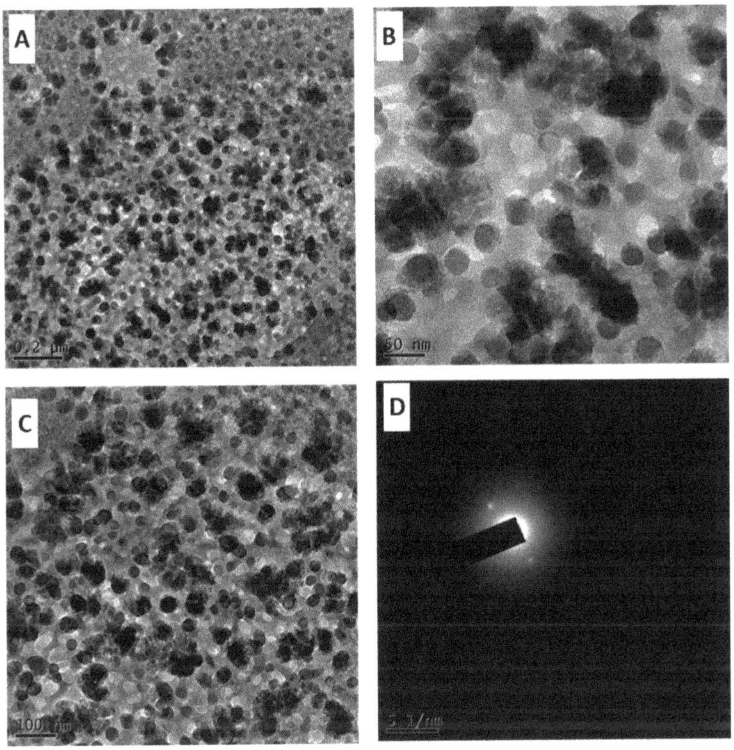


**Figure 8.** Transmission Electron Microscopy (TEM) images of AgNPs. Scale bars in the figures represent 0.2 μm (**A**), 20 nm (**B**), 100 nm (**C**), and 5 1/nm (**D**), respectively.

### 3.7. Antibacterial Assay

The disc diffusion assay was used in the current investigation to look into the antibacterial capabilities. It was evaluated how AgNPs affected the development of both Gram-positive (*Staphylococcus aureus* and *Escherichia coli*) and Gram-negative (*Klebsiella pneumoniae*) bacteria. Control plates loaded with plant extract (5 μL) exhibited a zone of inhibition (ZOI) measuring 7 mm. However, when filter papers infused with AgNP concentrations of 5 μL, 10 μL, and 15 μL were used, the diameter of the inhibition zone showed ranges of 10 mm, 15 mm, and 18 mm, respectively, for *K. pneumoniae* (as depicted in Figure 9). In the case of *S. aureus*, the ZOI for the control (5 μL) was 8 mm, while filter papers containing 5 μL, 10 μL, and 15 μL of AgNPs exhibited inhibition zone diameters of 10 mm, 14 mm, and 16 mm, respectively. As in the case of *E. coli*, the ZOI for the control shows a mortality of 5 mm and 7 mm, 9 mm, and 12 mm for 5 μL, 10 μL, and 15 μL of AgNPs. As the concentration of AgNPs rises, the antibacterial activity of AgNPs shows a dose-dependent response, with a stronger activity against *K. pneumoniae* as compared to *S. aureus* and *E. coli*. It shows that silver nanoparticles have a potent antibacterial impact, as numerous scientists have already found [17,25,30–39].

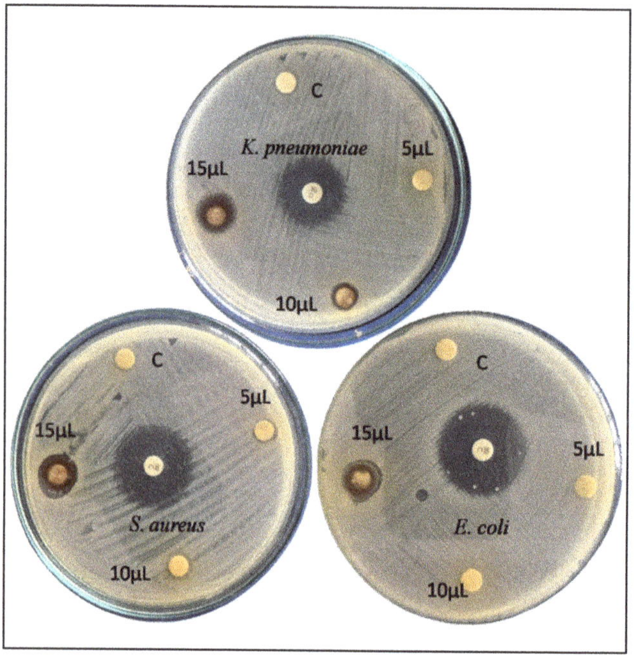


**Figure 9.** Antibacterial activity of AgNPs against *K. pneumoniae*, *S. aureus*, and *E. coli*. Three distinct concentrations (5 μL, 10 μL, and 15 μL) of AgNPs are evaluated, along with a control (C) that represents the leaf extract of *H. arnottiana*. The antibiotic, tetracycline, was applied at the central portion of the disc as a positive control.

The effects of green synthesized AgNPs in the *Allium cepa* root tip cells provide valuable insights into the potential cytological impacts of AgNP exposure [40–43]. The microscopic examination of the root tip cells revealed distinct effects on chromosomal behavior upon exposure to AgNPs (Figure 10). In the figure, the control group displayed regular nuclei, indicating the fundamental condition of chromosome (A) division when AgNPs were absent. Images (B and C) exhibit nuclei disintegration: the presence of AgNPs seemed to cause nuclei disintegration in certain cells (B and C). This occurrence implies that AgNPs might potentially disrupt the structural integrity of nuclei, resulting in irregular cellular states. (D) Typical Metaphase: in the control group, cells depicted a normal metaphase, signifying the correct alignment of chromosomes on the metaphase plate—a pivotal phase in chromosome separation. (E, F, and G) Delayed Metaphase: intriguingly, cells subjected to AgNP exposure presented lagging chromosomes during metaphase (E, F, G, and H). This observation implies a possible delay or hindrance in chromosome alignment, which could lead to the uneven distribution of genetic material during cell division. (I) Standard Anaphase: control cells exhibited regular anaphase, characterized by the accurate separation of sister chromatids towards opposite poles of the cell (J). This demonstrates a degree of similarity to other works [44–46]. A similar cytological analysis employing *H. arnottiana* leaf extract did not reveal significant alterations in the cell division of *Allium cepa* root tips (Supplementary Figure S2).

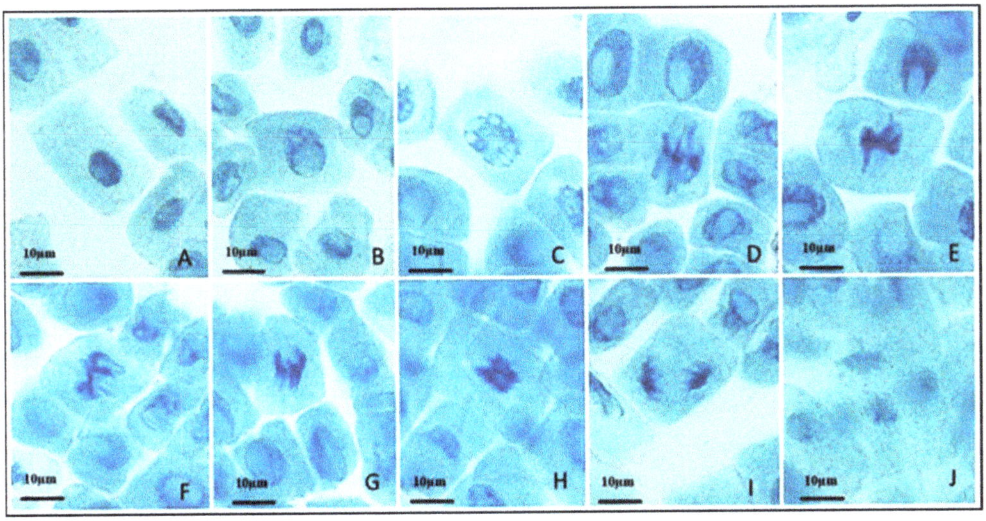


**Figure 10.** Chromosomal effect of green synthesized AgNPs. (**A**) Normal nuclei. (**B**,**C**) Disintegrated nuclei. (**D**) Normal metaphase. (**E–H**) Laggard metaphase. (**I**) Normal Anaphase. (**J**) Laggard anaphase.

### *3.8. Cytotoxicity Effect of AgNPs*

The anticancer activity of green synthesized AgNPs was estimated on DLA cells via a trypan blue assay (Figure 11). At a silver concentration of 12.5 µg, a 17.4% reduction in cell viability was observed, indicating a mild impact. As the AgNP concentration increased to 25 µL, 50 µL, and 100 µL, the cytotoxic effects escalated to 30.3%, 34.8%, and 48.7%, respectively. Notably, at AgNP concentrations of 150 µL and 200 µL, the cytotoxicity reached a significant level with a substantial reduction in cell viability at 57.4% and 100%. This result underscores the potential of AgNPs to induce strong cytotoxic responses at elevated concentrations, aligning with previous studies that have reported the dose-dependent cytotoxicity of nanoparticles [47]. The results of this study underscore the concentration-dependent cytotoxic effects of AgNPs on cell viability. The significant reduction in cell viability at higher AgNPs concentrations aligns with the existing understanding of nanoparticle toxicity trends [48–54]. These findings emphasize the importance of careful consideration and dose regulation when utilizing AgNPs in biomedical applications and highlight their potential for inducing cytotoxic responses at elevated concentrations. In cells exposed to AgNPs, a laggard anaphase was observed. This phenomenon implies that AgNPs might have interfered with the proper separation of chromatids during anaphase, potentially leading to unequal genetic distribution.

### *3.9. Antioxidant Activity*

Free radical scavenging activity was measured using the DPPH method, and various concentrations (20 µL, 40 µL, 60 µL, 80 µL, and 100 µL) of AgNPs were utilized for the analysis. The solution was exposed to UV spectroscopy for examination after an incubation time of 30 min in complete darkness. At 517 nm, the greatest amount of electrons was absorbed by DPPH-free radicals. The results show evidence of concentration-dependent scavenging activity. The sample concentration needed to provide 50% scavenging activity (EC50 value) was found to be 80 µL. The assessment of scavenging activity using *H. arnottiana* leaf extract displayed a notably higher EC50 value when compared to AgNPs synthesized through green methods, as depicted in Supplementary Figure S3. The current study confirms earlier findings by demonstrating a dose-dependent increase in the scavenging activity of AgNPs [25,39,55–59].

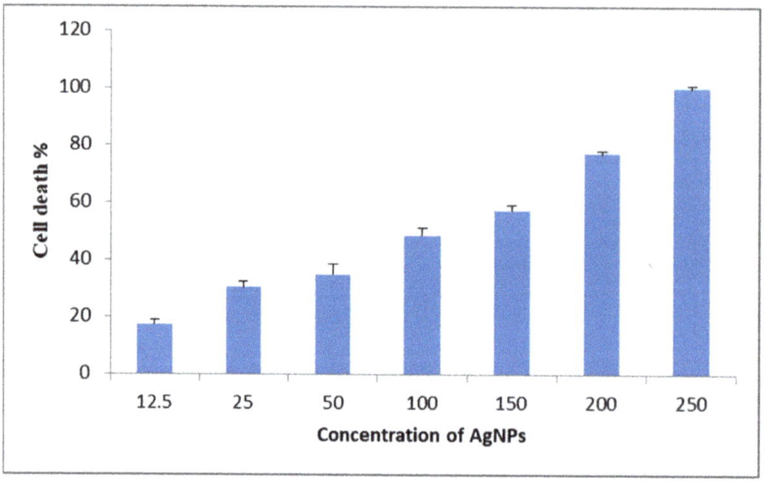


**Figure 11.** Anticancer property of AgNPs on DLA cancer cells.

## 4. Conclusions

In conclusion, the green synthesis of silver nanoparticles using *Holigarna arnottiana* leaf extract represents a promising approach that aligns with the principles of sustainability and eco-friendliness. The multifaceted bioactivities of these nanoparticles, including anticancer, antimicrobial, and scavenging properties, hold immense potential for future biomedical and environmental applications. This research aims to contribute valuable insights into the development of novel and sustainable nanomaterials with significant biomedical relevance.

**Supplementary Materials:** The following supporting information can be downloaded at: https://www.mdpi.com/article/10.3390/pharmaceutics15102468/s1, Figure S1: CV data of AgNPS synthesized using *H. arnottiana*; Figure S2: The Effect of Leaf Extract and Silver Nanoparticles on Mitotic Activity in Onion Root Tip Cells in the Context of Chromosomal Aberrations; Figure S3: DPPH activity of leaf extract of *H. arnottiana*.

**Author Contributions:** Conceptualization, A.P.A., V.S.B. and K.N.J.; methodology, A.P.A., O.S. and A.M.; software, S.R.V., K.N.J. and P.P.; validation, A.P.A., K.N.J., P.S. and V.V.; formal analysis, O.S., A.M., K.V.Z. and V.V.; investigation, O.S.; resources, V.S.B., S.R.V., P.A.J., P.P. and A.; data curation, A.P.A. and P.S.; writing—original draft, O.S., K.N.J. and P.A.J.; visualization, A.; supervision, A.P.A. and K.V.Z.; project administration, A.P.A. All authors have read and agreed to the published version of the manuscript.

**Funding:** This study did not receive any particular funding.

**Institutional Review Board Statement:** Not applicable.

**Informed Consent Statement:** Not applicable.

**Data Availability Statement:** The data generated and analyzed during the current study are available from the corresponding author upon reasonable request.

**Acknowledgments:** We thank the Sophisticated Test and Instrumentation Centre (STIC) Cochin, Kerala, India, for TEM analysis. We would like to extend our gratitude to Sudheer Kumar Paruthully of the Department of History at Sree Neelakanta Government Sanskrit College, Pattambi, Kerala, India, for his valuable assistance in collecting the *H. arnottiana* plant specimens.

**Conflicts of Interest:** The authors declare no conflict of interest.

## References


1. Subba Rao, Y.; Kotakadi, V.S.; Prasad, T.N.V.K.V.; Reddy, A.V.; Sai Gopal, D.V.R. Green Synthesis and Spectral Characterization of Silver Nanoparticles from Lakshmi Tulasi (*Ocimum sanctum*) Leaf Extract. *Spectrochim. Acta Part A Mol. Biomol. Spectrosc.* **2013**, *103*, 156–159. [CrossRef] [PubMed]
2. Rafique, M.; Sadaf, I.; Rafique, M.S.; Tahir, M.B. A Review on Green Synthesis of Silver Nanoparticles and Their Applications. *Artif. Cells Nanomed. Biotechnol.* **2017**, *45*, 1272–1291. [CrossRef] [PubMed]
3. Firdhouse, M.J.; Lalitha, P. Biosynthesis of Silver Nanoparticles and Its Applications. *J. Nanotechnol.* **2015**, *2015*, e829526. [CrossRef]
4. Chung, I.-M.; Park, I.; Seung-Hyun, K.; Thiruvengadam, M.; Rajakumar, G. Plant-Mediated Synthesis of Silver Nanoparticles: Their Characteristic Properties and Therapeutic Applications. *Nanoscale Res. Lett.* **2016**, *11*, 40. [CrossRef] [PubMed]
5. Holigarna Arnottiana—Black Varnish Tree. Available online: https://www.flowersofindia.net/catalog/slides/Black%20Varnish%20Tree.html (accessed on 27 March 2023).
6. Harsha, V.H.; Hebbar, S.S.; Hegde, G.R.; Shripathi, V. Ethnomedical Knowledge of Plants Used by Kunabi Tribe of Karnataka in India. *Fitoterapia* **2002**, *73*, 281–287. [CrossRef]
7. Pillai, S.S. Contact Dermatitis in Trivandrum. *Indian J. Dermatol. Venereol. Leprol.* **1989**, *55*, 297.
8. Mane, R.N.; Gholave, A.R.; Lekhak, M.M.; Yadav, S.R. Karyomorphological Analysis of *Holigarna arnottiana* (Anacardiaceae): An Endemic Species from the Western Ghats, India. *Cytologia* **2021**, *86*, 85–87. [CrossRef]
9. Ravi, A.; Oommen, P.S. Phytochemical characterization of *Holigarna arnottiana* Hook. F. *J. Pharm. Res.* **2012**, *5*, 3202–3203.
10. Manilal, A.; Idhayadhulla, A. Potential in Vitro Antimicrobial Efficacy of *Holigarna arnottiana* (Hook F). *Asian Pac. J. Trop. Biomed.* **2014**, *4*, 25–29. [CrossRef]
11. Pradeep, D.P.; Saj, O.P. Phytochemical and Antimicrobial Studies of Leaf and Bark Extracts of *Holigarna arnottiana* Hook F. *J. Pure Appl. Microbiol.* **2010**, *4*, 139–148.
12. Ravi, A.; Saj, O.P. Antioxidant and Cytotoxic Potential of the Plant *Holigarna arnottiana* Hook.F. Bark Ethanolic Extract. *World J. Pharm. Res.* **2015**, *4*, 1234–1244.
13. Yende, A.; Bhat, R.; Zainab, A.; Acharya, S.; Padyana, S. Evaluation of Antioxidant and Antimicrobial Activities of *Holigarna arnottiana* Hook.F. *J. Free Radic. Antioxidnats* **2013**, *139*, 278–288.
14. Jain, S.; Mehata, M.S. Medicinal Plant Leaf Extract and Pure Flavonoid Mediated Green Synthesis of Silver Nanoparticles and Their Enhanced Antibacterial Property. *Sci. Rep.* **2017**, *7*, 15867. [CrossRef] [PubMed]
15. Sadeghi, B.; Jamali, M.; Sh, K.; Amini Nia, A.; Ghafari, S. Synthesis and Characterization of Silver Nanoparticles for Antibacterial Activity. *R. Soc. Open Sci.* **2010**, *1*, 119–124.
16. Sadeghi, B.; Rostami, A.; Momeni, S.S. Facile Green Synthesis of Silver Nanoparticles Using Seed Aqueous Extract of *Pistacia atlantica* and Its Antibacterial Activity. *Spectrochim. Acta Part A Mol. Biomol. Spectrosc.* **2015**, *134*, 326–332. [CrossRef] [PubMed]
17. Ravichandran, V.; Vasanthi, S.; Shalini, S.; Ali Shah, S.A.; Harish, R. Green Synthesis of Silver Nanoparticles Using Atrocarpus Altilis Leaf Extract and the Study of Their Antimicrobial and Antioxidant Activity. *Mater. Lett.* **2016**, *180*, 264–267. [CrossRef]
18. Rout, Y.; Behera, S.; Ojha, A.K.; Nayak, P.L. Green synthesis of silver nanoparticles using *Ocimum sanctum* (Tulashi) and study of their antibacterial and antifungal activities. *J. Microbiol. Antimicrob.* **2012**, *4*, 103–109. [CrossRef]
19. Padalia, H.; Moteriya, P.; Chanda, S. Green Synthesis of Silver Nanoparticles from Marigold Flower and Its Synergistic Antimicrobial Potential. *Arab. J. Chem.* **2015**, *8*, 732–741. [CrossRef]
20. Patil, J.; Tatiya, A.; Wadekar, R.; Girase, T.; Patel, K. Phytochemistry of Uvaria Narum: A Multifaceted Perspective and Ethnopharmacological Potential. *Chem. Proc.* **2022**, *12*, 18. [CrossRef]
21. Francis, S.; Joseph, S.; Koshy, E.P.; Mathew, B. Microwave Assisted Green Synthesis of Silver Nanoparticles Using Leaf Extract of Elephantopus Scaber and Its Environmental and Biological Applications. *Artif. Cells Nanomed. Biotechnol.* **2018**, *46*, 795–804. [CrossRef]
22. Singh, C.; Baboota, R.K.; Kr Naik, P.; Singh, H. Biocompatible Synthesis of Silver and Gold Nanoparticles Using Leaf Extract of Dalbergia Sissoo. *Adv. Mater. Lett.* **2012**, *3*, 279–285. [CrossRef]
23. Mishra, K.; Ojha, H.; Chaudhury, N.K. Estimation of Antiradical Properties of Antioxidants Using DPPH Assay: A Critical Review and Results. *Food Chem.* **2012**, *130*, 1036–1043. [CrossRef]
24. Sathishkumar, M.; Sneha, K.; Won, S.W.; Cho, C.-W.; Kim, S.; Yun, Y.-S. Cinnamon Zeylanicum Bark Extract and Powder Mediated Green Synthesis of Nano-Crystalline Silver Particles and Its Bactericidal Activity. *Colloids Surf. B Biointerfaces* **2009**, *73*, 332–338. [CrossRef]
25. Ajaykumar, A.P.; Mathew, A.; Chandni, A.P.; Varma, S.R.; Jayaraj, K.N.; Sabira, O.; Rasheed, V.A.; Binitha, V.S.; Swaminathan, T.R.; Basheer, V.S.; et al. Green Synthesis of Silver Nanoparticles Using the Leaf Extract of the Medicinal Plant, Uvaria Narum and Its Antibacterial, Antiangiogenic, Anticancer and Catalytic Properties. *Antibiotics* **2023**, *12*, 564. [CrossRef]
26. Pilaquinga, F.; Amaguaña, D.; Morey, J.; Moncada-Basualto, M.; Pozo-Martínez, J.; Olea-Azar, C.; Fernández, L.; Espinoza-Montero, P.; Jara-Negrete, E.; Meneses, L.; et al. Synthesis of Silver Nanoparticles Using Aqueous Leaf Extract of *Mimosa albida* (Mimosoideae): Characterization and Antioxidant Activity. *Materials* **2020**, *13*, 503. [CrossRef]
27. Gardea-Torresdey, J.L.; Gomez, E.; Peralta-Videa, J.R.; Parsons, J.G.; Troiani, H.; Jose-Yacaman, M. Alfalfa Sprouts: A Natural Source for the Synthesis of Silver Nanoparticles. *Langmuir* **2003**, *19*, 1357–1361. [CrossRef]

28. Shankar, S.S.; Ahmad, A.; Sastry, M. Geranium Leaf Assisted Biosynthesis of Silver Nanoparticles. *Biotechnol. Prog.* **2003**, *19*, 1627–1631. [CrossRef] [PubMed]
29. Arokiyaraj, S.; Arasu, M.V.; Vincent, S.; Prakash, N.U.; Choi, S.H.; Oh, Y.-K.; Choi, K.C.; Kim, K.H. Rapid Green Synthesis of Silver Nanoparticles from Chrysanthemum Indicum L and Its Antibacterial and Cytotoxic Effects: An in Vitro Study. *Int. J. Nanomed.* **2014**, *9*, 379–388. [CrossRef]
30. Lokina, S.; Stephen, A.; Kaviyarasan, V.; Arulvasu, C.; Narayanan, V. Cytotoxicity and Antimicrobial Activities of Green Synthesized Silver Nanoparticles. *Eur. J. Med. Chem.* **2014**, *76*, 256–263. [CrossRef] [PubMed]
31. Loo, Y.Y.; Rukayadi, Y.; Nor-Khaizura, M.-A.-R.; Kuan, C.H.; Chieng, B.W.; Nishibuchi, M.; Radu, S. In Vitro Antimicrobial Activity of Green Synthesized Silver Nanoparticles Against Selected Gram-Negative Foodborne Pathogens. *Front. Microbiol.* **2018**, *9*, 1555. [CrossRef]
32. Zein, R.; Alghoraibi, I.; Soukkarieh, C.; Ismail, M.T.; Alahmad, A. Influence of Polyvinylpyrrolidone Concentration on Properties and Anti-Bacterial Activity of Green Synthesized Silver Nanoparticles. *Micromachines* **2022**, *13*, 777. [CrossRef] [PubMed]
33. Devasvaran, K.; Alallam, B.; Yunus, M.A.; Dewi, F.R.P.; Kamal, N.N.S.N.M.; Lim, V. Microwave-Assisted Green Synthesis of Silver Nanoparticles Using Alkaline Extracted Crude Polysaccharide of C. Nutans: Optimisation, Characterisation, Toxicity, Anticancer Potential and Antibacterial Studies. *J. Drug Deliv. Sci. Technol.* **2023**, *86*, 104688. [CrossRef]
34. Dinesh, D.; Murugan, K.; Madhiyazhagan, P.; Panneerselvam, C.; Mahesh Kumar, P.; Nicoletti, M.; Jiang, W.; Benelli, G.; Chandramohan, B.; Suresh, U. Mosquitocidal and Antibacterial Activity of Green-Synthesized Silver Nanoparticles from Aloe Vera Extracts: Towards an Effective Tool against the Malaria Vector Anopheles Stephensi? *Parasitol. Res.* **2015**, *114*, 1519–1529. [CrossRef] [PubMed]
35. Balakumaran, M.D.; Ramachandran, R.; Balashanmugam, P.; Mukeshkumar, D.J.; Kalaichelvan, P.T. Mycosynthesis of Silver and Gold Nanoparticles: Optimization, Characterization and Antimicrobial Activity against Human Pathogens. *Microbiol. Res.* **2016**, *182*, 8–20. [CrossRef] [PubMed]
36. Senthil, B.; Devasena, T.; Prakash, B.; Rajasekar, A. Non-Cytotoxic Effect of Green Synthesized Silver Nanoparticles and Its Antibacterial Activity. *J. Photochem. Photobiol. B Biol.* **2017**, *177*, 1–7. [CrossRef]
37. Saxena, A.; Tripathi, R.; Singh, R. Synthesis of Silver Nanoparticles Using Onion (*Allium cepa*) Extract and Their Antibacterial Activity. *Dig. J. Nanomater. Biostructures* **2010**, *5*, 427–432.
38. Ahmad, A.; Wei, Y.; Syed, F.; Tahir, K.; Rehman, A.U.; Khan, A.; Ullah, S.; Yuan, Q. The Effects of Bacteria-Nanoparticles Interface on the Antibacterial Activity of Green Synthesized Silver Nanoparticles. *Microb. Pathog.* **2017**, *102*, 133–142. [CrossRef]
39. Mathew, A. Microwave-Assisted Greener Synthesis of Silver Nanoparticles Using Entada Rheedii Leaf Extract and Investigation of Its Anticancer and Antimicrobial Properties. *Int. J. Nano Dimens.* **2022**, *13*, 329–334.
40. Kumari, M.; Mukherjee, A.; Chandrasekaran, N. Genotoxicity of Silver Nanoparticles in Allium Cepa. *Sci. Total Environ.* **2009**, *407*, 5243–5246. [CrossRef]
41. Pesnya, D.S. Cytogenetic Effects of Chitosan-Capped Silver Nanoparticles in the *Allium cepa* Test. *Caryologia* **2013**, *66*, 275–281. [CrossRef]
42. Debnath, P.; Mondal, A.; Hajra, A.; Das, C.; Mondal, N.K. Cytogenetic Effects of Silver and Gold Nanoparticles on *Allium cepa* Roots. *J. Genet. Eng. Biotechnol.* **2018**, *16*, 519–526. [CrossRef] [PubMed]
43. Sabira, O.; Vignesh, A.R.; Ajaykumar, A.P.; Varma, S.R.; Jayaraj, K.N.; Sebastin, M.; Nikhila, K.; Babu, A.; Rasheed, V.A.; Binitha, V.S.; et al. The Chemical Composition and Antimitotic, Antioxidant, Antibacterial and Cytotoxic Properties of the Defensive Gland Extract of the Beetle, Luprops Tristis Fabricius. *Molecules* **2022**, *27*, 7476. [CrossRef]
44. Ghosh, M.; J, M.; Sinha, S.; Chakraborty, A.; Mallick, S.K.; Bandyopadhyay, M.; Mukherjee, A. In Vitro and in Vivo Genotoxicity of Silver Nanoparticles. *Mutat. Res. Genet. Toxicol. Environ. Mutagen.* **2012**, *749*, 60–69. [CrossRef] [PubMed]
45. Prokhorova, I.M.; Kibrik, B.S.; Pavlov, A.V.; Pesnya, D.S. Estimation of Mutagenic Effect and Modifications of Mitosis by Silver Nanoparticles. *Bull. Exp. Biol. Med.* **2013**, *156*, 255–259. [CrossRef]
46. Pesnya, D.S.; Romanovsky, A.V. Comparison of Cytotoxic and Genotoxic Effects of Plutonium-239 Alpha Particles and Mobile Phone GSM 900 Radiation in the *Allium cepa* Test. *Mutat. Res. Genet. Toxicol. Environ. Mutagen.* **2013**, *750*, 27–33. [CrossRef] [PubMed]
47. Ahmed, S.; Saifullah; Ahmad, M.; Swami, B.L.; Ikram, S. Green Synthesis of Silver Nanoparticles Using Azadirachta Indica Aqueous Leaf Extract. *J. Radiat. Res. Appl. Sci.* **2016**, *9*, 1–7. [CrossRef]
48. Jeyaraj, M.; Rajesh, M.; Arun, R.; MubarakAli, D.; Sathishkumar, G.; Sivanandhan, G.; Dev, G.K.; Manickavasagam, M.; Premkumar, K.; Thajuddin, N.; et al. An Investigation on the Cytotoxicity and Caspase-Mediated Apoptotic Effect of Biologically Synthesized Silver Nanoparticles Using Podophyllum Hexandrum on Human Cervical Carcinoma Cells. *Colloids Surf. B Biointerfaces* **2013**, *102*, 708–717. [CrossRef]
49. Vasanth, K.; Ilango, K.; MohanKumar, R.; Agrawal, A.; Dubey, G.P. Anticancer Activity of Moringa Oleifera Mediated Silver Nanoparticles on Human Cervical Carcinoma Cells by Apoptosis Induction. *Colloids Surf. B Biointerfaces* **2014**, *117*, 354–359. [CrossRef]
50. Nakkala, J.R.; Mata, R.; Gupta, A.K.; Sadras, S.R. Biological Activities of Green Silver Nanoparticles Synthesized with Acorous Calamus Rhizome Extract. *Eur. J. Med. Chem.* **2014**, *85*, 784–794. [CrossRef]
51. Mittal, A.K.; Thanki, K.; Jain, S.; Banerjee, U.C. Comparative Studies of Anticancer and Antimicrobial Potential of Bioinspired Silver and Silver-Selenium Nanoparticles. *J. Mater. NanoSci.* **2016**, *3*, 22–27.

52. Sukirtha, R.; Priyanka, K.M.; Antony, J.J.; Kamalakkannan, S.; Thangam, R.; Gunasekaran, P.; Krishnan, M.; Achiraman, S. Cytotoxic Effect of Green Synthesized Silver Nanoparticles Using Melia Azedarach against in Vitro HeLa Cell Lines and Lymphoma Mice Model. *Process Biochem.* **2012**, *47*, 273–279. [CrossRef]
53. Chanthini, A.B.; Balasubramani, G.; Ramkumar, R.; Sowmiya, R.; Balakumaran, M.D.; Kalaichelvan, P.T.; Perumal, P. Structural Characterization, Antioxidant and in Vitro Cytotoxic Properties of Seagrass, *Cymodocea serrulata* (R.Br.) Asch. & Magnus Mediated Silver Nanoparticles. *J. Photochem. Photobiol. B Biol.* **2015**, *153*, 145–152. [CrossRef]
54. Ajaykumar, A.P.; Sabira, O.; Sebastian, M.; Varma, S.R.; Roy, K.B.; Binitha, V.S.; Rasheed, V.A.; Jayaraj, K.N.; Vignesh, A.R. A Novel Approach for the Biosynthesis of Silver Nanoparticles Using the Defensive Gland Extracts of the Beetle, *Luprops tristis* Fabricius. *Sci. Rep.* **2023**, *13*, 10186. [CrossRef]
55. Mousavi-Khattat, M.; Keyhanfar, M.; Razmjou, A. A Comparative Study of Stability, Antioxidant, DNA Cleavage and Antibacterial Activities of Green and Chemically Synthesized Silver Nanoparticles. *Artif. Cells Nanomed. Biotechnol.* **2018**, *46*, 1022–1031. [CrossRef] [PubMed]
56. Sreelekha, E.; George, B.; Shyam, A.; Sajina, N.; Mathew, B. A Comparative Study on the Synthesis, Characterization, and Antioxidant Activity of Green and Chemically Synthesized Silver Nanoparticles. *BioNanoScience* **2021**, *11*, 489–496. [CrossRef]
57. Lokapur, V.; Jayakar, V.; Shantaram, M. Phytochemical investigation, chemical composition and in vitro antioxidant activities of various crude extracts of *Holigarna ferrugenia* Marchand. *Med. Plants-Int. J. Phytomedicines Relat. Ind.* **2022**, *14*, 72–83. [CrossRef]
58. Reddy, N.J.; Nagoor Vali, D.; Rani, M.; Rani, S.S. Evaluation of Antioxidant, Antibacterial and Cytotoxic Effects of Green Synthesized Silver Nanoparticles by Piper Longum Fruit. *Mater. Sci. Eng. C* **2014**, *34*, 115–122. [CrossRef]
59. Yousaf, H.; Mehmood, A.; Ahmad, K.S.; Raffi, M. Green Synthesis of Silver Nanoparticles and Their Applications as an Alternative Antibacterial and Antioxidant Agents. *Mater. Sci. Eng. C* **2020**, *112*, 110901. [CrossRef]

*pharmaceutics*

MDPI

*Review*

# Plant Protease Inhibitors as Emerging Antimicrobial Peptide Agents: A Comprehensive Review

**Mónica G. Parisi [1], Brenda Ozón [2], Sofía M. Vera González [2], Javier García-Pardo [3,*] and Walter David Obregón [2,*]**

[1] Instituto de Ecología y Desarrollo Sustentable (INEDES, CONICET-UNLu) and Departamento de Ciencias Básicas, Universidad Nacional de Luján, Ruta 5 y Avenida Constitución, Luján B6700, Buenos Aires, Argentina; mparisi@unlu.edu.ar

[2] Centro de Investigación de Proteínas Vegetales (CIProVe) and Departamento de Ciencias Biológicas, Facultad de Ciencias Exactas, Universidad Nacional de La Plata, 47 y 115 s/N, La Plata B1900, Buenos Aires, Argentina; brendaozon@biol.unlp.edu.ar (B.O.); alu-sveragonzalez@exactas.unlp.edu.ar (S.M.V.G.)

[3] Institut de Biotecnologia i de Biomedicina (IBB) and Departament de Bioquímica i Biologia Molecular, Universitat Autònoma de Barcelona, 08193 Bellaterra, Barcelona, Spain

* Correspondence: javier.garcia.pardo@uab.cat (J.G.-P.); davidobregon@biol.unlp.edu.ar (W.D.O.); Tel.: +34-93-5868936 (J.G.-P.); +54-221-423-5333 (ext. 57) (W.D.O.)

**Abstract:** Antimicrobial peptides (AMPs) are important mediator molecules of the innate defense mechanisms in a wide range of living organisms, including bacteria, mammals, and plants. Among them, peptide protease inhibitors (PPIs) from plants play a central role in their defense mechanisms by directly attacking pathogens or by modulating the plant's defense response. The growing prevalence of microbial resistance to currently available antibiotics has intensified the interest concerning these molecules as novel antimicrobial agents. In this scenario, PPIs isolated from a variety of plants have shown potential in inhibiting the growth of pathogenic bacteria, protozoans, and fungal strains, either by interfering with essential biochemical or physiological processes or by altering the permeability of biological membranes of invading organisms. Moreover, these molecules are active inhibitors of a range of proteases, including aspartic, serine, and cysteine types, with some showing particular efficacy as trypsin and chymotrypsin inhibitors. In this review, we provide a comprehensive analysis of the potential of plant-derived PPIs as novel antimicrobial molecules, highlighting their broad-spectrum antimicrobial efficacy, specificity, and minimal toxicity. These natural compounds exhibit diverse mechanisms of action and often multifunctionality, positioning them as promising molecular scaffolds for developing new therapeutic antibacterial agents.

**Keywords:** plant protease inhibitors; antibacterial compounds; antimicrobial activity; antimicrobial peptides; antibiotic resistance; antifungal agents; cysteine-rich peptides

## 1. Introduction

Natural products (NPs) have long served as a source of active ingredients for a broad array of pharmaceutical applications, owing to their extensive chemical and structural diversity and generally low toxicity [1,2]. Predominantly sourced from plants, these compounds exhibit a wide range of biological mechanisms, offering novel activities and biological properties essential for the development of nutraceuticals, pharmacologically active substances, phytosanitary agents, and other economically valuable products [3–6].

The increasing emergence of multidrug-resistant (MDR) bacteria is a critical global problem [7–9]. From the One Health perspective, this can be considered one of the most significant challenges for the biomedical field, animal health, and phytosanitary services [10–13]. Indeed, it has recently been estimated that antimicrobial resistance (AMR) was directly responsible for 1.27 million global deaths and contributed to 4.95 million deaths, affecting countries in all regions and at all income levels, being higher in low- and middle-income countries, thus producing significant economic costs (WHO) [14].

In this scenario, the search for natural therapeutic agents, especially as alternatives to synthetic antibiotics, remains a fervent area of research and development within the scientific community. Notably, in the last 35 years, natural compounds have been the basis for over 35% of drugs approved by the Food and Drug Administration (FDA) (https://www.fda.gov/drugs) [15]. This trend is expected to continue, with predictions suggesting that natural and biopharmaceutical products, including small chemical compounds, peptides, and proteins, could form the majority of our pharmacotherapeutic tools in the near future [16].

In the plant kingdom, antimicrobial compounds act as the primary defense mechanism against pathogenic microorganisms. In this context, numerous plant antimicrobial molecules capable of inhibiting a number of human pathogens have been isolated from various plant sources [17–20]. These natural compounds include primary metabolites essential for plant growth, substance transport, photosynthesis, respiration, nutrient assimilation, flowering, and fruit ripening (i.e., proteins, lipids, and carbohydrates) and secondary plant metabolites (i.e., terpenes, polyphenols, and carotenoids), which do not participate in fundamental physiological processes, although they exhibit diverse bioactive chemical scaffolds. In addition to these small chemical compounds, a variety of proteins and peptides with promising antimicrobial properties have also been isolated and characterized so far. These include chitinases, β-1,3-glucanases, thaumatin-like proteins, endoproteases, peroxidases, ribonuclease-like proteins, γ-thionins, plant defensins, oxalate oxidases, oxalate oxidase-like proteins, proteases, and peptide protease inhibitors (PPIs) [21,22]. During the last two decades, up to eighteen PPIs with antimicrobial activity have been described so far. Among them, several PPIs have demonstrated broad-spectrum efficacy against a wide range of pathogens, including bacteria, fungi, and viruses [23–26]. These include Kunitz- and Trypsin-type inhibitors from plants belonging to the *Solanaceae*, *Fabaceae*, and *Moringaceae* families.

This work aims to offer a comprehensive review of plant-derived PPIs and their potential as novel antibacterial and antifungal agents, exploring their mechanisms of action, efficacy against various microbial strains, and the potential for their application as innovative antimicrobial peptide agents. Ultimately, we also discuss future directions of this field.

## 2. Antimicrobial Peptides (AMPs)

The extensive research has brought many novel antimicrobial candidates into clinical and pre-clinical development [27,28]. An important number of these molecules are antimicrobial peptides (AMPs), a diverse group of small molecules synthesized and ubiquitously distributed in plants and other living organisms. Due to antimicrobial potential and immunomodulatory capacities, research on AMPs has steadily increased in the last decade. Currently, over 3940 AMPs have been reported so far, including 3146 natural AMPs of different natural sources, including 383 bacteriocins/peptide antibiotics from bacteria; 5 from archaea; 8 from protists; 29 from fungi; 250 from plants; and 2463 from animals, according to the latest update of the Antimicrobial Peptide Database (APD3) (http://aps.unmc.edu/AP/) [29] (Figure 1). While the majority of AMPs have been identified in animals and plants, those isolated from archaea and protists include a set of relevant molecules, such as halocins, that have acquired specific adaptations to extreme environments. These adaptations could translate into robust stability and efficacy under conditions that are challenging for other antimicrobial compounds. The activity and mechanism of these and other AMPs have been thoroughly reviewed elsewhere [30–33]; therefore, this review will specifically focus on the recent developments in plant-derived AMPs, particularly those with protease inhibition activity.

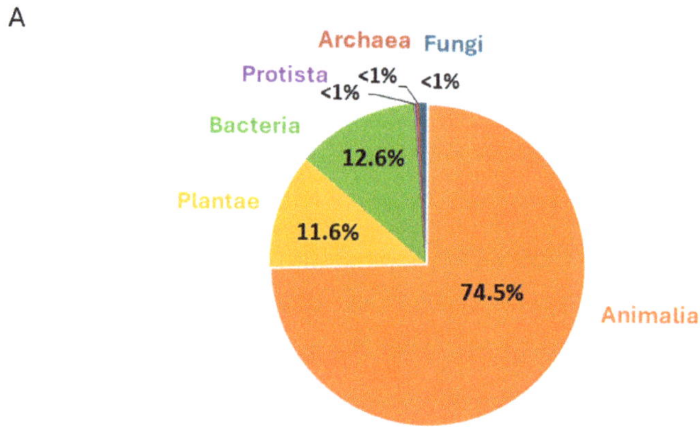


B

| Structural class | Peptide number | % |
|---|---|---|
| Helix | 568 | 15.27 |
| Beta structure | 93 | 2.51 |
| Helix and beta unpacked | 4 | 0.11 |
| Combined helix and beta packed | 119 | 3.19 |
| Neither helix nor beta structure | 24 | 0.65 |
| Rich in unusual amino acids | 117 | 3.15 |
| Disulfide bridge (but no 3D) | 548 | 14.74 |
| Unknown 3D structure | 2245 | 60.37 |

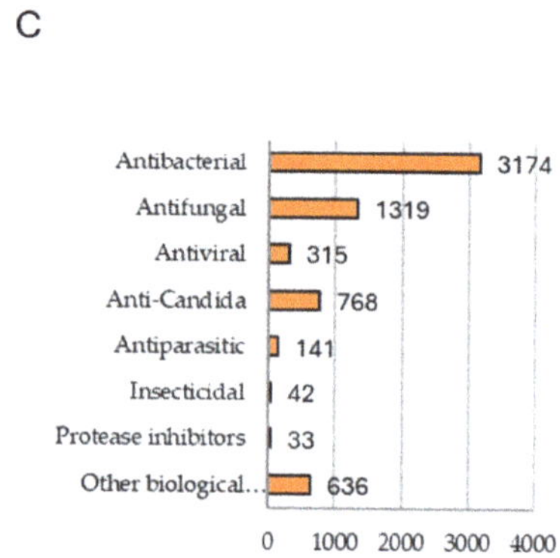


**Figure 1.** Composition and Diversity of Antimicrobial Peptides (AMPs). (**A–C**) These panels summarize the data from the Antimicrobial Peptide Database (APD3) (http://aps.unmc.edu/AP/) [29]. As of January 2024, APD3 has cataloged 3940 AMPs, categorized by (**A**) biological origin, (**B**) secondary structural characteristics, and (**C**) the spectrum of biological activity. In panel (**B**), Unknown 3D structure refers to AMPs for which the three-dimensional structures have not yet been determined.

Plant AMPs are generally classified based on their tridimensional structure, their sequence, and the presence of disulfide bonds, such as thionins, defensins, hevein-like peptides, knottins, stable-like peptides, lipid transfer proteins, snakins, and cyclotides [24,34–37]. Thionins are essential peptides found in monocots and dicots, with sizes ranging from 45 to 48 amino acid residues and a molecular weight of around 5 kDa [36]. They are classified into two groups: $\alpha/\beta$-thionins and $\gamma$-thionins, according to amino acid sequence homology, three-dimensional structural similarity, and disulfide bridge positions. Regarding the antimicrobial activity of thionins, antibacterial [38,39], antifungal, anti-larvicidal, and in vitro cytotoxic effects against mammal cell cultures have been reported [24,40].

Defensins constitute one of the largest groups of AMPs (Figure 2). They have been shown to disrupt cellular functions by binding extracellularly to the cell wall components, membrane targets, and/or to specific membrane lipids such as sphingolipids or phospholipids or binding to cell surface targets and membrane lipids [41]. An important aspect of AMP functionality involves the modulation of the host plant's defense mechanisms. In this regard, some AMPs have been found to prime the plant's immune response, preparing it for a more robust defense against subsequent pathogen attacks. This priming effect boosts the plant's innate immune system, enhancing its responsiveness through the modulation of various signaling cascades [42]. Hevein-like peptides are basic peptides containing 29–45 amino acid residues stabilized by 3–5 disulfide bonds, with a conserved chitin-binding domain (SXFGY/SXYGY), which is associated with antifungal activity [43]. These molecules were first identified in the latex of *Hevea brasiliensis* and are responsible for defense against a wide range of pathogenic fungi based on the interaction of the hevein domain with fungal chitin present in the cell, causing damage to the fungal cell.

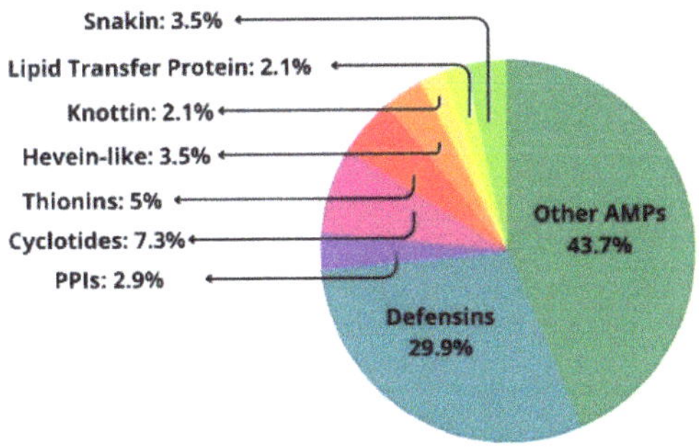


**Figure 2.** Relative abundance of AMPs available on the Antimicrobial Plant Database (https://aps.unmc.edu/) [29].

Plant knottins are peptides with approximately 30 amino acid residues, the smallest plant AMPs, involving three disulfide bonds (cysteine knot motifs) and two conformations, cyclic and linear, showing high thermal stability and resistance to chemicals and proteolytic action [44]. Knottins are known as "promiscuous peptides" based on their multiple biological functions: they can bind to cell membranes, $K^+$ and $Na^+$ channels in membranes, and acid-sense channels; are highly cytotoxic to human cells [45]; and have antimicrobial activity against bacteria, fungi, viruses, and insects via interacting with membranes to exert their activities.

The α-hairpinin family represents a class of Lys/Arg-rich plant defense peptides, with special Cys motifs (i.e., XnC1X3C2XnC3X3C4Xn, where X is any amino acid different from cysteine), to form a highly characteristic helix–loop–helix secondary structure with two antiparallel α-helices [46] stabilized by two disulfide bonds in the tertiary structure. This AMP family shows antibacterial, antifungal, trypsin-inactivating, and ribosome-inactivating activities [46,47].

Lipid transfer proteins (LTPs) are small cationic peptides (7 to 10 kDa), composed of a conservative pattern of eight Cys residues and four disulfide bonds that stabilize a tight tertiary fold with hydrophobic cavities. LTPs inhibit the growth of fungi and bacteria, thus participating in plant defense systems [48].

The snakin class contains 12 cysteine residues (cysteine-rich), the largest number of disulfide bonds and are constitutively or inducibly expressed via biotic or abiotic stress in different organs, such as the roots, stem, leaves, flowers, and seeds. Snakins inhibit both fungal (e.g., *Magnaporthe grisea*, *Fusarium solani*, and *Botrytis cinerea*) and bacterial (e.g., *Dickeya dadantii*, *Ralstonia solanacearum*, and *Sinorhizobium meliloti*) growth [49].

Cyclotides are long-chain cyclic peptides (28–37 amino acids) consisting of six loops with six conserved cysteine residues, stabilized via three disulfide bonds. They are anionic AMPs with high resistance to thermal and chemical denaturation as well as proteolytic degradation, making them potential therapeutic agents as antitumor, anti-HIV, insecticidal, and antibacterial agents [50,51].

In this context, it is known that around 3% of AMPs are plant PPIs (Figure 2) with proven effectiveness in inhibiting the growth of a variety of pathogenic bacterial and fungal strains [36]. Not surprisingly, many of these AMPs share common structural and functional properties with plant PPIs, including their ability to alter the integrity of microbial cell membranes, a mechanism crucial for their function [36,52–54]. This positions this set of natural peptides as promising molecules for developing novel therapeutics against MDR bacteria.

## 3. Mechanism of Antimicrobial Action of Plant AMPs

Typically, AMPs are characterized by their low molecular weight, displaying less than 100 amino acids with a positive net charge at physiological pH, due to the presence of lysine and arginine residues and a high proportion ($\geq$30%) of hydrophobic residues [32,55]. These characteristics render them highly cationic molecules that can interact with the phospholipid and lipopolysaccharide layers found in many types of bacteria, leading to pore formation, membrane destabilization, and rapid cell death [25,32,56–58]. At present, several mechanisms have been reported on cationic AMPs that cause cell membrane damage by interacting with the negatively charged phospholipid layer and lipopolysaccharides of the bacterial membrane, causing pores, membrane destabilization, and rapid cell death [25,56–58]. For Gram-negative bacteria, the AMP needs to permeabilize the outer membrane before it can access the cytoplasmic membrane. Conversely, for Gram-positive bacteria, the AMP can directly diffuse through the peptidoglycan layer via nano-sized pores.

AMPs can also mediate their action by acting as lipid transfer proteins. Through this mechanism, these proteins directly interact with fungal membrane lipids, effectively blocking pathogen penetration into the host cell and, thus, preventing infection [59]. Furthermore, many AMPs can specifically bind to the sphingolipid and phospholipid bilayers of the cell wall and/or the cell membrane of fungi, being able to enter or remain on the outside of the fungal cell. Regardless of their possible absorption, antifungal peptides can affect intracellular targets, causing, among other actions, the production of ROS, programmed cell death, mitochondrial dysfunction, the alteration of cation homeostasis, ATP efflux, cell cycle impairment, and autophagy, among others [29]. They can penetrate the target cell by breaking the membrane and interfering with protein or nucleic acid synthesis or cell division or by inhibiting protease activity [53,60]. For instance, heveins bind chitin in fungi [43] and some defensins interact with eukaryotic target proteins to target intracellular functions [41]. Another mechanism of action is priming the plant by inducing defense responses against pathogens due to the modulation of different cascades in their immune system (Figure 3) [61].

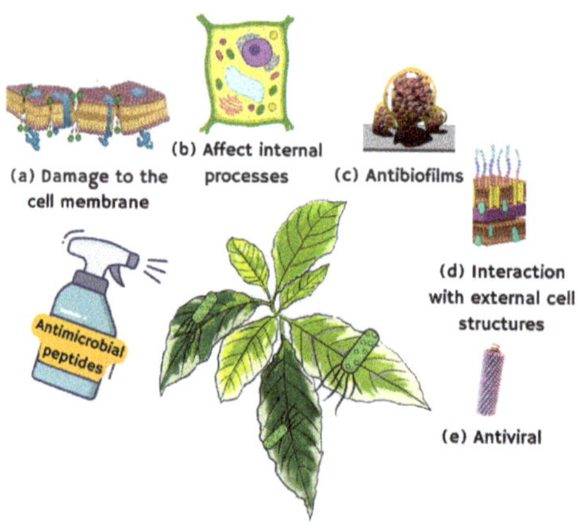


**Figure 3.** Mechanisms of action of antimicrobial peptides. (**a**) Damage of cell membranes, causing cell lysis. (**b**) Interference with internal cellular processes. (**c**) Inhibition of biofilm formation. (**d**) Interaction with external cell structures such as lipopolysaccharides, fimbriae, or flagella in bacteria or chitin in fungi. (**e**) Inhibition of virus attachment or replication [59].

## 4. Peptide Protease Inhibitors (PPIs) Derived from Plants

Protease inhibitors (PIs) are regulatory molecules of a proteinaceous or non-proteinaceous nature, ubiquitously distributed in animals, plants, and across microbial species. Those of a proteinaceous nature are typically known as PPIs. Based on the type of protease they inhibit, PPIs can be categorized into six main classes: serine protease inhibitors (SPIs), cysteine protease inhibitors (CPIs), aspartyl protease inhibitors (APIs), metalloprotease inhibitors (MPIs), glutamate protease inhibitors (GPIs), and threonine protease inhibitors (TPIs) [62–64].

PPIs play a central role in controlling the activity of target proteases, often inhibiting their excessive and uncontrolled activity in both normal and pathological conditions [65]. This regulation is essential for activating coenzymes and releasing biologically active polypeptides [66,67]. In the case of mammals, the presence of PPIs in plasma is closely associated with the regulation of critical proteolytic cascades, including blood coagulation and complement activation [66,68–70]. Moreover, proteolysis is involved in many vital biological functions, such as immunity, blood coagulation, cell cycle regulation, and tissue morphogenesis [71,72]. In this context, PPIs have been exploited by the pharmaceutical industry for different applications, including drug discovery and diagnosis [22].

Notably, several low-molecular PPIs from plants contain a high content of cysteine residues that form disulfide bridges that confer resistance to heat treatment, extreme pH or ionic forces, and proteolysis [62,73–77]. Owing to their multifunctionality and remarkable physicochemical properties, PPIs have found a wide array of applications in biotechnology and biomedicine, including their use as antimicrobial agents [21,78–80] (Figure 4). Despite their high stability, the presence of multiple cysteine residues involved in disulfide bridges often complicates the production of these molecules, requiring a comprehensive understanding of their folding mechanisms and three-dimensional structure [81,82].

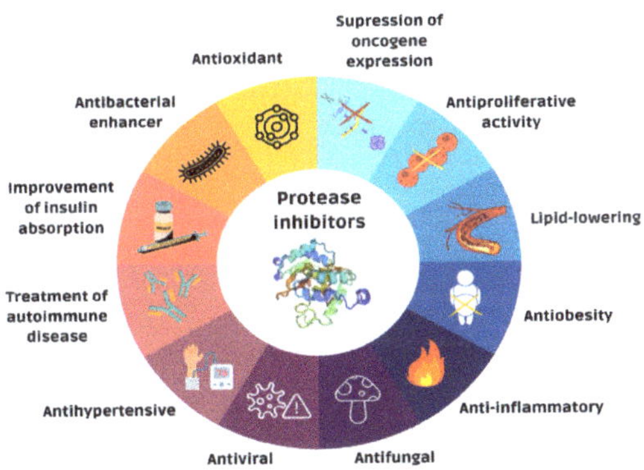


**Figure 4.** Plant peptide protease inhibitors and their potential applications in biomedicine.

## 5. Plant-Derived Peptide Protease Inhibitors with Antimicrobial Properties

Over the last 15 years, there has been a notable increase in the reports on novel PPIs with antimicrobial properties isolated from various plant sources [22,83–85]. An important number of these PPIs have been linked to antimicrobial roles in a physiological context. This is the case of the systemin hormone, an 18-amino acid peptide that plays a critical role in the defense against pathogens and is a key part of the jasmonic acid signaling pathway that triggers PPI expression [73]. In the *Solanaceae* family, the prosystemin mechanism involves the expression of genes predominantly associated with metallo carboxypeptidase inhibitors, cysteine PPIs, and trypsin PPIs, including Kunitz-type PPIs [74,86].

PPIs isolated from a variety of plants have shown potential in inhibiting the growth of pathogenic bacteria, protozoans, and fungal strains (Table 1) [21]. As introduced above, the antimicrobial activity of many plant PPIs is typically mediated through the ability of these molecules to alter the permeability of bacterial cytoplasmic membranes. This toxic action is facilitated via the cationic nature of these peptides, which bind to cell membranes through interactions between positive and negative charges, causing destabilization and the formation of temporary cavities or disruptions. Such damage leads to the leakage of cellular contents and ultimately results in bacterial death [67,75,85]. In addition to their antimicrobial actions, it is also common that these molecules simultaneously target diverse metabolic processes, which has fueled interest in applying these molecules for diverse biomedical applications.

**Table 1.** Peptide protease inhibitors derived from plants with antimicrobial activity.

| Plant Family | Plant Source | Name | Type of PPI | MW (kDa) | Activity | Target Microorganisms | Ref. |
|---|---|---|---|---|---|---|---|
| Fabaceae (Leguminosae) | *Inga edulis* | Inga Edulis Trypsin Inhibitor (IETI) | Kunitz | 19.7 | Antifungal | *Candida tropicalis* and *Candida buinensis* | [76] |
| | *Inga laurina* | Inga Laurina Trypsin Inhibitor (ILTI) | Kunitz | 20 | Antifungal | *C. tropicalis* and *C. buinensis* | [84] |
| | *Acacia plumosa* | Acacia plumosa Trypsin Inhibitor (ApTI A, B, C) | Kunitz | 20 | Antifungal | *Aspergillus niger, Thielaviopsis paradoxa,* and *Colletotrichum* spp. | [77] |
| | *Acacia nilotica* | Acacia nilotica Trypsin Inhibitor (AnTI) | Trypsin inhibitor | 21 | Antibacterial | G+: *Staphylococcus aureus and Bacillus subtilis,* G−: *Escherichia coli,* and *Pseudomonas aeruginosa* | [62] |
| | *Albizia amara* | Proteinaceous protease inhibitor (API) | Unknown | 49 | Antibacterial | G+: *B. subtilis* G−: *P. aeruginosa* | [87] |
| | | | | | Antifungal | *C. albicans* | |
| | *Luetzelburgia auriculata* | Luetzelburgia auriculata Bowman-Birk protease inhibitor (LzaBBI) | Bowman-Birk | 14.3 | Antibacterial | G+: *S. aureus* | [88] |
| | *Cassia fistula* | Fistulin | Trypsin inhibitor | 4 | Antibacterial | G+: *S. aureus*, and *B. subtilis* G−: *Klebsiella pneumoniae, E. coli,* and *P. aeruginosa* | [89] |
| | *Adenanthera pavonina* | Adepamycin | Trypsin inhibitor | 2.4 | Antibacterial | G+: *S. aureus* G−: *E. coli, Klebsiella oxytoca, K. pneumoniae,* and *P. aeruginosa* | [90,91] |
| | | | | | Antifungal | *C. albicans* and *C. tropicalis* | |
| Solanaceae | *Solanum tuberosum* L. Cv. Golden Valley | Potide-G | Kunitz | 5.57 | Antibacterial | G+: *S. aureus*, and *Listeria monocytogenes* G−: *E. coli* | [26] |
| | | | | | Antifungal | *C. albicans* and *Rhizoctonia solani* | |
| | *S. tuberosum* L. Cv. Gogu Valley | Peptide G2 (PG-2) | Kunitz | 3.2 | Antibacterial | G+: *S. aureus* | [92] |
| | | | | | Antifungal | *C. albicans* | |
| | *S. tuberosum* L. Cv. L. Jopung | Antifungal protein J (AFP-J) | Kunitz | 13.5 | Antifungal | *C. albicans, Trichosporon beigelii,* and *Saccharomyces cerevisiae* | [63] |
| | *Capsicum annuum* L. | Yellow Bell Pepper Trypsin Inhibitor (YBPTI) | Trypsin inhibitor | | Antifungal | *C. albicans* | [93] |

**Table 1.** *Cont.*

| Plant Family | Plant Source | Name | Type of PPI | MW (kDa) | Activity | Target Microorganisms | Ref. |
|---|---|---|---|---|---|---|---|
| *Cucurbitaceae* | *Coccinia grandis* (L.) *Voigt.* | *Coccinia grandis* (L.) inhibitor | Protease inhibitor | 14.3 | Antibacterial | G+: *S. aureus*, and *B. subtilis* G−: *K. pneumoniae*, *Proteus vulgaris*, and *E. coli* | [83] |
| | | | | | Antifungal | *C. albicans*, *Mucor indicus*, *Penicillium notatum*, *Aspergillus flavus Aspergillus oryzae*, and *Cryptococcus neoformans* | |
| *Rhamnaceae* | *Rhamnus frangula* | RflP-1 | Kunitz | 22.5 | Antibacterial | G+: *Bacillus* sp. and *Bacillus licheniformis* | [94,95] |
| *Euphorbiaceae* | *Jatropha curcas* | Jatropha curcas Trypsin inhibitor I (JcTI-I) | Trypsin inhibitor | 10.2 | Antibacterial | G+: *S. aureus* G−: *Salmonella enterica* | [96] |
| *Asteraceae* | *Helianthus annuus* | Sunflower Trypsin Inhibitor 1 (SFTI1) | Trypsin inhibitor | | Antibacterial | G+: *Staphylococcus epidermidis*, *S. aureus*, and *Enterococcus faecalis* G−: *E. coli*, *P. aeruginosa*, and *Salmonella typhimurium* | [97] |
| *Moringaceae* | *Moringa oleifera* | Moringa oleifera Protein Inhibitor (MoPI) | Phytocystatin | 19 | Antibacterial | G+: *S. aureus*, and *E. faecalis* | [98] |
| *Lamiaceae* | *Salvia hispanica* | Salvia hispanica Trypsin Inhibitor (ShTI) | Trypsin inhibitor | 11 | Antibacterial | G+: *Methicillin-resistant S. aureus* | [99] |

### 5.1. Plant Protease Inhibitors with Antibacterial Activity

Potatoes (*Solanum. tuberosum* spp.) have historically been a rich source of plant PPIs with antimicrobial (antifungal and antibacterial) activities against a wide variety of agricultural and clinical pathogens [100]. A prototypical example is the Kunitz-type serine PPI, Potide-G, isolated from the tubers of *S. tuberosum* L. cv. Golden Valley, which has been shown to be effective in inhibiting the growth of diverse human pathogens, including *S. aureus* and *L. monocytogenes*, through the regulation of extracellular enzymes related to nutrition. This inhibitor showed a strong potency with minimum inhibitory concentration (MIC) values lower than 30 μg/mL, demonstrating a potency comparable to established antibiotics [26]. Similarly, PG-2, a peptide isolated from the potato tubers of cv. Gogu Valley, exhibited antibacterial activity against *S. aureus* and other pathogens, displaying minimal cytotoxic effects against human red blood cells [92].

A significant number of studies have reported on the action mechanisms of trypsin inhibitors on the bacterial membrane. A trypsin PPI called API was purified from *A. amara* seeds, showing antimicrobial activity against *P. aeruginosa* and *B. subtilis*, being useful as a potential antibacterial component [87]. In 2018, Martins et al. [88] discovered a Bowman–Birk PPI (called LzaBBI) from *L. auriculata* seeds that exhibits robust antibacterial activity against *S. aureus*. This 14.3 kDa thermostable protein, similar to other Bowman–Birk inhibitors, demonstrates mixed-type inhibitory activity against trypsin and chymotrypsin. Scanning electron microscopy experiments unequivocally revealed that LzaBBI compromises *S. aureus* membrane integrity, possibly through oxidative stress and reactive oxygen species (ROS) production, leading to cell lysis. In a previous study, Satheesh and Murugan [83] identified a 14.3 kDa trypsin PPI from *C. grandis* leaves that effectively inhibits several pathogenic bacteria, including *S. aureus*. Similarly, a 4 kDa trypsin inhibitor from *C. fistula* leaves, called Fistulin, demonstrated significant antibacterial activity, paralleling the efficacy of streptomycin sulfate by inhibiting microbial proteases [89]. Additionally, a new

Kunitz-type inhibitor was isolated from *R. frangula* leaves. This inhibitor displayed a potent inhibition against several proteases from Bacillus sp, such as sd *B. licheniformis* [94,95].

In 2020, Almeida and coworkers [90] synthesized Adepamycin, a peptide based on a trypsin inhibitor from *A. pavonina* seeds, showing antimicrobial activity by damaging the membrane integrity of various pathogens without displaying toxicity against human red blood cells. Additionally, Costa et al. [96] explored the action of a peptide from *J. curcas* seeds, JcTI-I, on proteases from *S. enterica* and *S. aureus*, suggesting an alternative antimicrobial mechanism of trypsin inhibitors. Further studies, such as those by Mehmood et al. [62], Rodrigues et al. [91], and Wang et al. [97], have further uncovered the bacteriostatic properties of trypsin inhibitors and their efficacy against bacterial pathogens.

Notably, two recent studies have unveiled novel PPIs with significant antimicrobial potential from Moringa and Salvia plants. In 2021, Cotabarren et al. [98] reported the purification and characterization of the first phytocystatin isolated from *M. oleifera* seeds (MoPI). MoPI, with a molecular mass of 19 kDa, shows higher physicochemical stability against acidic pHs and high temperatures. The study highlighted MoPI as one of the most potent cysteine PPI identified to date, exhibiting Ki and IC50 values within the nanomolar range. Additionally, MoPI has been recognized as a multifunctional molecule, showing robust antimicrobial activity against human pathogens, including *E. faecalis* and *S. aureus*, as well as significant anticoagulant activity. More recently, a trypsin inhibitor named ShTI was isolated and characterized from mucilage and fat-free chia seeds (*S. hispanica* L.). This inhibitor demonstrated exceptional thermostability and wide pH tolerance, likely due to its disulfide bridge structure [99]. ShTI exhibited antibacterial activity against *S. aureus*, including strains resistant to methicillin. However, it showed no significant effects against the Gram-negative bacteria tested. As such, ShTI presents itself as a promising candidate for standalone use or in combination therapy with oxacillin for managing *S. aureus* infections.

### 5.2. Plant Protease Inhibitors with Antifungal Activity

Phytopathogenic fungi produce extracellular proteinases that play an active role in the development of plant diseases [101]. In response to the attack, plants synthesize inhibitors to inactivate these proteinases [21]. This phenomenon was observed for the first time in tomatoes infected by *Phytophthora infestans* [102], in which it was found that high levels of trypsin and chymotrypsin inhibitors correlated with the resistance of the plants to the pathogen [103]. A similar strategy is also used by *S. tuberosum* spp., which produce various PPIs with strong antifungal activity [100]. Importantly, these bioactive molecules extend their fungicidal efficacy beyond phytopathogenic strains to include numerous human pathogens, highlighting their broad-spectrum antimicrobial potential.

Among the inhibitors mentioned in the previous section, Potide-G and PG-2, both Kunitz-type serine PPIs, have been shown to effectively inhibit the growth of several pathogenic fungal strains, such as *C. albicans* [26,92]. Similarly, AFP-J, isolated from *S. tuberosum* cv. L. Jopung [63], targets chymotrypsin, pepsin, and trypsin, exhibiting significant antifungal activity against *C. albicans*, *T. beigelii*, and *S. cerevisiae*. Additional Kunitz-type trypsin and chymotrypsin inhibitors with antifungal properties have been isolated and characterized. Among these, ApTIA, ApTIB, and ApTIC were extracted from the seeds of the Brazilian plant *A. plumosa*, demonstrating notable antifungal activity against pathogenic fungi, including *A. niger*, *T. paradoxa*, and *Colletotrichum* spp. Their mechanism of action is believed to involve the inhibition of serine proteases secreted by these fungi [77]. Moreover, Kunitz-type inhibitors such as RfIP-1 and PDInhibitor have been identified from *R. frangula* and *Conyza dioscoridis*, respectively. These inhibitors have exhibited strong antifungal activity against various fungal strains, including the commercial aerobic filamentous fungus *A. oryzae*. Furthermore, the protease inhibitor from *Coccinia grandis* has demonstrated an antifungal effect on *C. albicans*, *Mucor indicus*, *Penicillium notatum*, *Aspergillus flavus*, *Aspergillus oryzae*, and *Cryptococcus neoformans* [83].

From the Fabaceae family, the tropical trees *I. edulis* and *I. laurina* have also been explored as sources of novel bioactive molecules with antifungal properties. Two distinct Kunitz-type trypsin inhibitors, ILTI and IETI, exhibiting anticandidal activity, were isolated, and their activities characterized [76,84]. ILTI demonstrated potent inhibitory enzymatic activity against trypsin and significantly inhibited the growth of *C. tropicalis* and *C. buinensis*, causing changes in the membrane and permeabilization of yeast cells, as well as the production of reactive oxygen species (ROS). However, this inhibitor did not show efficacy against human pathogenic bacteria such as *E. coli*, *S. aureus*, and *K. pneumoniae*. IETI, characterized by a single trypsin-reactive site stabilized by a disulfide bridge and notable pH and thermal stability, displayed growth-inhibitory activity towards *Candida* spp., including *C. buinensis* and *C. tropicalis*. The yeast inhibitory action of these inhibitors involves multiple mechanisms, including protease inhibition, plasma membrane disruption affecting cell viability, oxidative stress targeting mitochondria, and inducing apoptosis in yeast by blocking critical serine peptidases (e.g., metacaspases) and a nuclear apoptosis mediator [104]. It was also observed that API from *Albizia amara* had an inhibitory effect on *C. albicans* [87], and Adepamycin from *Adenanthera pavonina* inhibited the growth of *C. albicans* and *C. tropicalis* [87,88]. Only recently, Cotabarren et al. [93] reported the purification and characterization of YBPTI, a thermostable trypsin inhibitor from yellow pepper (*C. annuum* L.) seeds with dual antifungal and hypoglycemic properties. The purified inhibitor showed potent and specific in vitro activity against *C. albicans* and showed no cytotoxicity against human Hela cells (ATCC CCL-2). Furthermore, the inhibitor also exhibited α-1,4-glucosidase inhibition activity, positioning this inhibitor as one of the first plant-derived molecules with such a particular dual combination of biological activities, demonstrating that this new inhibitor could be potentially used as an antifungal agent in pharmaceutical preparations to prevent invasive *Candida* infections.

In summary, plant-derived PPIs represent a valuable resource for the development of novel antimicrobial agents. The inherent stability, broad-spectrum antimicrobial efficacy, multifaceted functionality, and minimal toxicity to human cells position these natural compounds as promising candidates for advancing the treatment of microbial infections. Additionally, some of these inhibitors combine both strong antifungal activity and antibacterial properties, potentially serving as dual-purpose antimicrobial molecules.

## 6. Concluding Remarks and Future Perspectives

The worldwide rise of AMR and its impacts on plant, animal, and human health represent a formidable challenge to global health, with the rate of resistance development outpacing the introduction of new antibiotics. The widespread and sometimes indiscriminate use of antimicrobials in medicine, agriculture, and industry has accelerated this process, driving the emergence and spread of AMR as one of the world's most pressing public health issues. Currently, infections by AMR pathogens are causing over 700,000 deaths per year. It is estimated that by 2050, AMR infections will be responsible for 10 million deaths annually worldwide, significantly impacting the global economy [13,105,106]. Due to its great complexity, it must be approached from different disciplines to frame it within the One Health approach. Thus, several countries have implemented national action plans to combat antibiotic-resistant microbes, following the guidelines of the FAO [107] and the WHO [14]. In this context, natural products, particularly from plant sources, offer a vast repository of chemical diversity that could lead to the discovery of novel antimicrobial agents [2].

Over the last 20 years, a number of natural PPIs with antibacterial and antifungal properties have been isolated from diverse plant sources. In this review, we aim to provide a comprehensive analysis of these molecules and insight into their specific activities, with a special focus on exploring PPIs as relevant agents against biomedical pathogens. The review has described the classification and the diverse mechanisms of action of these proteinaceous molecules, which exhibited a wide range of activities, ranging from disrupting microbial cell membranes to inhibiting vital enzymatic pathways. Taken together, these mechanisms provide novel opportunities for combating pathogens that have evolved resistance to current treatments.

However, to date, only a limited number of antimicrobial peptides (AMPs), such as nisin, gramicidin, polymyxins, and daptomycin have been adopted for clinical use [27]. There remains a significant need for extensive research to identify and enhance potential plant-derived PPIs with therapeutic significance. In this context, efforts have been made to develop novel computational tools aimed at predicting antimicrobial peptide sequences [108]. Moreover, such in silico approaches can also be utilized to improve molecules already known for their antimicrobial activity. Structure-based drug discovery strategies, for instance, might employ the known three-dimensional structures of target proteins to design molecules with enhanced biological activities [109]. This rational design approach will facilitate the development of novel, engineered molecules with superior antimicrobial properties.

In summary, this review illustrates the enormous potential of the PPIs from plant species in medicine. Their broad-spectrum antimicrobial activities, combined with their specificity and minimal toxicity towards non-target organisms, highlight their value as a sustainable and effective alternative to conventional synthetic chemicals. This, coupled with their multifunctionality, positions these molecules as integral components of the next generation of antimicrobial agents.

**Author Contributions:** Conceptualization, M.G.P., J.G.-P. and W.D.O.; writing—review and editing, M.G.P., B.O., S.M.V.G., J.G.-P. and W.D.O. All authors have read and agreed to the published version of the manuscript.

**Funding:** W.D.O. (Researcher Career Program) and B.O. (Doctoral Fellow) are members of CONICET, Argentina. The authors acknowledge the support from Universidad Nacional de La Plata (Argentina) under grants UNLP I+D 2020 project Code X-920, and the ANPCyT under Project PICT-2021-00955. J.G.-P. was supported by the Spanish Ministry of Science and Innovation with a Juan de la Cierva Incorporacion (IJC2019-041039-I).

**Institutional Review Board Statement:** Not applicable.

**Informed Consent Statement:** Not applicable.

**Data Availability Statement:** Not applicable.

**Conflicts of Interest:** The authors declare no conflicts of interest.

## References

1. Lorsbach, B.A.; Sparks, T.C.; Cicchillo, R.M.; Garizi, N.V.; Hahn, D.R.; Meyer, K.G. Natural products: A strategic lead generation approach in crop protection discovery. *Pest Manag. Sci.* **2019**, *75*, 2301–2309. [CrossRef]
2. Bergonzi, M.C.; Heard, C.M.; Garcia-Pardo, J. Bioactive Molecules from Plants: Discovery and Pharmaceutical Applications. *Pharmaceutics* **2022**, *14*, 2116. [CrossRef] [PubMed]
3. Harvey, A.L. Natural products in drug discovery. *Drug Discov. Today* **2008**, *13*, 894–901. [CrossRef] [PubMed]
4. Harvey, A.L.; Edrada-Ebel, R.; Quinn, R.J. The re-emergence of natural products for drug discovery in the genomics era. *Nat. Rev. Drug Discov.* **2015**, *14*, 111–129. [CrossRef] [PubMed]
5. Haider, R. harmaceutical and Biopharmaceuticals Industries: Revolutionizing Healthcare. *Asian J. Nat. Sci.* **2023**, *2*, 11.
6. Clemons, P.A.; Bodycombe, N.E.; Carrinski, H.A.; Wilson, J.A.; Shamji, A.F.; Wagner, B.K.; Koehler, A.N.; Schreiber, S.L. Small molecules of different origins have distinct distributions of structural complexity that correlate with protein-binding profiles. *Proc. Natl. Acad. Sci. USA* **2010**, *107*, 18787–18792. [CrossRef] [PubMed]
7. Despotovic, M.; de Nies, L.; Busi, S.B.; Wilmes, P. Reservoirs of antimicrobial resistance in the context of One Health. *Curr. Opin. Microbiol.* **2023**, *73*, 102291. [CrossRef] [PubMed]
8. Tang, K.W.K.; Millar, B.C.; Moore, J.E. Antimicrobial Resistance (AMR). *Br. J. Biomed. Sci.* **2023**, *80*, 11387. [CrossRef] [PubMed]
9. Mudenda, S.; Daka, V.; Matafwali, S.K. World Health Organization AWaRe framework for antibiotic stewardship: Where are we now and where do we need to go? An expert viewpoint. *Antimicrob. Steward. Healthc. Epidemiol.* **2023**, *3*, e84. [CrossRef] [PubMed]
10. Collignon, P.J.; McEwen, S.A. One Health-Its Importance in Helping to Better Control Antimicrobial Resistance. *Trop. Med. Infect. Dis.* **2019**, *4*, 22. [CrossRef] [PubMed]
11. McEwen, S.A.; Collignon, P.J. Antimicrobial Resistance: A One Health Perspective. *Microbiol. Spectr.* **2018**, *6*. [CrossRef] [PubMed]
12. White, A.; Hughes, J.M. Critical Importance of a One Health Approach to Antimicrobial Resistance. *Ecohealth* **2019**, *16*, 404–409. [CrossRef] [PubMed]

13. Velazquez-Meza, M.E.; Galarde-Lopez, M.; Carrillo-Quiroz, B.; Alpuche-Aranda, C.M. Antimicrobial resistance: One Health approach. *Vet. World* **2022**, *15*, 743–749. [CrossRef] [PubMed]
14. WHO. Antimicrobial Resistance. Available online: https://www.who.int/news-room/fact-sheets/detail/antimicrobial-resistance (accessed on 2 April 2024).
15. FDA. Drugs. Available online: https://www.fda.gov/drugs (accessed on 2 April 2024).
16. Newman, D.J.; Cragg, G.M. Natural Products as Sources of New Drugs over the Nearly Four Decades from 01/1981 to 09/2019. *J. Nat. Prod.* **2020**, *83*, 770–803. [CrossRef] [PubMed]
17. Boman, H.G. Antibacterial peptides: Basic facts and emerging concepts. *J. Intern. Med.* **2003**, *254*, 197–215. [CrossRef] [PubMed]
18. Bulet, P.; Stocklin, R.; Menin, L. Anti-microbial peptides: From invertebrates to vertebrates. *Immunol. Rev.* **2004**, *198*, 169–184. [CrossRef] [PubMed]
19. van Loon, L.C.; Rep, M.; Pieterse, C.M. Significance of inducible defense-related proteins in infected plants. *Annu. Rev. Phytopathol.* **2006**, *44*, 135–162. [CrossRef] [PubMed]
20. Parisi, M.G.; Moreno, S.; Fernandez, G. Isolation and characterization of a dual function protein from Allium sativum bulbs which exhibits proteolytic and hemagglutinating activities. *Plant Physiol. Biochem.* **2008**, *46*, 403–413. [CrossRef] [PubMed]
21. Kim, J.Y.; Park, S.C.; Hwang, I.; Cheong, H.; Nah, J.W.; Hahm, K.S.; Park, Y. Protease inhibitors from plants with antimicrobial activity. *Int. J. Mol. Sci.* **2009**, *10*, 2860–2872. [CrossRef]
22. Shamsi, T.N.; Parveen, R.; Fatima, S. Characterization, biomedical and agricultural applications of protease inhibitors: A review. *Int. J. Biol. Macromol.* **2016**, *91*, 1120–1133. [CrossRef]
23. Srivastava, S.; Dashora, K.; Ameta, K.L.; Singh, N.P.; El-Enshasy, H.A.; Pagano, M.C.; Hesham, A.E.; Sharma, G.D.; Sharma, M.; Bhargava, A. Cysteine-rich antimicrobial peptides from plants: The future of antimicrobial therapy. *Phytother. Res.* **2021**, *35*, 256–277. [CrossRef] [PubMed]
24. Lima, A.M.; Azevedo, M.I.G.; Sousa, L.M.; Oliveira, N.S.; Andrade, C.R.; Freitas, C.D.T.; Souza, P.F.N. Plant antimicrobial peptides: An overview about classification, toxicity and clinical applications. *Int. J. Biol. Macromol.* **2022**, *214*, 10–21. [CrossRef] [PubMed]
25. Zhang, Q.Y.; Yan, Z.B.; Meng, Y.M.; Hong, X.Y.; Shao, G.; Ma, J.J.; Cheng, X.R.; Liu, J.; Kang, J.; Fu, C.Y. Antimicrobial peptides: Mechanism of action, activity and clinical potential. *Mil. Med. Res.* **2021**, *8*, 48. [CrossRef] [PubMed]
26. Kim, M.H.; Park, S.C.; Kim, J.Y.; Lee, S.Y.; Lim, H.T.; Cheong, H.; Hahm, K.S.; Park, Y. Purification and characterization of a heat-stable serine protease inhibitor from the tubers of new potato variety "Golden Valley". *Biochem. Biophys. Res. Commun.* **2006**, *346*, 681–686. [CrossRef] [PubMed]
27. Dijksteel, G.S.; Ulrich, M.M.W.; Middelkoop, E.; Boekema, B. Review: Lessons Learned From Clinical Trials Using Antimicrobial Peptides (AMPs). *Front. Microbiol.* **2021**, *12*, 616979. [CrossRef] [PubMed]
28. Mazurkiewicz-Pisarek, A.; Baran, J.; Ciach, T. Antimicrobial Peptides: Challenging Journey to the Pharmaceutical, Biomedical, and Cosmeceutical Use. *Int. J. Mol. Sci.* **2023**, *24*, 9031. [CrossRef]
29. Wang, G.; Li, X.; Wang, Z. APD3: The antimicrobial peptide database as a tool for research and education. *Nucleic Acids Res.* **2016**, *44*, D1087–D1093. [CrossRef]
30. Kundu, R. Cationic Amphiphilic Peptides: Synthetic Antimicrobial Agents Inspired by Nature. *ChemMedChem* **2020**, *15*, 1887–1896. [CrossRef] [PubMed]
31. Kumar, V.; Singh, B.; van Belkum, M.J.; Diep, D.B.; Chikindas, M.L.; Ermakov, A.M.; Tiwari, S.K. Halocins, natural antimicrobials of Archaea: Exotic or special or both? *Biotechnol. Adv.* **2021**, *53*, 107834. [CrossRef] [PubMed]
32. Huan, Y.; Kong, Q.; Mou, H.; Yi, H. Antimicrobial Peptides: Classification, Design, Application and Research Progress in Multiple Fields. *Front. Microbiol.* **2020**, *11*, 582779. [CrossRef] [PubMed]
33. Saeed, S.I.; Mergani, A.; Aklilu, E.; Kamaruzzman, N.F. Antimicrobial Peptides: Bringing Solution to the Rising Threats of Antimicrobial Resistance in Livestock. *Front. Vet. Sci.* **2022**, *9*, 851052. [CrossRef] [PubMed]
34. Ramazi, S.; Mohammadi, N.; Allahverdi, A.; Khalili, E.; Abdolmaleki, P. A review on antimicrobial peptides databases and the computational tools. *Database* **2022**, *2022*, baac011. [CrossRef] [PubMed]
35. Bin Hafeez, A.; Jiang, X.; Bergen, P.J.; Zhu, Y. Antimicrobial Peptides: An Update on Classifications and Databases. *Int. J. Mol. Sci.* **2021**, *22*, 11691. [CrossRef] [PubMed]
36. Li, J.; Hu, S.; Jian, W.; Xie, C.; Yang, X. Plant antimicrobial peptides: Structures, functions, and applications. *Bot. Stud.* **2021**, *62*, 5. [CrossRef] [PubMed]
37. Dos Santos-Silva, C.A.; Zupin, L.; Oliveira-Lima, M.; Vilela, L.M.B.; Bezerra-Neto, J.P.; Ferreira-Neto, J.R.; Ferreira, J.D.C.; de Oliveira-Silva, R.L.; Pires, C.J.; Aburjaile, F.F.; et al. Plant Antimicrobial Peptides: State of the Art, In Silico Prediction and Perspectives in the Omics Era. *Bioinform. Biol. Insights* **2020**, *14*, 1177932220952739. [CrossRef] [PubMed]
38. Taveira, G.B.; Carvalho, A.O.; Rodrigues, R.; Trindade, F.G.; Da Cunha, M.; Gomes, V.M. Thionin-like peptide from Capsicum annuum fruits: Mechanism of action and synergism with fluconazole against *Candida* species. *BMC Microbiol.* **2016**, *16*, 12. [CrossRef] [PubMed]
39. Taveira, G.B.; Mello, E.O.; Carvalho, A.O.; Regente, M.; Pinedo, M.; de La Canal, L.; Rodrigues, R.; Gomes, V.M. Antimicrobial activity and mechanism of action of a thionin-like peptide from *Capsicum annuum* fruits and combinatorial treatment with fluconazole against *Fusarium solani*. *Biopolymers* **2017**, *108*, e23008. [CrossRef] [PubMed]

40. Hong, K.; Austerlitz, T.; Bohlmann, T.; Bohlmann, H. The thionin family of antimicrobial peptides. *PLoS ONE* **2021**, *16*, e0254549. [CrossRef] [PubMed]
41. Fu, J.; Zong, X.; Jin, M.; Min, J.; Wang, F.; Wang, Y. Mechanisms and regulation of defensins in host defense. *Signal. Transduct. Target. Ther.* **2023**, *8*, 300. [CrossRef]
42. Tiwari, P.; Srivastava, Y.; Sharma, A.; Vinayagam, R. Antimicrobial Peptides: The Production of Novel Peptide-Based Therapeutics in Plant Systems. *Life* **2023**, *13*, 1875. [CrossRef]
43. Odintsova, T.; Shcherbakova, L.; Slezina, M.; Pasechnik, T.; Kartabaeva, B.; Istomina, E.; Dzhavakhiya, V. Hevein-Like Antimicrobial Peptides Wamps: Structure-Function Relationship in Antifungal Activity and Sensitization of Plant Pathogenic Fungi to Tebuconazole by WAMP-2-Derived Peptides. *Int. J. Mol. Sci.* **2020**, *21*, 7912. [CrossRef] [PubMed]
44. Molesini, B.; Treggiari, D.; Dalbeni, A.; Minuz, P.; Pandolfini, T. Plant cystine-knot peptides: Pharmacological perspectives. *Br. J. Clin. Pharmacol.* **2017**, *83*, 63–70. [CrossRef] [PubMed]
45. Postic, G.; Gracy, J.; Perin, C.; Chiche, L.; Gelly, J.C. KNOTTIN: The database of inhibitor cystine knot scaffold after 10 years, toward a systematic structure modeling. *Nucleic Acids Res.* **2018**, *46*, D454–D458. [CrossRef] [PubMed]
46. Slavokhotova, A.A.; Rogozhin, E.A. Defense Peptides From the alpha-Hairpinin Family Are Components of Plant Innate Immunity. *Front. Plant Sci.* **2020**, *11*, 465. [CrossRef] [PubMed]
47. Tam, J.P.; Wang, S.; Wong, K.H.; Tan, W.L. Antimicrobial Peptides from Plants. *Pharmaceuticals* **2015**, *8*, 711–757. [CrossRef] [PubMed]
48. Souza, A.A.; Costa, A.S.; Campos, D.C.O.; Batista, A.H.M.; Sales, G.W.P.; Nogueira, N.A.P.; Alves, K.M.M.; Coelho-de-Souza, A.N.; Oliveira, H.D. Lipid transfer protein isolated from noni seeds displays antibacterial activity in vitro and improves survival in lethal sepsis induced by CLP in mice. *Biochimie* **2018**, *149*, 9–17. [CrossRef] [PubMed]
49. Berrocal-Lobo, M.; Segura, A.; Moreno, M.; Lopez, G.; Garcia-Olmedo, F.; Molina, A. Snakin-2, an antimicrobial peptide from potato whose gene is locally induced by wounding and responds to pathogen infection. *Plant Physiol.* **2002**, *128*, 951–961. [CrossRef]
50. Gould, A.; Camarero, J.A. Cyclotides: Overview and Biotechnological Applications. *Chembiochem* **2017**, *18*, 1350–1363. [CrossRef]
51. Jacob, B.; Vogelaar, A.; Cadenas, E.; Camarero, J.A. Using the Cyclotide Scaffold for Targeting Biomolecular Interactions in Drug Development. *Molecules* **2022**, *27*, 6430. [CrossRef]
52. Slezina, M.P.; Odintsova, T.I. Plant Antimicrobial Peptides: Insights into Structure-Function Relationships for Practical Applications. *Curr. Issues Mol. Biol.* **2023**, *45*, 3674–3704. [CrossRef]
53. Sharma, P.; Kaur, J.; Sharma, G.; Kashyap, P. Plant derived antimicrobial peptides: Mechanism of target, isolation techniques, sources and pharmaceutical applications. *J. Food Biochem.* **2022**, *46*, e14348. [CrossRef] [PubMed]
54. Kumar, P.; Kizhakkedathu, J.N.; Straus, S.K. Antimicrobial Peptides: Diversity, Mechanism of Action and Strategies to Improve the Activity and Biocompatibility In Vivo. *Biomolecules* **2018**, *8*, 4. [CrossRef] [PubMed]
55. Mahlapuu, M.; Hakansson, J.; Ringstad, L.; Bjorn, C. Antimicrobial Peptides: An Emerging Category of Therapeutic Agents. *Front. Cell Infect. Microbiol.* **2016**, *6*, 194. [CrossRef] [PubMed]
56. Monroc, S.; Badosa, E.; Feliu, L.; Planas, M.; Montesinos, E.; Bardaji, E. De novo designed cyclic cationic peptides as inhibitors of plant pathogenic bacteria. *Peptides* **2006**, *27*, 2567–2574. [CrossRef] [PubMed]
57. Guha, S.; Ghimire, J.; Wu, E.; Wimley, W.C. Mechanistic Landscape of Membrane-Permeabilizing Peptides. *Chem. Rev.* **2019**, *119*, 6040–6085. [CrossRef] [PubMed]
58. Lee, H.; Lim, S.I.; Shin, S.H.; Lim, Y.; Koh, J.W.; Yang, S. Conjugation of Cell-Penetrating Peptides to Antimicrobial Peptides Enhances Antibacterial Activity. *ACS Omega* **2019**, *4*, 15694–15701. [CrossRef]
59. Montesinos, E. Functional Peptides for Plant Disease Control. *Annu. Rev. Phytopathol.* **2023**, *61*, 301–324. [CrossRef] [PubMed]
60. Le, C.F.; Fang, C.M.; Sekaran, S.D. Intracellular Targeting Mechanisms by Antimicrobial Peptides. *Antimicrob. Agents Chemother.* **2017**, *61*, e02340-16. [CrossRef] [PubMed]
61. Tiwari, M.; Pati, D.; Mohapatra, R.; Sahu, B.B.; Singh, P. The Impact of Microbes in Plant Immunity and Priming Induced Inheritance: A Sustainable Approach for Crop protection. *Plant Stress* **2022**, *4*, 100072. [CrossRef]
62. Mehmood, S.; Imran, M.; Ali, A.; Munawar, A.; Khaliq, B.; Anwar, F.; Saeed, Q.; Buck, F.; Hussain, S.; Saeed, A.; et al. Model prediction of a Kunitz-type trypsin inhibitor protein from seeds of *Acacia nilotica* L. with strong antimicrobial and insecticidal activity. *Turk. J. Biol.* **2020**, *44*, 188–200. [CrossRef] [PubMed]
63. Park, Y.; Choi, B.H.; Kwak, J.S.; Kang, C.W.; Lim, H.T.; Cheong, H.S.; Hahm, K.S. Kunitz-type serine protease inhibitor from potato (*Solanum tuberosum* L. cv. Jopung). *J. Agric. Food Chem.* **2005**, *53*, 6491–6496. [CrossRef] [PubMed]
64. Clemente, M.; Corigliano, M.G.; Pariani, S.A.; Sanchez-Lopez, E.F.; Sander, V.A.; Ramos-Duarte, V.A. Plant Serine Protease Inhibitors: Biotechnology Application in Agriculture and Molecular Farming. *Int. J. Mol. Sci.* **2019**, *20*, 1345. [CrossRef] [PubMed]
65. Bode, W.; Huber, R. Natural protein proteinase inhibitors and their interaction with proteinases. *Eur. J. Biochem.* **1992**, *204*, 433–451. [CrossRef] [PubMed]
66. Laskowski, M., Jr.; Kato, I. Protein inhibitors of proteinases. *Annu. Rev. Biochem.* **1980**, *49*, 593–626. [CrossRef] [PubMed]
67. Cotabarren, J.; Lufrano, D.; Parisi, M.G.; Obregon, W.D. Biotechnological, biomedical, and agronomical applications of plant protease inhibitors with high stability: A systematic review. *Plant Sci.* **2020**, *292*, 110398. [CrossRef] [PubMed]
68. Dzik, S. Complement and Coagulation: Cross Talk Through Time. *Transfus. Med. Rev.* **2019**, *33*, 199–206. [CrossRef] [PubMed]

69. Bonturi, C.R.; Silva Teixeira, A.B.; Rocha, V.M.; Valente, P.F.; Oliveira, J.R.; Filho, C.M.B.; Fatima Correia Batista, I.; Oliva, M.L.V. Plant Kunitz Inhibitors and Their Interaction with Proteases: Current and Potential Pharmacological Targets. *Int. J. Mol. Sci.* **2022**, *23*, 4742. [CrossRef] [PubMed]
70. Varkoly, K.; Beladi, R.; Hamada, M.; McFadden, G.; Irving, J.; Lucas, A.R. Viral SERPINS-A Family of Highly Potent Immune-Modulating Therapeutic Proteins. *Biomolecules* **2023**, *13*, 1393. [CrossRef] [PubMed]
71. Udvardy, A. The role of controlled proteolysis in cell-cycle regulation. *Eur. J. Biochem.* **1996**, *240*, 307–313. [CrossRef] [PubMed]
72. Heutinck, K.M.; ten Berge, I.J.; Hack, C.E.; Hamann, J.; Rowshani, A.T. Serine proteases of the human immune system in health and disease. *Mol. Immunol.* **2010**, *47*, 1943–1955. [CrossRef] [PubMed]
73. Pearce, G. Systemin, hydroxyproline-rich systemin and the induction of protease inhibitors. *Curr. Protein Pept. Sci.* **2011**, *12*, 399–408. [CrossRef]
74. Coppola, M.; Corrado, G.; Coppola, V.; Cascone, P.; Martinelli, R.; Digilio, M.C.; Pennacchio, F.; Rao, R. Prosystemin Overexpression in Tomato Enhances Resistance to Different Biotic Stresses by Activating Genes of Multiple Signaling Pathways. *Plant Mol. Biol. Rep.* **2015**, *33*, 1270–1285. [CrossRef] [PubMed]
75. Huang, Y.; Huang, J.; Chen, Y. Alpha-helical cationic antimicrobial peptides: Relationships of structure and function. *Protein Cell* **2010**, *1*, 143–152. [CrossRef]
76. Dib, H.X.; de Oliveira, D.G.L.; de Oliveira, C.F.R.; Taveira, G.B.; de Oliveira Mello, E.; Verbisk, N.V.; Chang, M.R.; Correa Junior, D.; Gomes, V.M.; Macedo, M.L.R. Biochemical characterization of a Kunitz inhibitor from Inga edulis seeds with antifungal activity against *Candida* spp. *Arch. Microbiol.* **2019**, *201*, 223–233. [CrossRef] [PubMed]
77. Lopes, J.L.; Valadares, N.F.; Moraes, D.I.; Rosa, J.C.; Araujo, H.S.; Beltramini, L.M. Physico-chemical and antifungal properties of protease inhibitors from *Acacia plumosa*. *Phytochemistry* **2009**, *70*, 871–879. [CrossRef] [PubMed]
78. Cid-Gallegos, M.S.; Corzo-Rios, L.J.; Jimenez-Martinez, C.; Sanchez-Chino, X.M. Protease Inhibitors from Plants as Therapeutic Agents—A Review. *Plant Foods Hum. Nutr.* **2022**, *77*, 20–29. [CrossRef] [PubMed]
79. de Freitas, M.A.G.; Amaral, N.O.; Alvares, A.; de Oliveira, S.A.; Mehdad, A.; Honda, D.E.; Bessa, A.S.M.; Ramada, M.H.S.; Naves, L.M.; Pontes, C.N.R.; et al. Blood pressure-lowering effects of a Bowman-Birk inhibitor and its derived peptides in normotensive and hypertensive rats. *Sci. Rep.* **2020**, *10*, 11680. [CrossRef] [PubMed]
80. Gitlin-Domagalska, A.; Maciejewska, A.; Debowski, D. Bowman-Birk Inhibitors: Insights into Family of Multifunctional Proteins and Peptides with Potential Therapeutical Applications. *Pharmaceuticals* **2020**, *13*, 421. [CrossRef] [PubMed]
81. Bronsoms, S.; Villanueva, J.; Canals, F.; Querol, E.; Aviles, F.X. Analysis of the effect of potato carboxypeptidase inhibitor pro-sequence on the folding of the mature protein. *Eur. J. Biochem.* **2003**, *270*, 3641–3650. [CrossRef] [PubMed]
82. Arolas, J.L.; Ventura, S. Protease inhibitors as models for the study of oxidative folding. *Antioxid. Redox Signal.* **2011**, *14*, 97–112. [CrossRef] [PubMed]
83. Satheesh, L.S.; Murugan, K. Antimicrobial activity of protease inhibitor from leaves of *Coccinia grandis* (L.) Voigt. *Indian J. Exp. Biol.* **2011**, *49*, 366–374. [PubMed]
84. Macedo, M.L.; Ribeiro, S.F.; Taveira, G.B.; Gomes, V.M.; de Barros, K.M.; Maria-Neto, S. Antimicrobial Activity of ILTI, a Kunitz-Type Trypsin Inhibitor from *Inga laurina* (SW.) Willd. *Curr. Microbiol.* **2016**, *72*, 538–544. [CrossRef] [PubMed]
85. de Souza Nascimento, A.M.; de Oliveira Segundo, V.H.; Felipe Camelo Aguiar, A.J.; Piuvezam, G.; Souza Passos, T.; Florentino da Silva Chaves Damasceno, K.; de Araujo Morais, A.H. Antibacterial action mechanisms and mode of trypsin inhibitors: A systematic review. *J. Enzyme. Inhib. Med. Chem.* **2022**, *37*, 749–759. [CrossRef] [PubMed]
86. Coppola, M.; Lelio, I.D.; Romanelli, A.; Gualtieri, L.; Molisso, D.; Ruocco, M.; Avitabile, C.; Natale, R.; Cascone, P.; Guerrieri, E.; et al. Tomato Plants Treated with Systemin Peptide Show Enhanced Levels of Direct and Indirect Defense Associated with Increased Expression of Defense-Related Genes. *Plants* **2019**, *8*, 395. [CrossRef] [PubMed]
87. Dabhade, A.R.; Mokashe, N.U.; Patil, U.K. Purification, characterization, and antimicrobial activity of nontoxic trypsin inhibitor from *Albizia amara* Boiv. *Process Biochem.* **2016**, *51*, 659–674. [CrossRef]
88. Martins, T.F.; Vasconcelos, I.M.; Silva, R.G.G.; Silva, F.D.A.; Souza, P.F.N.; Varela, A.L.N.; Albuquerque, L.M.; Oliveira, J.T.A. A Bowman-Birk Inhibitor from the Seeds of Luetzelburgia auriculata Inhibits *Staphylococcus aureus* Growth by Promoting Severe Cell Membrane Damage. *J. Nat. Prod.* **2018**, *81*, 1497–1507. [CrossRef] [PubMed]
89. Arulpandi, I.; Sangeetha, R. Antibacterial Activity of Fistulin: A Protease Inhibitor Purified from the Leaves of *Cassia fistula*. *ISRN Pharm.* **2012**, *2012*, 584073. [CrossRef] [PubMed]
90. Almeida, L.H.O.; Oliveira, C.F.R.; Rodrigues, M.S.; Neto, S.M.; Boleti, A.P.A.; Taveira, G.B.; Mello, E.O.; Gomes, V.M.; Santos, E.L.D.; Crusca, E., Jr.; et al. Adepamycin: Design, synthesis and biological properties of a new peptide with antimicrobial properties. *Arch. Biochem. Biophys.* **2020**, *691*, 108487. [CrossRef] [PubMed]
91. Rodrigues, M.S.; de Oliveira, C.F.R.; Almeida, L.H.O.; Neto, S.M.; Boleti, A.P.A.; Dos Santos, E.L.; Cardoso, M.H.; Ribeiro, S.M.; Franco, O.L.; Rodrigues, F.S.; et al. Adevonin, a novel synthetic antimicrobial peptide designed from the *Adenanthera pavonina* trypsin inhibitor (ApTI) sequence. *Pathog. Glob. Health* **2018**, *112*, 438–447. [CrossRef] [PubMed]
92. Kim, J.Y.; Gopal, R.; Kim, S.Y.; Seo, C.H.; Lee, H.B.; Cheong, H.; Park, Y. PG-2, a Potent AMP against Pathogenic Microbial Strains, from Potato (*Solanum tuberosum* L. cv. Gogu Valley) Tubers Not Cytotoxic against Human Cells. *Int. J. Mol. Sci.* **2013**, *14*, 4349–4360. [CrossRef] [PubMed]

93. Cotabarren, J.; Ozon, B.; Claver, S.; Geier, F.; Rossotti, M.; Garcia-Pardo, J.; Obregon, W.D. A Multifunctional Trypsin Protease Inhibitor from Yellow Bell Pepper Seeds: Uncovering Its Dual Antifungal and Hypoglycemic Properties. *Pharmaceutics* **2023**, *15*, 781. [CrossRef] [PubMed]
94. Bacha, A.B.; Jemel, I.; Moubayed, N.M.S.; Abdelmalek, I.B. Purification and characterization of a newly serine protease inhibitor from *Rhamnus frangula* with potential for use as therapeutic drug. *3 Biotech* **2017**, *7*, 148. [CrossRef] [PubMed]
95. Manojlovic, N.T.; Solujic, S.; Sukdolak, S.; Milosev, M. Antifungal activity of Rubia tinctorum, *Rhamnus frangula* and *Caloplaca cerina*. *Fitoterapia* **2005**, *76*, 244–246. [CrossRef] [PubMed]
96. Costa, H.P.; Oliveira, J.T.; Sousa, D.O.; Morais, J.K.; Moreno, F.B.; Monteiro-Moreira, A.C.; Viegas, R.A.; Vasconcelos, I.M. JcTI-I: A novel trypsin inhibitor from *Jatropha curcas* seed cake with potential for bacterial infection treatment. *Front. Microbiol.* **2014**, *5*, 5. [CrossRef] [PubMed]
97. Wang, C.; Shao, C.; Fang, Y.; Wang, J.; Dong, N.; Shan, A. Binding loop of sunflower trypsin inhibitor 1 serves as a design motif for proteolysis-resistant antimicrobial peptides. *Acta Biomater.* **2021**, *124*, 254–269. [CrossRef] [PubMed]
98. Cotabarren, J.; Claver, S.; Payrol, J.A.; Garcia-Pardo, J.; Obregon, W.D. Purification and Characterization of a Novel Thermostable Papain Inhibitor from Moringa oleifera with Antimicrobial and Anticoagulant Properties. *Pharmaceutics* **2021**, *13*, 512. [CrossRef] [PubMed]
99. de Souza, A.A.; Lima, A.M.; Dede Oliveira BezerraSousa, D.; Nogueira, F.C.; do Sacramento Neto, J.C.; Dias, L.P.; Araujo, N.M.S.; Nagano, C.S.; Junior, H.V.N.; da Silva, C.R.; et al. Chia (*Salvia hispanica* L.) Seeds Contain a Highly Stable Trypsin Inhibitor with Potential for Bacterial Management Alone or in Drug Combination Therapy with Oxacillin. *Probiotics. Antimicrob. Proteins* **2023**, *15*, 1221–1233. [CrossRef] [PubMed]
100. Bartova, V.; Barta, J.; Jarosova, M. Antifungal and antimicrobial proteins and peptides of potato (*Solanum tuberosum* L.) tubers and their applications. *Appl. Microbiol. Biotechnol.* **2019**, *103*, 5533–5547. [CrossRef] [PubMed]
101. Jashni, M.K.; Mehrabi, R.; Collemare, J.; Mesarich, C.H.; de Wit, P.J. The battle in the apoplast: Further insights into the roles of proteases and their inhibitors in plant-pathogen interactions. *Front. Plant Sci.* **2015**, *6*, 584. [CrossRef]
102. Woloshuk, C.P.; Meulenhoff, J.S.; Sela-Buurlage, M.; van den Elzen, P.J.; Cornelissen, B.J. Pathogen-induced proteins with inhibitory activity toward *Phytophthora infestans*. *Plant Cell* **1991**, *3*, 619–628. [CrossRef] [PubMed]
103. Vernekar, J.V.; Tanksale, A.M.; Ghatge, M.S.; Deshpande, V.V. Novel bifunctional alkaline protease inhibitor: Protease inhibitory activity as the biochemical basis of antifungal activity. *Biochem. Biophys. Res. Commun.* **2001**, *285*, 1018–1024. [CrossRef] [PubMed]
104. Carmona-Gutierrez, D.; Frohlich, K.U.; Kroemer, G.; Madeo, F. Metacaspases are caspases. Doubt no more. *Cell Death Differ.* **2010**, *17*, 377–378. [CrossRef] [PubMed]
105. Aslam, B.; Khurshid, M.; Arshad, M.I.; Muzammil, S.; Rasool, M.; Yasmeen, N.; Shah, T.; Chaudhry, T.H.; Rasool, M.H.; Shahid, A.; et al. Antibiotic Resistance: One Health One World Outlook. *Front. Cell Infect. Microbiol.* **2021**, *11*, 771510. [CrossRef] [PubMed]
106. Woods, M.; McAlister, J.A.; Geddes-McAlister, J. A One Health approach to overcoming fungal disease and antifungal resistance. *WIREs Mech. Dis.* **2023**, *15*, e1610. [CrossRef] [PubMed]
107. FAO. Antimicrobial Resistance. Available online: https://www.fao.org/antimicrobial-resistance/key-sectors/animal-production/en/ (accessed on 2 April 2024).
108. Barcenas, O.; Pintado-Grima, C.; Sidorczuk, K.; Teufel, F.; Nielsen, H.; Ventura, S.; Burdukiewicz, M. The dynamic landscape of peptide activity prediction. *Comput. Struct. Biotechnol. J.* **2022**, *20*, 6526–6533. [CrossRef] [PubMed]
109. Yue, L.; Song, L.; Zhu, S.; Fu, X.; Li, X.; He, C.; Li, J. Machine learning assisted rational design of antimicrobial peptides based on human endogenous proteins and their applications for cosmetic preservative system optimization. *Sci. Rep.* **2024**, *14*, 947. [CrossRef] [PubMed]

MDPI AG
Grosspeteranlage 5
4052 Basel
Switzerland
Tel.: +41 61 683 77 34

*Pharmaceutics* Editorial Office
E-mail: pharmaceutics@mdpi.com
www.mdpi.com/journal/pharmaceutics

www.ingramcontent.com/pod-product-compliance
Lightning Source LLC
Chambersburg PA
CBHW062039220926
51850CB00010B/2602
* 9 7 8 3 7 2 5 8 7 8 6 1 1 *